# INFRARED AND MILLIMETER WAVES

VOLUME 9   MILLIMETER COMPONENTS
                  AND TECHNIQUES, PART I

# CONTRIBUTORS

EDWARD E. ALTSHULER

J. W. M. BAARS

ANDERS BONDESON

G. BOUCHER

PH. BOULANGER

P. CHARBIT

G. FAILLON

A. HERSCOVICI

TATSUO ITOH

E. KAMMERER

MARVIN B. KLEIN

WALLACE M. MANHEIMER

K. MIYAUCHI

G. MOURIER

EDWARD OTT

JUAN RIVERA

# INFRARED AND MILLIMETER WAVES

## VOLUME 9 MILLIMETER COMPONENTS AND TECHNIQUES, PART I

*Edited by* **KENNETH J. BUTTON**

NATIONAL MAGNET LABORATORY
MASSACHUSETTS INSTITUTE OF TECHNOLOGY
CAMBRIDGE, MASSACHUSETTS

1983

*ACADEMIC PRESS*

*A Subsidiary of Harcourt Brace Jovanovich, Publishers*

New York   London
Paris   San Diego   San Francisco   São Paulo   Sydney   Tokyo   Toronto

ACADEMIC PRESS, INC.
111 Fifth Avenue, New York, New York 10003

*United Kingdom Edition published by*
ACADEMIC PRESS, INC. (LONDON) LTD.
24/28 Oval Road, London NW1 7DX


Library of Congress Cataloging in Publication Data

Main entry under title:

Infrared and millimeter waves.

   Includes bibliographies and indexes.
   Contents: v. 1. Sources of radiation.--v. 2. Instrumentation.--[etc.]--v. 9. Millimeter components
and techniques. Part 1.
   1. Infra-red wave devices.  2. Millimeter wave
devices.  I. Button, Kenneth J.
TA1570.I52      621.36'2              79-6949
ISBN 0-12-147709-6 (v. 9)


PRINTED IN THE UNITED STATES OF AMERICA

83 84 85 86      9 8 7 6 5 4 3 2 1

D
621.362
INF

# CONTENTS

Chapter 7    **Multimode Analysis of Quasi-Optical**
             **Gyrotrons and Gryoklystrons**

*Anders Bondeson, Wallace M. Manheimer, and Edward Ott*

# LIST OF CONTRIBUTORS

Numbers in parentheses indicate the pages on which the authors' contributions begin.

EDWARD E. ALTSHULER (177), *Rome Air Development Center, Electromagnetic Sciences Division, Hanscom Air Force Base, Massachusetts 01731*

J. W. M. BAARS (241), *Max-Planck-Institut für Radioastronomie, 5300 Bonn, Germany*

ANDERS BONDESON[1] (309), *Laboratory for Plasma and Fusion Energy Studies, University of Maryland, College Park, Maryland 20742*

G. BOUCHER (283), *THOMSON-CSF, Division Tubes Électroniques, 92102 Boulogne Billancourt, France*

PH. BOULANGER (283), *THOMSON-CSF, Division Tubes Électroniques, 92102 Boulogne Billancourt, France*

P. CHARBIT (283), *THOMSON-CSF, Division Tubes Électroniques, 92102 Boulogne Billancourt, France*

G. FAILLON (283), *THOMSON-CSF, Division Tubes Électroniques, 92102 Boulogne Billancourt, France*

A. HERSCOVICI (283), *THOMSON-CSF, Division Tubes Électroniques, 92102 Boulogne Billancourt, France*

TATSUO ITOH (95), *Department of Electrical Engineering, The University of Texas at Austin, Austin, Texas 78712*

E. KAMMERER (283), *THOMSON-CSF, Division Tubes Électroniques, 92102 Boulogne Billancourt, France*

MARVIN B. KLEIN (133), *Hughes Research Laboratories, Malibu, California 90265*

WALLACE M. MANHEIMER (309), *Plasma Theory Branch, Plasma Dynamics Division, Naval Research Laboratory, Washington, D.C. 20375*

K. MIYAUCHI (1), *Research and Development Bureau, Nippon Telegraph and Telephone Public Corporation, Yokosuka, Japan 238-03*

G. MOURIER (283), *Division Tubes Électroniques, THOMSON-CSF, 92102 Boulogne Billancourt, France*

---

[1]Present address: Institute for Electromagnetic Field Theory, Chalmers University of Technology, 412 96 Göteborg, Sweden.

EDWARD OTT (309), *Laboratory for Plasma and Fusion Energy Studies, University of Maryland, College Park, Maryland 20742, and Plasma Theory Branch, Plasma Dynamics Division, Naval Research Laboratory, Washington, D.C. 20375*

JUAN RIVERA[2] (95), *Department of Electrical Engineering, The University of Texas at Austin, Austin, Texas 78712*

[2]Present address: TRW Inc., Redondo Beach, California 90278.

# PREFACE

This is the first of a series of books that will adhere closely to the theme "Millimeter Components and Techniques." We have not emphasized millimeter waves as the strict theme of a book since Volume 4: "Millimeter Systems" published in 1981. We are correcting our neglect of this emerging technology by publishing three books in rapid succession. This first one of the subseries opens with a chapter on the general topic of "Millimeter-Wave Communications" by Dr. K. Miyauchi. Then we address one of the most important problems in millimeter-wave components: the transmission line. We shall have five or six chapters on this subject eventually, but we start, of course, with Professor Tatsuo Itoh and Dr. Juan Rivera on "Comparative Studies of Millimeter-Wave Transmission Lines," which includes some important examples of devices. Chapter 3 deals with the emerging science and technology of "Dielectric Waveguide Electrooptic Devices" by Dr. Marvin B. Klein. There will be more on transmission lines and resonators in Volume 10. In the meantime, we shall open up three more aspects of this theme, namely, "Millimeter-Wave Propagation and Remote Sensing of the Atmosphere" by Edward E. Altshuler, "Technology of Large Radio Telescopes for Millimeter and Submillimeter Wavelengths" by J. W. M. Baars, and, finally, two chapters on gyrotrons.

The next book, Volume 10: "Millimeter Components and Techniques, Part II," will continue with the development of this theme. We shall have a chapter, "Microwave Open Resonator Techniques," by Professor A. L. Cullen and a chapter, "Microwave Open Resonators in Gyrotrons," by Professors Cheng-he Xu and Le-zhu Zhou. Dr. C. W. Roberson and colleagues have given us "A Free-Electron Laser Driven by a Long-Pulse Induction Linac." We are sure that we shall include "Integrated-Circuit Antennas" by David B. Rutledge and colleagues and "Near-Millimeter Imaging with Integrated Planar Receptors" by Professor K. S. Yngvesson. Finally, as promised, we have tried to get extensive reviews of emerging component and transmission techniques, and we shall not go into production until we have "Properties and Capabilities of Millimeter-Wave IMPATT Devices" by R. K. Mains and G. I. Haddad. Concerning detectors, we have a chapter on "$^3$He Refrigerators and Bolometers for Infrared and Millimeter-Wave Observations" by Dr. G. Chanin and Dr. J. P. Torre.

In the next volumes we have hopes for the long-awaited chapter on "Groove Guide for Short Millimetric Waveguide Systems" by Professor Douglas Harris and Yat Man Choi, "The Modified H Guide" by Professor Frederick Tischer, "Millimeter-Wave Hybrid Integrated Circuit Techniques" by A. G. Cardiosmenos, and "Photoconductive Detectors" by D. K. Shivanandan. For additional emphasis on components, we expect "Millimeter-Wave Integrated Circuits" from Charles Seashore, "InP and GaAs Devices at Millimeter Wavelengths" by I. G. Eddison, "Dielectric-Based Active and Passive Millimeter-Wave Components" by N. Deo, and "Integrated Fin-Line Components for Radar and Radiometer Applications" by W. Menzel. The chapters in the planning stages are literally too numerous to mention here so we refer you to the Preface of Volume 10.

# CONTENTS OF OTHER VOLUMES

xiii

CHAPTER 1

# Millimeter-Wave Communications

*K. Miyauchi*

*Research and Development Bureau*
*Nippon Telegraph and Telephone Public Corporation*
*Yokosuka, Japan*

## I. Introduction

The exploitation of new frequency regions has always led to technological advances in the history of radio communication. The millimeter-wave region is a new frontier; it borders on the microwave region, which has been developed in the past 30 years and is now widely applied in various fields in modern society.

Improvements in microwave technology have made advanced and inexpensive equipment available. The steadily increasing demand for communications has resulted in the installation of numerous microwave systems and a critical shortage of frequencies. Millimeter waves are quite attractive because their information-carrying capabilities are far greater than those of microwaves.

Millimeter-wave technology, in comparison with that of microwaves, has disadvantages such as large rainfall attenuation, large circuit loss, low efficiency of receivers, and low transmitter output power. Recent intensive research has solved many problems associated with these disadvantages and made it possible to provide systems that feature good performance in the millimeter-wave region. We are now able to produce devices and components that satisfy the performance, reliability, and productibility requirements necessary to build practical millimeter-wave communication equipment up to 100 GHz.

In Section II of this chapter, we shall describe devices and components for millimeter-wave communication systems. Active devices for the generation, amplification, and detection of signals are of primary importance. Only electron tubes and point-contact diodes were available for these purposes in the past. We shall not describe the electron tubes because they have a relatively narrow area of application owing to the large size and high voltage of their power supply, although they are still quite useful for high-power amplifiers and oscillators. The point-contact diodes have had substantial problems in terms of electrical performance, long-term stability, and reproducibility. It recently became possible to build practical millimeter-wave communication equipment employing sophisticated solid-state devices such as IMPATT, Schottky-barrier, $p-n$ junction, and $p-i-n$ diodes.

The most remarkable breakthrough was the invention of impact-ionization avalanche and transit-time (IMPATT) diodes and the improvement of gallium arsenide diodes. The IMPATT diode utilizes the avalanche effect of a reversely biased $p-n$ junction and is capable of oscillation and amplification in a frequency range of up to several hundred gigahertz. Although some cognates and different versions of the IMPATT diode, such as the TRAPATT, LSA, BARITT, and TUNNETT, have been proposed, the IMPATT diode is virtually the only device for carrier generation with sufficient power in the millimeter-wave region. Today, we can safely attain an oscillation power of about 8.0 mW at 100 GHz.

Gallium arsenide is a III–V compound and has an electron mobility approximately six times greater than that of silicon. The dielectric constants of gallium arsenide and silicon being equal, gallium arsenide provides us with diodes whose cutoff frequency is six times higher than that of silicon diodes. As a result of the improvement of the impurity doping and surface treatment technology of semiconductors, we can now obtain good diodes whose cutoff frequency is higher than 1000 GHz and whose current slope factor is very close to the theoretical limit.

Passive components are also indispensable in building communication equipment. We shall discuss various kinds of filters, including the band-splitting, channel-dropping, bandpass, and low-pass types, and nonreciprocal devices such as Y-junction circulators and self-resonant isolators. The design principles of most of these millimeter-wave components are similar to those of microwave components, except for some circuits that are designed with an oversize waveguide or according to quasi-optical principles to reduce heat loss. It is well known to specialists that poor machinery causes a large heat loss in passive waveguide components in the millimeter-wave region. Heat loss can be reduced to very close to the theoretical limit with advanced fabrication technology such as electroforming, photolithographic etching, and cold hobbing.

In Section III, guided millimeter-wave communication systems are introduced. A low-loss and wide-band transmission medium operating with hyperfrequencies has been sought since the beginning of the development of microwave technology. In the early 1930s the transmission characteristics of the $TE_{01}$ mode created interest in the fact that the theoretical attentuation constant of this mode decreases without limit with an increase in operating frequency. It was not easy to realize the expected low-loss characteristic experimentally. It was found after intensive studies that mode conversions due to mechanical imperfections of waveguides prevented low-loss transmission. Practical transmission media such as helix and dielectric-lined waveguides were finally developed.

It is possible to install a waveguide line along a practical route with many curved sections and undulations if low-loss, small-diameter flexible waveguides and mirror corners are utilized together with helix and dielectric-lined waveguides. Studies were carried out to develop a large-capacity communication system by utilizing the wide-band characteristics of millimeter-wave waveguide lines. Research engineers also succeeded in developing ultrahigh-speed digital repeaters and low-loss multiplexing and demultiplexing networks in the millimeter-wave region.

Integrating these techniques, full-scale field trials on large-capacity guided millimeter-wave, long-distance communication systems were carried out in some countries very successfully in the mid-1970s to confirm the system's feasibility. We include, in Section III, the example of a system that can transmit a few hundred thousand two-way telephone channels in a single waveguide line. In Sections IV and V, millimeter-wave terrestrial radio and satellite communication systems are presented.

Intensive research has been carried out to overcome difficulties due to severe rainfall attenuation in the design of practical communication systems that feature good performance and reasonable cost by utilizing advantages inherent in millimeter waves, such as a broad frequency bandwidth and a small-aperture antenna with a sharp directivity. As for large-capacity terrestrial communications, microwave systems operating at several gigahertz are widely used for long-distance circuits. The spacing of repeater stations is typically 50 km, but the spacing becomes as short as 3 km at 20 GHz when we require the same order of outage time due to atmospheric phenomena as that of microwave systems. This suggests that the feasibility of the systems will be established only when we can lower the cost and the electronics failure rate of repeaters as well as improve the reliability of the power supply by more than one order of magnitude.

20L-P1 and DR 18 are large-capacity and long-distance digital radio systems operating near 20 GHz that have recently been developed and introduced into commercial use. They are typical examples of a new technology that has overcome difficulties due to severe rainfall attenuation at frequencies near the millimeter-wave range. We shall describe these systems in some detail because practical large-capacity and long-distance systems at higher frequencies have not yet been developed.

Many short-distance small- and medium-capacity radio systems are being researched for distribution network and subscriber line applications above 25 GHz, and some have already been introduced into commercial use. Large-capacity inter- and intracity radio systems with new installation concepts have also been proposed. These systems are briefly described in Section IV.

In Section V, satellite communications are described. In the past, satellite communications were mainly put to international use. However, rapid progress has been made during the past few years in utilizing them in domestic and regional systems. This situation has stimulated the development of techniques for exploiting new frequencies, such as millimeter waves, owing to the radio spectrum shortage. Millimeter waves offer almost the same advantages in satellite communications as in terrestrial radio communications: a large-capacity transmission capability due to a wide frequency spectrum and the possibility of small earth stations with small antennas. We can also expect other advantages, such as opportunities for frequency reuse employing a multibeam satellite antenna of moderate size. Another advantage is less interference between terrestrial and satellite systems than with microwaves because of the small number of millimeter-wave stations.

Rainfall attenuation in the millimeter-wave region also results in a large degradation of the performance of satellite communication systems. Satellite communication systems now in operation mostly employ microwaves such as 4 and 6 GHz; however, much experimentation has been done to evaluate the effects of rainfall attenuation on system performance at higher frequencies.

1. MILLIMETER-WAVE COMMUNICATIONS

Today, several systems in practical use operate at 12 and 14 GHz. It is expected that systems operating at 20 and 30 GHz will be introduced into practical use in the near future. In Section V we describe some examples of satellite communication systems in the millimeter-wave region as well as the state of the art in equipment, spacecraft, and earth station facilities. This chapter is thus intended to cover millimeter-wave applications in various communication fields. Most of these developments, however, are still in the beginning stages.

## II.   Devices and Components for Millimeter-Wave Communication Systems

### A.   SOLID-STATE DEVICES

### 1.   *IMPATT Devices*

Since the first proposal of fundamental operation (Read, 1958) and the succeeding oscillation experiment (Johnston *et al.*, 1964), impact-ionization avalanche and transit-time (IMPATT) devices have secured an important position as powerful solid-state devices in the microwave and millimeter-wave regions. IMPATT diodes employ the impact-ionization and transit-time properties of semiconductors. When a reverse voltage is applied to a $p-n$ junction beyond its breakdown voltage, avalanche current flows and the carriers transit to the electrode with a time delay. A negative resistance arises at frequencies corresponding to the thickness of the transit layer.

Because the IMPATT diode has a nonlinearity in its rf voltage and current characteristics, its field of application is very wide, and almost all conceivable functions, even frequency conversion and multiplication, are realized. Among them, oscillation and amplification are the two most important uses.

a.   *IMPATT Diodes.*   The doping profile first proposed by Read was $p^+-n-i-n^+$ or $n^+-p-i-p^+$. However, the practical structure presently used in the millimeter-wave region is $p^+-n$ or $p^+-p-n-n^+$. The $p^+-n$ profile is called the single-drift region (SDR) type and the $p^+-p-n-n^+$ profile is called the double-drift region (DDR) type. Although the DDR type has a more complex profile, it is capable of delivering more power and higher efficiency (Scharfetter *et al.*, 1970; Seidel *et al.*, 1971). Silicon is commonly used as the semiconductor material.

Figure 1 shows a typical structure of an IMPATT diode. The junction, from which heat is generated, is positioned close to the heat sink. Because the efficiency is less than 10% in the millimeter-wave region, good heat

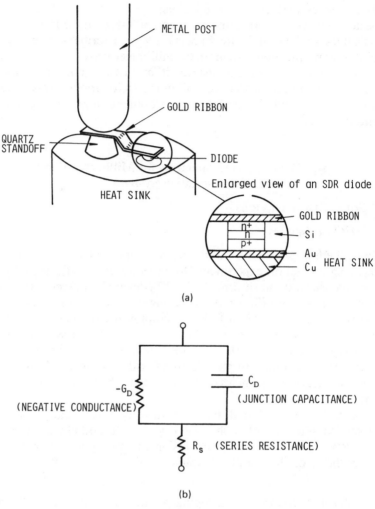

FIG. 1   (a) Diode structure and (b) equivalent circuit of an IMPATT diode. (From Suzuki, 1975.)

transfer is required. The heat sink is made of copper or metalized type-IIA diamond. The quartz standoff is for mechanical protection. The electrical contact is made by a gold ribbon between the diode and the standoff and by a metal post between the standoff and the external power supply. The diameter of the quartz standoff and the width and length of the gold ribbon are below 100 $\mu$m for millimeter-wave use. The equivalent circuit is expressed as a parallel connection of $C_D$ (the junction capacitance) and $-G_D$ (the negative conductance arising from the impact-ionization and transit-time effect), which is followed by $R_s$ (the series resistance). The resistance within the semiconductor substrate and metal surface and other imperfections of electrical contacts are included in $R_s$.

b. *IMPATT Oscillators*

(1) General description. IMPATT oscillators are used for local oscillators of frequency converters, transmitting carrier sources, and sweep oscillators. In designing IMPATT oscillators for practical communication systems, one should pay careful attention to the maximum obtainable power level under a given reliability, i.e., under a certain limitation of temperature rise of the junction, because the junction temperature is directly related to the mean time of failure (MTTF) of the device. Moreover, the maximum obtainable power level is closely related to the mechanical dimensions and electrical parameters of the diode.

Under a simplified equivalent circuit (Fig. 1), the maximum obtainable power $P_{opt}$, where the load impedance is optimized, is calculated as follows for an SDR diode (Suzuki, 1975):

$$P_{opt} = \frac{4D\kappa \, \Delta T_J}{27\alpha^2 V_{dc}} \left( 1 - \frac{R_s\omega^2\varepsilon^2\pi^2D^3V_{dc}}{32\gamma\kappa \, \Delta T_J \; W^2} \right)^3. \tag{1}$$

TABLE I

CONSTANTS IN EQ. (1)

| Symbol | Quantity |
|---|---|
| $D$ | Diode diameter |
| $R_s$ | Series resistance |
| $\Delta T_J$ | Junction temperature rise |
| $V_{dc}$ | Operating voltage[a] |
| $W$ | Active layer width |
| $\alpha$ | Constant[b] |
| $\gamma$ | Constant[c] |
| $\varepsilon$ | Dielectric constant of semiconductor |
| $\kappa$ | Effective thermal conductivity from junction to heat sink |
| $\omega$ | Operating angular frequency |

[a] $V_{dc}$ is determined from the breakdown voltage $V_B$ and dc bias.

[b] $G_D = G_m(1 - \alpha V_{rf}^2)$, where $G_m$ is the maximum value of $G_D$ and $V_{rf}$ the rf voltage across the diode.

[c] $G_m = \gamma \left(\frac{1}{4}\pi D^2\right)I_{dc}$, where $I_{dc}$ is the dc bias current.

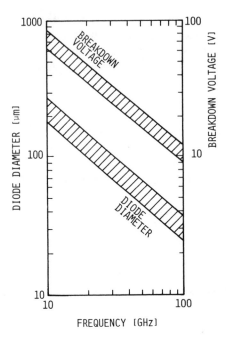

FIG. 2 Optimum breakdown voltage and diode diameter for an SDR diode.

The constants in this equation are listed in Table I. Equation (1) shows us that the oscillation power is determined from electrical constants (such as breakdown voltage, thickness of active layer, and series resistance), mechanical dimensions (such as the junction diameter and diode height), thermal parameters (such as thermal resistance and junction temperature rise), and material constants (such as electrical conductivity, thermal conductivity, and the dielectric constant). Some of these, of course, are related.

The breakdown voltage $V_B$ and junction diameter $D$ are of primary importance, because they determine the oscillation frequency range. Based on the many experimental results reported so far, one can find the appropriate relationship of $V_B$, $D$, and their oscillation frequency shown in Fig. 2. Figure 2 gives the results for an SDR diode. For the DDR type, diodes with approximately the same diameter but with a breakdown voltage of approximately $1.5 V_B$ are used.

Local oscillators for frequency converters and transmitter carrier sources are required to have high frequency stability. Frequency stabilization is achieved by means of a waveguide cavity resonator attached to the external circuit. The $TE_{01n}$ mode of the cylindrical cavity is used for obtaining a high quality factor ($Q$). To minimize the change in oscillation frequency with the variation in ambient temperature, the cavity is made of Invar (the thermal expansion coefficient is typically less than $10^{-5}/°C$) or the thermal expansion of the cavity is compensated for by a dielectric rod inserted into the cavity. A frequency stability of $\pm 7 \times 10^{-5}$ (0–50°C) has been obtained by an Invar cavity without compensation (Akaike *et al.*, 1975).

IMPATT oscillators can also be operated as variable-frequency oscillators. Oscillation frequency tuning is summarized as follows:

(a)  varying IMPATT diode impedance: bias-current tuning;
(b)  varying external circuit impedance: mechanical tuning, electrical tuning (varactor, yttrium iron garnet).

In the first of the two methods, the impedance of the diode is varied while the external circuit is held fixed. In the second, the impedance of the external circuit is changed. Current- and varactor-tuned oscillators have been used as

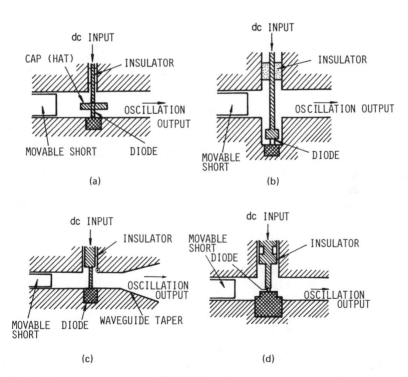

FIG. 3  Diode mount structures of IMPATT oscillators: (a) cap (hat) type, (b) coaxial-waveguide type, (c) reduced-height-waveguide type, (d) platform type. (From IECE Japan, 1979.)

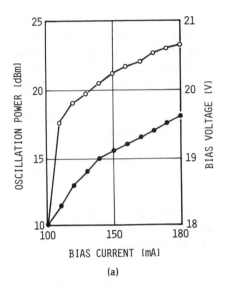
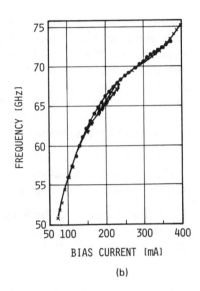

(a)                                            (b)

Fig. 4   Oscillation characteristics of IMPATT oscillators. (a) Fixed-frequency DDR oscillator. Oscillation frequency = 80 GHz; $V_B$ = 15.9 V; o, output power; •, bias voltage. Thermal
resistance of the diode = 60.1°C/W. A diamond heat sink is used. (From Nagao et al. 1978. ©
1978 IEEE.) (b) Bias-current-tuned oscillator for distances between the diode and movable
short of 1.5 mm (×), 2.0 mm (•), and 2.5 mm (▼). $V_B$ = 13.3 V. (From Akaike et al., 1976b. ©
1976 IEEE.)

part of the measuring equipment of the W-40G guided millimeter-wave
communication system (Akaike et al., 1977).

(2)   Experimental results. Various kinds of diode mounts have been
described — typical structures are shown in Fig. 3. In these structures, the
diode is mounted where the characteristic impedance is low and impedance
matching between the waveguide load and diode is performed by a trans-
former.

Figure 4 shows the oscillation characteristics of IMPATT oscillation. In
Fig. 4a, the power and frequency of 80-GHz DDR oscillator (Nagao et al.,
1978) are shown. The heat sink is made of type IIA diamond. The input dc
power is 4 W when the output rf power is 200 mW. The junction diameter is
50 μm, and the estimated junction temperature is 265°C (for an ambient
temperature of 25°C). An MTTF of 4.6 × 10⁴ h is expected. The oscillation
frequency versus dc bias current of a 60-GHz bias-current-tuned oscillator
(Akaike et al., 1976b) is shown in Fig. 4b. The swept frequency covers
almost the entire frequency bandwidth of the R-620 waveguide (50–75
GHz).

c. *IMPATT Amplifiers*

(1)   General description. Two types of amplifiers are possible: negative-resistance amplifiers and injection-locked amplifiers. In negative-resistance amplifers, the negative resistance of the diode is utilized. When the input signal is reflected by a negative resistance, the reflection coefficient is greater than unity and therefore amplification is possible. In injection-locked amplifiers, the oscillator frequency is locked by an injected signal. When an input signal is injected into a free-running oscillator, the frequency of the oscillator synchronizes with the frequency of the injected signal. This is known as injection locking. The frequency of the synchronized oscillator is varied with the injected frequency. The oscillator therefore amplifies the input signal.

The main difference between these two types of amplifiers is that in negative-resistance amplifiers the reflection coefficient (i.e., the gain) is not affected by the input level, and therefore this type of amplifier is linear, whereas in injection-locked amplifiers the output power is always determined by the oscillation power, whether or not the input signal exists. Injection-locked amplifiers, therefore, are used only for FM or PM signals. Both types are used as power amplifiers and not as low-level amplifiers because in negative-resistance amplifiers the noise figure is fairly large (typically greater than 20 dB) and in injection-locked amplifiers the gain – bandwidth product is constant [see (3)] and sufficient bandwidth cannot be obtained when the input level is too low.

(2)   Negative-resistance amplifiers. The negative conductance $G_D$ of the diode is a function of the rf voltage $V_{rf}$ across the diode and is approximately expressed as (Kuno and English, 1973)

$$G_D = G_m(1 - \alpha V_{rf}^2), \qquad (2)$$

where $G_m$ is the small-signal conductance and $\alpha$ a constant. When the input level is relatively low, the output power increases linearly with the input power. When the input power increases, however, the gain begins to saturate; i.e., output compression occurs. The relationship between gain and output power calculated from Eq. (2) (Akaike *et al.*, 1976a) is shown in Fig. 5. When compression is low and gain high, the amplifier output power is below the oscillation power, but as compression becomes higher or gain lower, the amplifier output power becomes comparable to the oscillation power and then exceeds it. In general, higher output power is available if one allows greater compression. However, when compression increases, distortion of the amplified waveform may increase. Therefore, allowable compression selection is an important factor in designing IMPATT negative-resistance amplifiers.

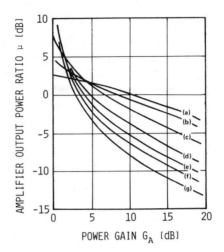

FIG. 5  Amplifier output ratio as a function of power gain of an IMPATT diode. Amplifier output ratio = (amplifier outpower power)/(oscillator output power). Measurement was made at 80 GHz. Measured data compression: (a) 10 dB, (b) 6 dB, (c) 3 dB, (d) 1.5 dB, (e) 1.0 dB, (f) 0.75 dB, (g) 0.5 dB. (From Akaike *et al.*, 1976a. © 1976 IEEE.)

The characteristics of an 86-GHz amplifier manufactured for amplifying 800 Mbit/sec phase-shift keying (PSK) signals are shown in Fig. 6 (Ando *et al.*, 1978). The quality of the amplified waveform for various compression values was evaluated by measuring the bit error rate of the amplified PSK signals. No significant carrier-to-noise ratio (C/N) degradation (less than 0.7 dB) was observed for compressions up to 2 dB.

(3)  Injection-locked amplifiers. In injection-locked amplifiers, the output frequency or phase varies with the input FM or PM signal. How fast the oscillator can follow the frequency (or phase) change of the injected signal depends upon the locking bandwidth $B$ of the oscillator and the gain (the ratio of oscillation power to injected power) $G_D$ and is expressed as follows:

$$\sqrt{G_D}\ B = Q \text{ (constant).}$$

Less distorted waveforms are obtained for oscillators with larger values of $Q$.

An experiment using a 47-GHz amplifier was performed. The input signal was an 800-Mbit/sec PSK signal and no distortion was observed (Fukatsu *et al.*, 1971). The value of $Q$ in this oscillator was 12 GHz.

### 2.  Schottky-Barrier Devices

a.  *Schottky-Barrier Diodes.*  The Schottky-barrier diode is a general term for metal–semiconductor contact diodes. The use of gallium arsenide

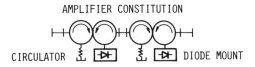

AMPLIFIER CONSTITUTION

CIRCULATOR        DIODE MOUNT

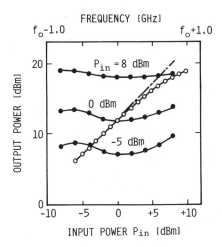

FIG. 6 Input–output relationship and frequency response of an 86-GHz DDR negative-resistance amplifier. The breakdown voltage of the diodes is 16 V; $f_0 = 86.35$ GHz; ●, frequency response; ○, amplifier output. (From Ando *et al.*, 1978. © 1978 IEEE.)

as a semiconductor material is common in the millimeter-wave region. Typical diode structures and an equivalent circuit are shown in Fig. 7. The junction diameter usually used lies in the range 2–20 $\mu$m. For downconverters, a diameter of a few micrometers is preferred because the handling power is very low and the junction temperature rise presents no problem. For upconverters, however, the optimum diameter is selected depending upon power-handling capability and matching between the diode and external circuit. Diameters of approximately 15 $\mu$m at 40 GHz and 8 $\mu$m at 100 GHz are used.

The current and voltage relationship is given as follows:

$$I = I_s[\exp(eV_B/nkT) - 1] + C_0(1 - V_B/\phi)^{-\gamma} \, dV_B/dt,$$

$$V_B = V_T - IR_s,$$

where $I_s$ is the saturation current, $e$ the electronic charge, $k$ the Boltzmann constant, $T$ the temperature of the diode, $\phi$ the diffusion voltage (0.85 V for gallium arsenide), $V_T$ the voltage across the Schottky junction, and $n$ and $\gamma$ the slope factors of current and junction capacitance, respectively. For ideal

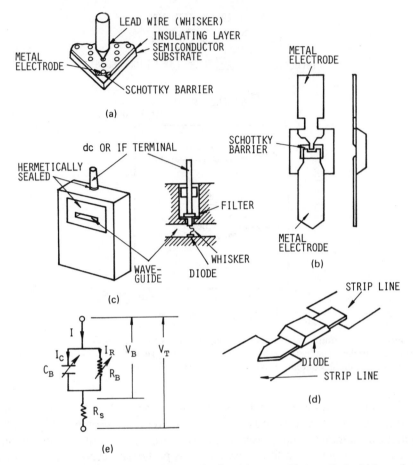

FIG. 7   Diode structure and equivalent circuit: (a) honeycomb-type diode, (b) beam-lead diode, (c) waveguide-wafer-type package, (d) beam-lead diode mounted on a strip line, (e) equilvalent circuit. $I_R = I_s [\exp(AV_B) - 1]$, $I_C = C_B \, dV_B/dt$, $V_T = V_B + IR_s$, $C_B = C_0(1 - V_B/\phi)^{-\gamma}$. [(a) From Young and Irvin, 1965. © 1965 IEEE. (b) From Nippon Electric Company Ltd., 1979. (c) From Akaike et al., 1975. (e) From Akaike and Onishi, 1977. © 1977 IEEE.]

metal–semiconductor contact diodes, $n$ is equal to unity. In practically available Schottky-barrier diodes, however, $n$ is about 1.1. The quantity $\gamma$ is 0.5 for ideal, one-dimensional Schottky-barrier diodes. In actual three-dimensional diodes, $\gamma$ is not constant and is less than 0.5, owing to the three-dimensional effect and stray capacitances around the junction.

In the microwave region, diodes encapsulated in pill packages are commonly used. However, in the millimeter-wave region, special care in diode packaging is necessary because the wavelength becomes shorter and the parasitic effect of the package determines the upper limit of available frequency. One way to extend this limit is to make smaller packages whose resonant frequency is relatively high. Experimentally, a package with a 1-mm diameter and 0.5-mm height works up to 70 GHz (Kawasaki *et al.*, 1977). Another and more efficient method is shown in Fig. 7c, in which all parasitic elements in the close vicinity of the diode are removed. The diode is directly mounted within a "waveguide wafer." This structure has, in principle, no frequency-limiting factors except diode $Q = 1/\omega C_0 R_s$ and can operate at any high frequency when a proper waveguide size is selected. The two waveguide ports are hermetically sealed to ensure the reliability of the diode.

The beam-lead type has a planar structure (Fig. 7b,d) and can be mounted on a strip line. Because the surface of the semiconductor is covered with silicon oxide, it is free from environmental contamination and promises a high reliability.

b. *Downconverters.* Downconverters are widely used in heterodyne receivers not only for communications but also for physical studies and other measurements. There have been, therefore, a great number of theoretical and experimental studies of downconverters. The fundamental characteristics have already been discussed in many textbooks and theoretical and experimental papers. We shall therefore limit our discussion to broadband downconverters for high-speed digital communications.

Downconverters for communication equipment are required to have a low noise figure and a sufficient bandwidth. The receiver noise figure $F_T$ is expressed as

$$F_T = L_c(t - 1 + F_{IF}),$$

where $L_c$ is the conversion loss, $t$ is the effective noise temperature ratio of the diode, and $F_{IF}$ is the noise figure of the IF amplifier. For good Schottky-barrier diodes, $t$ being approximately unity (experimentally about 1.1), $F_T$ is approximately expressed as the product (the sum in decibels) of $L_c$ and $F_{IF}$. Consequently, the conversion loss is a convenient quality factor of downconverters.

The signal bandwidth of digital transmission is usually broad — on the order of several hundred megahertz, especially in guided millimeter-wave transmission. Choosing an appropriate IF and obtaining a sufficient IF bandwidth are necessary.

Typical structures of downconverters are shown in Fig. 8 (Barnes *et al.*, 1977). In Fig. 8a, the waveguide – wafer-type package is used (a single-ended mixer). The signal passes through a bandpass filter and a circulator. The

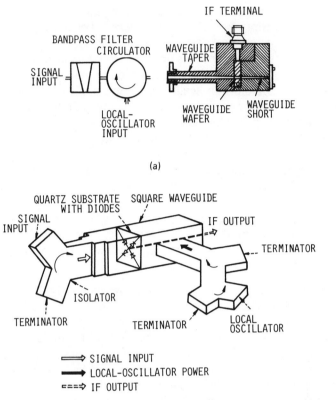

FIG. 8  Typical downconverter structures: (a) using a waveguide wafer, (b) using a beam-lead diode. [(a) From Kanmuri and Kawasaki, 1976. © 1976 IEEE. (b) From Barnes *et al.*, 1977. © 1977, AT&T. Reprinted from *The Bell System Technical Journal* by permission.]

local-oscillator power is fed from another port of the circulator and reflected by the bandpass filter and then enters the mixer. The waveguide height where the diode is mounted is reduced to obtain a good matching between the diode and waveguide. Typical conversion loss is given in Table II (Akaike *et al.*, 1975). The IF is 1.7 GHz. The variation in frequency response is less than ±0.5 dB for 1.7 ± 0.4 GHz.

In Fig. 8b, a double-balanced mixer is shown. The downconverter employs four beam-lead diodes mounted on a quartz substrate. Signal and local-oscillator power are orthogonally fed to the diodes. The IF signal is then taken out of the center of the four diodes. The conversion loss is typically 5 dB at 51.5 GHz (Dunn *et al.*, 1977).

TABLE II

CONVERSION LOSS OF WAVEGUIDE–WAFER-TYPE
DOWNCONVERTERS

| Center frequency (GHz) | Diode junction diameter ($\mu$m) | Conversion loss (dB) |
|---|---|---|
| 50.28 | 2.5 | 5.3 |
| 63.48 | ~ 3–3.5 | 5.8 |
| 65.85 | ~ 3–3.5 | 5.0 |
| 80.67 | 2.5 | 7.0 |
| 86.35 | ~ 3–4 | 5.5 |

An ultrabroadband mixer has been constructed (Kanmuri and Kawasaki, 1975). In this mixer, a waveguide-wafer diode holder is used, and the wave absorber is stuffed in the IF coaxial circuit. The rf bandwidth covers the entire rectangular waveguide bandwidth (Fig. 9).

An attempt has been made to extend the microwave integrated circuit (MIC) technique to the millimeter-wave region (Carlson *et al.*, 1978). The MIC technique is promising from the point of view of circuit miniaturization, economy, and reliability.

c. *Upconverters.* For upconverters, two kinds of diodes are used: Schottky-barrier diodes and *p–n* junction varactor diodes. In varactor upconverters, frequency conversion is achieved by the nonlinearity of junction capacitance. In Schottky-barrier upconverters, however, the forward conduction current flows across the barrier for a period of time even though the diode is reversely biased, and the nonlinearities of barrier resistance and junction capacitance both contribute to frequency conversion. Recently, a theoretical treatment of Schottky-barrier upconverters has been reported and some design parameters of these devices have been discussed (Akaike and Ohnishi, 1977).

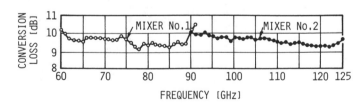

FIG. 9 Conversion loss of ultrabroadband mixers. The circuit is fixed for input rf frequencies. The local-oscillator frequency is varied so that a constant IF (= 1.7 GHz) is obtained. (From Kanmuri and Kawasaka, 1976. © 1976 IEEE.)

A comparison of Schottky-barrier diode upconverters and $p-n$ junction varactor upconverters was made at 50 GHz, and it is reported that Schottky-barrier diodes exhibit better characteristics with respect to ease in circuit adjustment and suppression of unwanted spurious oscillation (Kita *et al.*, 1973). In this section Schottky-barrier upconverters are discussed. Varactor upconverters will be briefly reviewed in the next section.

The waveguide mount for upconverters is almost the same as that for downconverters. Because the upconverter is a device for transmitters and the diode handles a fairly high level of power, its reliability—that is, the junction temperature rise of the diode—should be carefully checked.

Figure 10 shows the input–output characteristics of an 86-GHz upconverter (single ended) (Seki *et al.*, 1974). The diode is reversely biased by the rectified current through a resistor connected in the dc circuit. Resistance is chosen to give minimum conversion loss. The diode (honeycomb type) junction diameter is 7 $\mu$m. When the incident local-oscillator power and IF power are 18 and 16.7 dBm, respectively, an output power of 11.9 dBm is obtained with an output compression of 1.5 dB. The estimated junction temperature rise is 58.7°C. Therefore, this upconverter is expected to have a failure rate of less than 100 fits when the ambient temperature is 25°C.

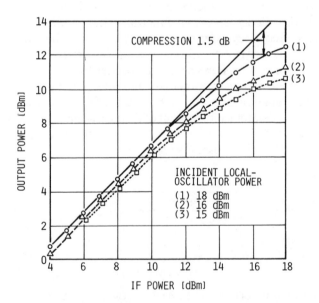

FIG. 10  Output vs input characteristics of an 86-GHz upconverter. The IF is 1.7 GHz. The IF bandwidth is greater than 800 MHz. (From Akaike *et al.*, 1976a. © 1976 IEEE.)

### 3. *p–n Junction Varactors*

*p–n* junction varactors have the same construction as Schottky-barrier diodes, except that the doping profile of the semiconductor is a $p^+–n$ one-sided abrupt junction. Gallium arsenide is commonly used in the millimeter-wave region. For usual operation, the diode is reversely biased and the equivalent circuit is expressed as a series connection of a variable capacitance and a series resistance. Because *p–n* junction varactors are used in relatively high-power devices (multipliers and upconverters), diode diameters in the range 10–30 μm are commonly used. The junction capacitance has the same form as in Schottky-barrier diodes, except that $\gamma$ is approximately 0.5, because the three-dimensional effect does not occur for larger diameter values. Theoretical treatment of varactor devices is well established, and sufficient information is available in textbooks.

Cross-coupled waveguide structures are commonly used for multipliers. The diode is mounted at the junction of the two waveguides. For triplers, an idler circuit (the second-harmonic circuit) is necessary. Figure 11 is a conversion-loss plot for doublers and triplers (Akaike *et al.*, 1975). Typically, the conversion loss is 4 dB at 50 GHz and 8 dB at 80 GHz.

Varactors can also be used for upconverters. An output power of 13 dBm is obtained at 39 GHz when the local-oscillator input is 23 dBm and the IF power is 9 dBm (Mahien, 1973).

### 4. *p–i–n Diode Modulator*

In a *p–i–n* diode, the semiconductor wafer has a *p* region (usually a heavily doped *p* region) and an *n* region (usually a heavily doped *n* region)

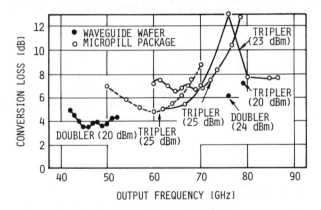

FIG. 11 Measured conversion loss of multipliers. Values in parentheses show input power levels. (From Akaike *et al.*, 1975.)

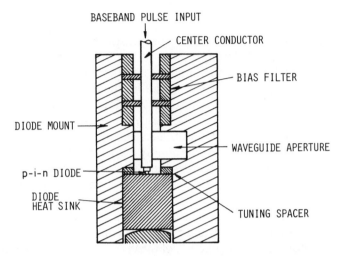

FIG. 12    $p-i-n$ diode modulator mount. (From Clemetson *et al.*, 1971. © 1971, AT&T. Reprinted from *The Bell System Technical Journal* by permission.)

separated by an intrinsic layer. Silicon is used in the millimeter-wave region. The Ohmic contact is made to the two heavily doped regions. The impedance of a $p-i-n$ diode is expressed as $R_s + 1/j\omega C_D$ for reverse bias and $R_s$ for forward bias, where $R_s$ is the series resistance and $C_D$ is the capacitance determined by the $i$ layer thickness. Unlike varactors and Schottky-barrier diodes, capacitance $C_D$ hardly varies with variation of reverse bias. This contact capacitance property is used for switching the rf carrier (pulse modulation). Thus the electrical property is determined by the intrinsic region, so that the determination of the $i$ layer is important in design.

TABLE III

$p-i-n$ Diode Path-Length 2-Phase Modulator Characteristics

| Frequency (GHz) | Input power[a] (oscillator output power) (dBm) | Output power (dBm) | Loss (dB) $p-i-n$ switch | Circulator |
|---|---|---|---|---|
| 56.4 | 21.3 | 20 | 0.5 | 0.5 |
| 57.4 | 21.4 | 20 | 0.6 | 0.5 |
| 107.2 | 14.9 | 12.0 | 0.8 | 0.7 |
| 108.3 | 15.7 | 13.4 | 0.8 | 0.5 |

[a] Net input power to the modulator is about 0.3dB (50-GHz range) and about 0.7dB (100-GHz range) lower than these values. The loss is due to an isolator connected after the oscillator.

Because the $p-i-n$ diode handles a fairly high level of rf and pulse driver power, the $i$ layer thickness and junction diameter must be properly chosen. The $i$ layer thickness is determined by the peak voltages of rf and pulse inputs and the electric field breakdown of the material. The diameter is determined by the matching of the diode to the external circuit. For millimeter-wave usage, an $i$ layer thickness of less than a few micrometers and a diameter of less than several tens of micrometers are commonly used.

A detailed discussion of the diode and circuit design and experimental results are given by Clemetson et al. (1971). The mount structure is shown in Fig. 12, and obtained characteristics are given in Table III (Clemetson et al., 1971; Clemetson et al., 1972).

## B. MULTIPLEXING AND DEMULTIPLEXING NETWORKS

A multiplexing and demultiplexing network (MDN) consists of band-splitting and channel-dropping filters. Specifications for the frequency reso-nances of these filters are determined mainly by the overall rf frequency bandwidth, guard band, rf channel spacing, and desired signal spectrum. In guided millimeter-wave communications, these filters should be optimally combined so that the overall loss-versus-frequency characteristic of the MDN matches that of the installed waveguide line and repeater gain. In addition, the size and arrangement of the repeaters and other equipment impose mechanical requirements on these filter arrays.

To achieve low thermal loss, the $TE_{01}$ mode of circular (or semicircular) waveguides is employed in band-splitting filters. For band-splitting filters, a semicircular type, a figure-8 type (Ohtomo et al., 1975), a Michelson-inter-ferometer type (Harkless et al., 1977), and a center-excited band-branching type (Nicholson and Watson, 1976) have been proposed. The first three types are described here. For channel-dropping filters, which extract one rf channel signal, a ring-resonator type, a commutating hybrid type, a band-pass filter plus circulator type, and a semicircular waveguide plus rectangu-lar cavity type have been proposed. The ring-resonator type is described here. For the other types, see the digest of the International Conference on Millimetric Waveguide Systems, held in London in 1976.

Recently, filters using quasi-optical techniques have been reported. Al-though comparatively larger than the wavelength type, this type of filter is expected to have lower loss at higher frequencies.

### 1. Band-Splitting Filters

a. *Semicircular Type.* The semicircular waveguide hybrid is made of two semicircular waveguides coupled through multiholes on a thin common flat wall. The structure is shown in Fig. 13a (Ohtomo et al., 1975). When a signal with a frequency band between $f_L$ and $f_H$ enters the upper waveguide

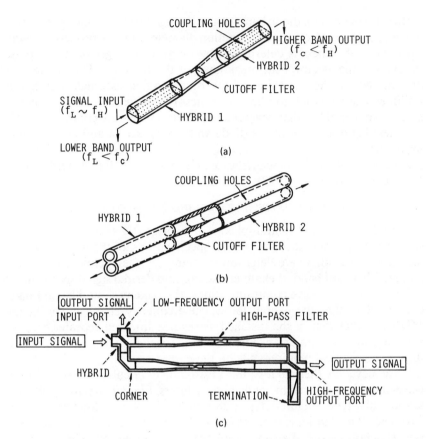

FIG. 13 Structures of band-splitting filters: (a) semicircular type, (b) figure-8 type, (c) structures of band-splitting filters. [(a), (b) From Ohtomo *et al.,* 1975. (c) From Harkless *et al.,* 1977. © 1977, AT&T. Reprinted from *The Bell System Technical Journal* by permission.]

of hybrid 1, it is divided between the two output ports of hybrid 1. The lower band between $f_L$ and $f_c$ is reflected by the cutoff filters with a cutoff frequency $f_c$ and emerges from the lower waveguide after again passing through hybrid 1. The higher band between $f_c$ and $f_H$ gets through the cutoff filters and emerges from the lower waveguide of hybrid 2.

To obtain a broad bandwidth, the diameter and distribution of coupling holes are optimally designed. A relative bandwidth (the ratio of overall bandwidth to cutoff frequency) of 70% has been obtained. For connecting it to the circular waveguide line or to the figure-8-type filters (see the next section), circular–semicircular transducers are necessary. A frequency bandwidth $\sim 43-87$ GHz is divided into two bands ($\sim 43-65$ and

~ 65 – 87 GHz). The insertion loss is 0.8 – 1.45 dB for the lower band (~ 43 – 65 GHz) and ~ 1.15 – 1.55 dB for the higher band (~ 65 – 87 GHz), including the loss of the mode transducers. The minimum loss of 0.8 dB is composed of 0.25 dB for the hybrids, 0.1 dB for the cutoff filters, 0.1 dB for additional waveguides (including semicircular helix waveguides), and 0.35 dB for the mode filters.

The cutoff frequency deviation from $f_c$ (= 65.025 GHz) is −45 MHz. The guard band (defined as the difference between frequencies where the transmission loss of the higher and lower bands is 20 dB) is 280 MHz. A narrow guard band has been achieved by using cubed cosine cutoff filters. The voltage standing-wave ratio (VSWR) is less than 1.28 for the input port and lower-band output port and less than 1.12 for the higher-band output port.

b. *Figure-8 Type.* The figure-8-type filter (Ohtomo *et al.*, 1975) (named after its cross-sectional shape) consists of two parallel $TE_{01}$-mode circular waveguides, coupled by an array of small holes (Fig. 13b). Its relative bandwidth is about 40%, narrower than that of the semicircular filter, because the coupling is achieved only by the axial magnetic field. The figure-8-type filter has three merits: low loss, ease of fabrication with high accuracy, and sharp cutoff response. It is best used for dividing – combining relatively narrow subbands. The bandwidth is determined by the diameter and spacing of the coupling holes. The total length (i.e., the number of holes) determines the cutoff response. Figure 14 shows the measured frequency

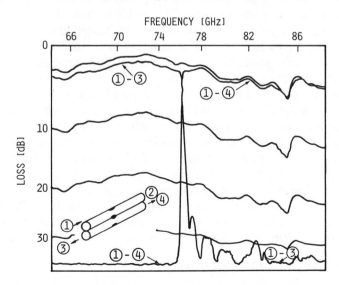

FIG. 14 Frequency response of a figure-8-type band-splitting filter. (From Ohtomo *et al.*, 1975).

response used for handling a frequency band of 65 to 87 GHz. The loss of the filter is $\sim 0.7-0.8$ dB. The measured cutoff frequency, 76.083 GHz, coincides well with the design value of 76.060 GHz. The guard band is 300 MHz.

One can also utilize two rectangular waveguides and the $TE_{10}$ mode. Because the rectangular $TE_{10}$ mode is the fundamental mode, the length can be reduced to one-third, and hence the loss is comparable to that of the figure-8-type. However, in the rectangular type, the attenuation constant is relatively large, so that the cutoff response is less sharp.

c.   *Michelson-Interferometer Type.*   Figure 13c shows the Michelson-interferometer-type filter (Harkless *et al.*, 1977), which consists of two hybrids, two high-pass filters, four tapers, and two 90° corners. The signals entering the input port are split equally between two output ports without coupling to the fourth port. The high-frequency component gets through the high-pass filter, combined in phase with the second hybrid, and is taken through the high-frequency output port. In the taper and high-pass filter, a sufficiently low level of conversion to the $TE_{02}$ mode is required (less than $-40$ dB). For the contour of the high-pass filter, the function raised to the cosine $n$th power is usually used. The contour is realized by numerical control machining.

### 2.   *Channel-Dropping Filters*

The structure of a two-cavity ring-type filter is shown in Fig. 15a. Two ring-shaped traveling-wave cavities are coupled to each other by a multihole coupler and to upper and lower waveguides by branch-guide couplers. The portion that forms two cavities is made of Invar to minimize resonant frequency variation with temperature change. The metal surface is plated with silver. A dielectric rod is inserted in each cavity to make fine adjustments of the resonant frequency and to cancel backward waves caused by imperfect coupler directivity. When signals are fed into port 1, the component that is resonant with these cavities gets through port 3.

Figure 15b shows typical filter characteristics. The channel-dropping loss in ports 1–3 at a resonant frequency of 84.13 GHz is 0.76 dB, and the passband loss is about 0.18 dB. These values include the loss of the attached waveguide. The measured loss in decibels is $\sim 1.8-2.1$ times larger than the theoretical value calculated from the dc conductivity of silver. The number of cavities and the loaded $Q$ are determined by the rf channel allocation, the waveform requirements, and the cavities and allowable thermal loss.

### 3.   *Quasi-Optical Branching Filters*

Branching filters using quasi-optical techniques have recently been developed. The ring-type branching filter is composed of two ring resonators, two

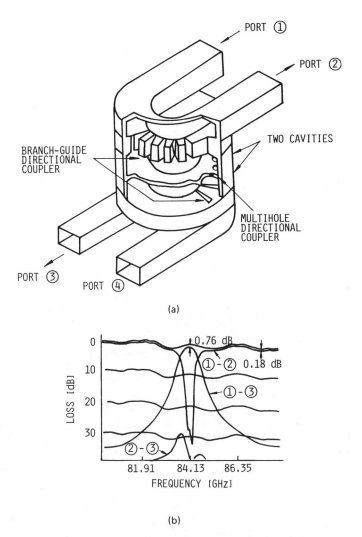

(a)

(b)

FIG. 15 (a) Structure and (b) frequency response of a ring-resonator-type channel-dropping filter. The 3-dB bandwidth is 820 MHz. [(a) From Suzuki *et al.*, 1972. (b) From Ohtomo *et al.*, 1975.]

mica-plate beam splitters, and a wire-grid beam splitter (Fig. 16a) Nakajima, 1978. The incident beam couples with a resonator through the mica-plate beam splitter. The reflecting surface of the mirror represents a section of the equatorial zone of a prolate rotational ellipsoid so that the ring resonator will show no astigmatism. The coupling of the two resonators is achieved by the metal wire grid. Excitation of the Gaussian beam is achieved by a conical

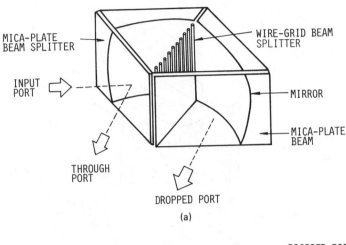

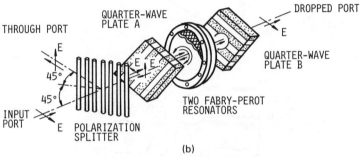

FIG. 16 Structures of quasi-optical branching filters: (a) ring type, (b) Fabry–Perot type. [(a) From Nakajima, 1978. (b) From Watanabe and Nakajima, 1978. © 1978, IEE.]

dielectric horn, which has a hemispherical end to match the radiated mode with the resonant mode of the resonator. The measured channel-dropping loss and through-channel loss are 0.5 and 0.2 dB, respectively, at 107 GHz, excluding the launching loss (0.7 dB per one beam mode exciter). The size of the filter is $20 \times 27 \times 46$ mm$^2$.

The Fabry–Perot-type branching filter consists of two parallel-plane Fabry–Perot resonators, one polarization splitter, and two quarter-wave plates (Fig. 16b) (Watanabe and Nakajoma, 1979). The polarization splitter is composed of wire grids. The quarter-wave plates are made of artificial anisotropic dielectric materials, which consist of alternate sheets of Teflon and formed polystyrene. The Fabry–Perot resonators consist of one plane mesh with square cells for the inner partially transparent mirror and two dielectric plates for the outer partially transparent mirrors.

A linearly polarized incident Gaussian beam passes through the polarization splitter and is converted into a circularly polarized Gaussian beam by quarter-wave plate A. A resonant circularly polarized Gaussian beam passes through the Fabry–Perot resonator and is reconverted into a linearly polarized Gaussian beam by another quarter-wave plate B. The nonresonant circularly polarized Gaussian beam is reflected by the Fabry–Perot resonator and reconverted into a linearly polarized Gaussian beam by quarter-wave plate A. Its polarization direction is perpendicular to that of the incident beam polarization, and therefore the nonresonant beam is reflected by the polarization splitter.

Measured channel-dropping loss is about 0.5 dB at 99.25 GHz. Minimum through-channel loss is about 0.3 dB.

## C.  Filters and Nonreciprocal Circuits

### 1.  *Filters*

Filters are used in repeaters for eliminating spurious interferences and local-oscillator noise. As in the microwave region, a waveguide type of filter

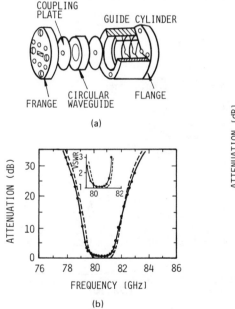

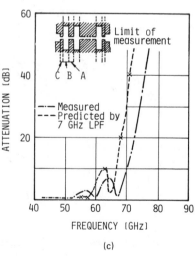

Fig. 17  Structures and frequency responses of waveguide filters: (a) structure of bandpass filter using TE$_{111}$-mode cavities, (b) frequency response of a four-section bandpass filter, (——, measured; ---, calculated), (c) internal structure and frequency response of a waffle-iron low-pass filter. [(a), (b) From Nakagami and Takenaka, 1975. © 1975 IEEE. (c) From Kitazume and Ishihara, 1975. © 1975 IEEE.]

is commonly used, and the basic design principle and fundamental structure are almost the same. However, thermal loss increases with the decrease in skin depth and wavelength. To achieve high machining precision, photolithographic etching, electroforming, and hobbing have been introduced.

A direct-coupled bandpass filter and waffle-iron low-pass filter are shown in Fig. 17 (Kitazume and Ishihara, 1975; Nakagami and Takenaka, 1975). In the bandpass filter, inductive windows are photolithographically etched (the film thickness is 50 $\mu$m) and waveguide spacers and inductive windows are mechanically pressed and adjoined. The waffle-iron low-pass filter was made by electroforming. Both frequency responses nearly coincide with the theoretically predicted values. To minimize the roughness of the waveguide internal surface and reduce thermal loss, hobbing techniques have been successfully introduced (Ishii and Oki, 1975; Mills *et al.*, 1976).

## 2. *Circulators and Isolators*

Nonreciprocal circuits are another essential element for eliminating reflections among components in repeaters. For circulators, Y-junction circulators are commonly preferred for their size and handling ease. The struc-

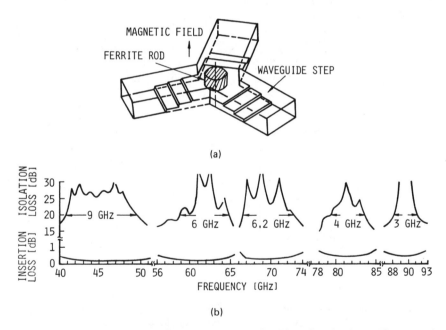

FIG. 18   (a) Structure and (b) frequency response of a Y-junction circulator. The external magnetic field is 5000–6000 Oe. The VSWR inside the band is less than 1.2. [(b) From Nakagami *et al.*, 1974.]

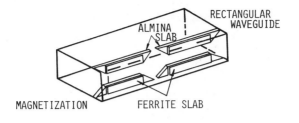

(a)

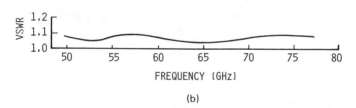

(b)

FIG. 19 (a) Structure and (b) frequency response of a self-resonance isolator. Curves A–F show the resonance of one ferrite slab. VSWR: Voltage standing-wave ratio. [(a) From Ohara *et al.*, 1976. (b) From Suzuki and Ohara, 1975.]

ture and frequency response are shown in Fig. 18 (Nakagami *et al.*, 1974). Three waveguide ports are joined in a Y configuration on the same plane. A ferrite rod is placed at the joint, and the magnetic field is applied perpendicularly to the plane. To obtain good matching and a broader bandwidth, an impedance matching circuit (inserting dielectric and/or waveguide steps) is placed around the ferrite. The fractional 20-dB bandwidth is typically 12% at 50 GHz and 4% at 90 GHz.

One can construct an isolator by terminating one port of a Y-junction circulator with a matched load. Because the bandwidth of one repeater rf channel is less than 1 GHz in most cases, most isolators used in repeaters are of this type. But there is another type of isolator that provides a very broad bandwidth: a self-resonance type (Ohara *et al.*, 1976; Suzuki and Ohara,

1975), which has been developed for use in broadband measuring equipment.

The structure and frequency responses of the self-resonance isolator are shown in Fig. 19. Ferrite slabs, each of which has been magnetized and has a different resonance frequency, are placed on the $H$ plane of the rectangular waveguide where the circular polarization exists. Alumina slabs are used for impedance matching. Even when the resonance bandwidth of one ferrite slab is narrow, one can obtain a broader bandwidth by cascading ferrites with different resonance frequencies as shown in Fig. 19. Although the insertion loss increases as the number of ferrites increases, the value is not significant. Figure 19b shows the frequency response of a 60-GHz-band isolator in which six ferrite slabs with different resonance frequencies are mounted. The 20-dB bandwidth covers the bandwidth of the R-620 waveguide (50–75 GHz). The insertion loss is less than 1.8 dB and the VSWR is less than 1.07.

This type of isolator has the following merits.

(1)  The structure is simple and fairly small.
(2)  The in-band VSWR is very low.
(3)  The out-band VSWR is almost the same as the in-band VSWR.
(4)  Any broad bandwidth can be realized at the expense of insertion loss.
(5)  No external magnetic field is necessary.

## III.   Guided Millimeter-Wave Communication Systems

### A.   Transmission Medium — Circular Waveguides

#### 1.   Transmission Modes and Transmission Loss

The attenuation constant of a conventional rectangular waveguide increases with frequency according to a relationship involving (frequency)$^{1/2}$. Typical measured values are 1000 dB/km at 30 GHz for the R-320 waveguide and 4000 dB/km at 90 GHz for the R-740 waveguide. To achieve a good transmission medium, the loss should be more than two orders of magnitude less than those values. The $TE_{01}$ mode of an oversized circular waveguide provides very low loss. Moreover, the theoretical attenuation constant of the $TE_{01}$ mode decreases with increasing frequency.

The electric and magnetic field patterns are shown in Fig. 20. In a family of $TE_{0n}$ circularly symmetrical modes, the electric lines of force are closed circles and have no component perpendicular to the waveguide wall. This is the most remarkable difference from other TM and TE modes. As is evident from Fig. 20, there is no surface current component of axial direction on the guide wall. Hence, the Ohmic loss of the $TE_{01}$ mode is expected to be lower

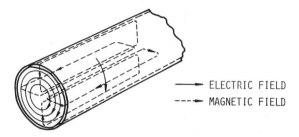

→ ELECTRIC FIELD
---→ MAGNETIC FIELD

FIG. 20   Field patterns of the $TE_{01}$ mode in a circular waveguide.

than that of other modes. The attenuation constant $\alpha$ of the $TE_{01}$ mode is given by

$$\alpha = \frac{\sqrt{2\pi f \epsilon \rho}}{D} \frac{(f_c/f)^2}{\sqrt{1 - (f_c/f)^2}}, \tag{3}$$

where $f$ and $f_c$ are the operating and cutoff frequencies, $D$ the diameter of the circular waveguide and $\rho$ the resistivity of the guide material. From Eq. (3), one knows that $\alpha$ is inversely proportional to $D^3$ and $f^{3/2}$ for $f_c/f \ll 1$; i.e., lower loss is obtained for a larger diameter and loss decreases with increasing frequency. Figure 21 shows the attenuation constants as a function of frequency for typical modes in a circular waveguide. The loss of the $TE_{0n}$ mode decreases monotonically with frequency. For other modes, there is minimum attenuation at frequencies not very far above the cutoff frequency. The $TE_{01}$ mode is apparently the most desirable from the point of view of low-loss characteristics.

Two problems arise. First, the $TE_{01}$ mode is degenerate with the $TM_{11}$ mode, and the $TM_{11}$ mode has a fairly large attenuation constant. Therefore, to achieve a low-loss transmission—i.e., pure $TE_{01}$-mode transmission—one must break the degeneracy of these two modes. Imperfect breaking results in excess thermal loss. Second, because the waveguide usually used carries several hundred transmission modes (in most cases, the waveguide diameter is selected to be approximately 10 or more times the wavelength chosen to obtain sufficiently low thermal loss), signal distortion may be caused by conversion from the $TE_{01}$ mode to other spurious modes and reconversion to the $TE_{01}$ mode. In other words, the power of spurious modes converted from the $TE_{01}$ mode should be absorbed to obtain a low-distortion waveguide line. This absorbed power is regarded as another transmission loss.

For practical circular waveguides, therefore, the transmission characteristics deviate from the theoretical value owing mainly to excess thermal loss caused by the increase in guide attenuation constant and signal distortion

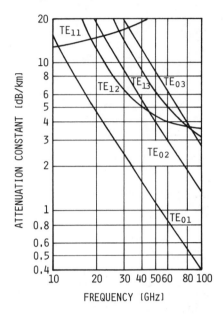

FIG. 21 Theoretical attenuation constant for the TE modes in circular waveguides. The waveguide inside diameter is 51 mm. (From Oguchi, 1968).

caused by the conversion–reconversion effect or mode conversion loss. Both of these phenomena are caused by the dimensional imperfections of waveguides and their joints. The practical achievement of low-loss transmission of the $TE_{01}$ mode is thus governed mainly by the mechanical precision of the installed circular waveguide. For a given diameter of the circular waveguide, thermal loss dominates in the lower-frequency region and mode conversion loss increases in the higher-frequency region, so that the overall loss versus frequency shows a broad bandpass characteristic.

## 2. Dielectric-Lined and Helix Waveguides

Several forms of waveguide structure have been proposed for minimizing the excess thermal loss and mode conversion–reconversion effect. Among these the dielectric-lined and helix waveguides (Miller, 1954; Unger, 1957, 1958) are of practical importance. Their fundamental structure is shown in Fig. 22. In the dielectric-lined waveguide, the inner surface of the waveguide is coated with a thin dielectric layer whose thickness is of the order of 100 $\mu$m. The $TE_{01}$-mode electric field component that is perpendicular to the wall being zero as shown in the preceding section, the electric field within the dielectric film is very weak. On the other hand, the $TM_{11}$ mode has the perpendicular component. Hence this thin dielectric layer changes the propagation constant of the $TM_{11}$ mode without any significant change in

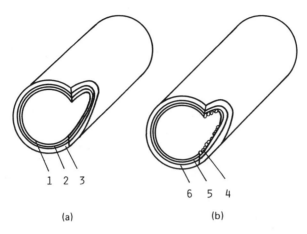

FIG. 22  Basic structure of practical circular waveguides: (a) dielectric-lined waveguide, (b) helix waveguide. 1, dielectric thin film; 2, plated copper; 3, reinforcement tube (steel); 4, thin-wire helix; 5, dielectric stuffing; 6, reinforcement tube.

that of the $TE_{01}$ mode. Thus the degeneracy between the $TE_{01}$ and $TM_{11}$ modes is broken.

In the helix waveguide, the innner surface is lined with a thin insulated copper-wire helix, the outside of which is stuffed with dielectric material. The outer surface of the guide is protected by mechanical reinforcement material. Because the $TE_{0n}$ mode has no surface current component of wave transmission direction, the helix causes no significant loss for these modes. On the other hand, other modes having a surface current component suffer large loss owing to the helix and the dielectric.

For practical communication systems, the route inevitably has curves and bends. The diameter of the waveguide is mainly determined by the operation frequency bandwidth and route condition. Waveguides with different internal diameters have been developed in different countries: 50- or 51-mm-i.d. waveguides in England, France, Italy, and Japan, a 60-mm-i.d. waveguide in the United States and the USSR, and a 70-mm-i.d. waveguide in the Federal Republic of Germany.

A 50-mm-i.d. fiberglass–epoxy-resin helix waveguide was developed in England (Moore *et al.,* 1976; Ritchie and Childs, 1976). The helix is covered by a fiberglass–epoxy resin (for structural support and absorption of spurious modes) and then by an aluminum tube (for gas and water vapor resistance). The outer surface is wound with a resin-impregnated tape. It is constructed on precision mandrels, and each layer is applied by a winding action that enables cheap and continuous production; this form of waveguide was preferred in England mainly for economy. The measured attenuation constant, for a substantially straight section 1177 m long, falls to a minimum of 1.25 dB/km around 80 GHz and rises to 2 dB/km at 100 GHz. Two loss peaks at 56 and 86 GHz are observed.

In France, 50- and 60-mm-i.d. helix waveguides were developed (Allanic, 1976; Bendayan *et al.*, 1974; Carratt, 1976). They were made by a "continuous" or "semi-continuous" process. In the continuous process, dielectric tape (ethylene terephtalate, 0.1 mm thick) with a thermal setting adhesive on one side, a half-hard aluminum tube 1.5 mm thick, and a steel tube ~ 2 – 3 mm thick are successively applied to an enameled wire (0.5 mm in diameter) helix. A damping filter is inserted between the aluminum and steel tubes. In the semicontinuous process, after polymerization of the epoxy resin the mandrel is water cooled and the waveguide is drawn, and the guide is then pushed into an 82.5-mm-i.d. steel tube. The measured attenuation constant, for an installed waveguide line 1.2 km long, is less than 2.1 dB/km at 40 – 85 GHz, with a minimum of 1.4 dB/km at 64 GHz.

Steel-jacketed dielectric-lined and helix waveguides with internal diameters of 51 mm were developed in Japan (Yamaguchi *et al.*, 1974). In the dielectric-lined waveguide, plated copper (20 $\mu$m), oxidized copper (~ 1 – 2 $\mu$m), adhesive (~ 1 – 2 $\mu$m), and polyethylene (200 $\mu$m) are successively coated on the inner surface of a 51-mm steel tube. In the helix waveguide, the inner dielectric layer consists of three layers (a medium-loss layer, a high-loss layer, and a reinforcing layer, with a total thickness of 2.5 mm). The thickness of the polyethylene lining is determined by a waveguide loss calculation taking into account a manufactured waveguide straightness of about 1000 m, a correlation distance of 0.1 m, and an installation straightness of about 300 m. The attenuation loss is about 1.5 dB/km at 50 GHz and 1.2 dB/km at 80 GHz. The helix waveguide is used for mode filtration in the installed waveguide line. The filter portion is 20% (a 5-m helix waveguide at 20-m intervals). The dimensional tolerances for manufactured waveguides are as follows: the standard deviation of the diameter is about 5 $\mu$m, the rms ellipticity is less than 20 $\mu$m, and the straightness is about 1000 m. The average $TE_{11}$- and $TE_{12}$-mode attenuation is about 3.1 – 3.4 and 5.6 – 6.6 dB/m, respectively, and the $TE_{01}$-mode attenuation is about 1.34 – 1.39 dB/km at 50 GHz.

In the United States, 60-mm-i.d. steel-jacketed waveguides were developed (Boyd *et al.*, 1977). The steel tube has an inner diameter of 60.070 mm and a thickness of 3.68 mm. The inner surface of the steel tube is plated with high-conductivity copper (5 $\mu$m thick) to which a thin polyethylene layer (180 $\mu$m thick) is bonded. This structure is determined so as to balance heat loss and mode conversion loss over the 40 – 110-GHz frequency band. The materials and manufacturing processes are chosen to give extremely low thermal loss, minimum geometric distortion, and high long-term reliability. The helix waveguide is made of an insulated copper-wire helix on the inner surface, the outside of which is stuffed with high-loss iron-loaded epoxy with a thickness of approximately 4.1 mm. The helix and epoxy are protected by a 3.8-mm-thick steel tube. Individual sections of the waveguides are jointed by fusion welding. To provide high Ohmic dissipation to spurious mode energy and minimize geometrical distortion, special care is taken with the high-loss iron-loaded epoxy and the precision temperature control during the epoxy cure cycle. The installed waveguide is a 99% dielectric-lined and 1% helix waveguide. The measured loss of the installed waveguide line with this combination is less than 1 dB/km at 40–110 GHz.

In the Federal Republic of Germany, 70-mm-i.d. aluminum-jacketed dielectric-lined waveguides were developed (Krahn, 1976; Schnitger and Unger, 1974). The tube material is an AlMgSi alloy (aluminum with a small addition to magnesium and silicon). An aluminum block is drawn into the tube and then the inner surface is oxidized. The aluminum oxide forms the dielectric lining. Although the measured $\tan \delta$ is $3 \times 10^{-2}$ and is larger than that of commonly used dielectrics (e.g., $\tan \delta \simeq 2 \times 10^{-4}$ for polyethylene), because the dielectric constant is as high as 8 and the thickness is less than $100\ \mu\text{m}$ the lining provides sufficient breaking of the degeneracy between the $TE_{01}$ and $TM_{11}$ modes and the attenuation constant is sufficiently low. The oxide layer shows a very strong resistance to peeling, which is often a problem in dielectric-lined waveguides. Electrical measurements and calculations show that the attenuation constant should be less than 1 dB/km in the frequency range $\sim 35–110$ GHz.

Work in design and development was carried out in Italy (Bernardi et al., 1976; Corazza et al., 1974) and the USSR (Suchkov et al., 1973). The 60-mm-i.d. helix waveguide developed in the USSR has an attenuation of 2.1 dB/km at 35 to 40 GHz and about 3.0 dB/km at 80 GHz.

### 3. Circular Waveguide Components

a. *Corner Waveguides.* A single-corner waveguide provides a 90° change in the direction of the waveguide line and a double-corner waveguide provides an arbitrary direction change. Corner waveguides utilize a quasi-optical mirror reflection principle (Marcatili, 1964). The measured characteristic is close to the theoretical value calculated from Marcatili's formula. The measured curve shows a fluctuation of approximately 0.1 dB p.–p. This fluctuation is estimated to be affected by a higher-order circular mode (Yamaguchi et al., 1974).

In a double-corner waveguide, the fluctuation becomes greater because conversion–reconversion is twice that of the single-corner waveguide. There is an optimum distance between the two corners of a double-corner waveguide. Based on a theoretical calculation, the distance is approximately 10 cm for a 51-mm waveguide.

b. *Small-Diameter Flexible Waveguide.* A flexible waveguide is also important for providing an arbitrary-angle bend section in the waveguide line. An optimum design for improving loss characteristics was obtained (Inada *et al.,* 1975) in which low-loss, flexible waveguides were achieved by proper selection of the inner diameter and optimum wall impedance. A small-diameter, dielectric-lined waveguide was investigated in England (South, 1976).

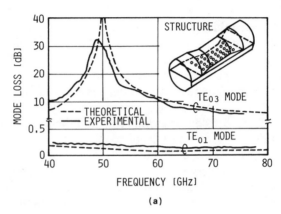

(a)

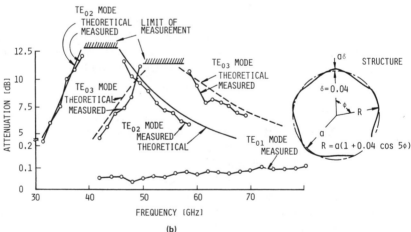

(b)

FIG. 23 Structure and performance of mode filters: (a) phase-inversion-type mode filter, (b) circular-arc–polygonal-type mode filter. P = 5 and 8, $a = 25.5$ mm, $\delta_5 = 0.02$, $\delta_8 = 0.01$, $L = 1000$ mm. [(a) From Hashimoto, 1974. (b) From Inada *et al.,* 1975. © 1975 IEEE.]

c. *Mode Filters.* When $TE_{0n}$ modes are significant in the waveguide line, mode filtration components other than a helix waveguide are necessary. The following are mode filters for filtering the $TE_{0n}$-mode family.

(1) Phase-inversion type (Hashimoto, 1976). The structure of this mode filter is shown in Fig. 23a. A circular waveguide is divided into two semicircular guides by a metallic sheet with resistive layers on both sides. One of the semicircular guides has a smaller radius, which causes a phase difference in the $TE_{0n}$ modes propagating in the two semicircular guides. When one selects a proper length of semicircular section, an opposite phase difference for the undesired $TE_{0n}$ modes and a zero difference for the signal $TE_{01}$ mode occur at the output of the filter. The undesired modes are offset and converted into unsymmetric modes (mainly $TE_{1n}$, $TE_{3n}$, and $TM_{1n}$). Experimental results for the $TE_{03}$-mode filter are shown in Fig. 23a. The insertion loss of the $TE_{01}$ mode is as small as 0.2 dB in the 40–80-GHz range.

(2) Circular-arc–polygonal type (Inada *et al.*, 1975). The structure of this filter is similar to that of the conventional circular waveguide. The cross section is slightly deformed to have a circular-arc–polygonal shape, as shown in Fig. 23b (pentagonal type). When the $TE_{0n}$ mode enters this polygonal waveguide, mode conversion occurs owing to the sectional deformation, e.g., $TE_{02}$ into $TE_{51}$ in the pentagonal case and $TE_{03}$ into $TE_{81}$ in the octagonal case. The converted modes are absorbed in the helix waveguide connected next to the mode filter. In Fig. 23b the measured and theoretical performances of the $TE_{02}$–$TE_{03}$ mode filter are shown. This structure gives no significant $TE_{01}$-mode transmission loss.

## B. W-40G SYSTEM

### 1. *System Outline*

The W-40G system was developed in the Electrical Communication Laboratories, NTT, Japan (Miyauchi, 1975; Miyauchi *et al.*, 1975a). The hypothetical reference circuit of the W-40G consists of nine supervisory-switching sections, each of which includes a pair of protection switches, transmitting and receiving code converters, a chain of transmitters and receivers, multiplexing and demultiplexing networks, and a waveguide line (Fig. 24). The protection switches and code converters are installed at each terminal station, and the transmitters and receivers and multiplexing and demultiplexing networks are installed at each repeater station. The repeater station is not manned on principle. The main features of the system are shown in Table IV.

Regenerative repeaters transmit an 800-Mbit/sec digital signal and maintain an error rate of less than $10^{-7}$ for the 2500-km circuit. The total transmission capacity of 300,000 telephone channels (both ways) is consid-

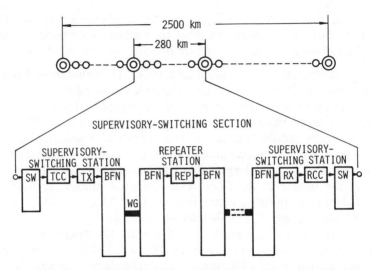

FIG. 24  Hypothetical reference circuit of the W-40G system. SW, protection switch; TCC, transmitting code converter, TX, transmitter; BFN, multiplexing and demultiplexing network; WG, waveguide line; REP, regenerative repeater, RX, receiver, RCC, receiving code converter. (From Miyauchi *et al.*, 1975a.)

ered appropriate for the first large-capacity, guided millimeter-wave transmission system because this capacity is about three times that of the existing 60-MHz analog coaxial cable system installed in an 18-tube coaxial cable.

Curved sections and corners along the waveguide route cause a decrease

TABLE IV

W-40G SYSTEM FEATURES

| Feature | Dimension |
|---------|-----------|
| Number of millimeter-wave channels | 28 (both E–W and W–E) |
| Overall transmission band | 43–87 GHz |
| Transmission capacity | ~ 21 Gbit/sec, two way; 800 Mbit/sec per millimeter-wave channel |
| Transmission line | |
|   Standard waveguide | Hybrid–tandem (four dielectric and one helix), 51 mm i.d. |
|   Conduit | 150-mm-bore steel pipe |
| Transmitter | Up converter |
| Intermediate frequency | 1.7 GHz |
| Modulation and demodulation | Four-phase PSK, coherent detection |
| Error rate | $< 10^{-7}$ per 2500 km |
| Average repeater spacing | 15 km |

in both the available frequency bandwidth and the repeater spacing. The system parameters have been chosen to match the Japanese terrain. The frequency bandwidth is minimized by employing four-phase phase shift keying (PSK), an appropriate frequency allocation design, and an optimum waveform transmission (the approximate zero-crossing waveform equalization). As a result the necessary frequency bandwidth for the transmission capacity of 30,000 telephone channels is about 43 GHz. The average repeater spacing is 15 km when the waveguide route has a radius of curvature similar to that of Japan's existing coaxial cable routes.

A frequency band from 43.40 to 86.75 GHz is divided into transmitting and receiving bands, each of which is further divided into four frequency blocks. One frequency block consists of seven millimeter-wave channels. Guard bands with appropriate bandwidths are arranged between each frequency block. The frequency separation between adjacent channels is 740 MHz, which corresponds to 1.85 times the clock rate.

The system consists of 28 millimeter-wave channels for both the east-to-west and west-to-east directions. Two of the 28 are standby channels to ensure system reliability. Supervisory and control channels are allocated between the transmitting and receiving bands.

Low-loss and compact branching filters are employed in the multiplexing and demultiplexing network, e.g., a semicircular waveguide filter for separating the transmitting and receiving bands, figure-8 or rectangular waveguide filters for dividing frequency blocks, and ring filters for dropping individual channels. These branching filters are installed at the top of repeater bays and are connected to the waveguide transmission line and the repeaters by a tapered waveguide and rectangular waveguide links, respectively.

### 2. *Repeaters*

The repeater has five parts (panels): a transmitter–receiver panel, a delay-equalizer panel, a modulator–demodulator panel, an alarm-control panel, and a power-supply panel (Seki *et al.*, 1975). The power consumption is about 100 W per repeater. The repeater station is built on the ground. A block diagram of a repeater is shown in Fig. 25.

The transmitter–receiver panel has the function of transforming between rf and IF (1.7 GHz) and has two frequency converters (an upconverter and a downconverter), a common local oscillator, rf filters for suppressing spurious interference, and IF amplifiers. Both frequency converters employ GaAs Schottky-barrier diodes. The transmitting power and noise figure are typically 12.0 dBm and 10.1 dB at 50 GHz and 8.4 dBm and 11.4 dB at 87 GHz. To extend the repeater station spacing, a negative-resistance IMPATT amplifier is connected after the upconverter. Recently an output power of

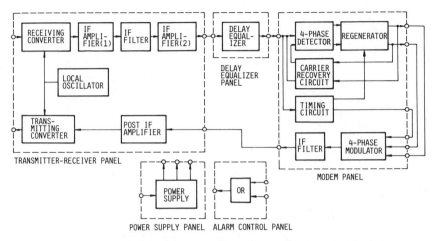

FIG. 25   W-40G system repeater block diagram. (From Miyauchi *et al.,* 1975a.)

23 dBm at 87 GHz was obtained. The frequency of the local oscillator is stabilized by means of a waveguide cavity. A frequency variation of less than $\pm 7 \times 10^{-5}$ for an ambient temperature of $0-40°C$ was specified from the obtainable pull-in range of the carrier recovery circuit of the modulator–demodulator panel and the capability of the waveguide cavity attached to the IMPATT local oscillator.

The delay-equalizer panel consists of waveform-shaping filters and delay equalizers in the IF band. Thomson-type filters realize approximate zero-crossing waveforms, and delay equalizers, which are made of wide-band tape meander line circuits embedded in Teflon, compensate for the linear delay of the waveguide transmission line. The linear delay is caused by the dispersion effect. Equalization is achieved to an accuracy of $\pm 0.8$ nsec/GHz by replacing mop-up delay equalizers.

The modulator–demodulator panel provides IF carrier recovery, coherent detection, timing-wave recovery, decision, pulse reshaping, and quarternary phase-shift keying of an IF carrier. Quarternary phase $\pi/2$-shift keying is performed by making use of a pair of miniaturized ring modulators (silicon Schottky-barrier diode) in the IF band at a pulse rate of 800 Mbit/sec. An amplitude variation of less than 0.2 dB and a phase error of less than $\pm 1°$ (for an ambient temperature of $\sim 0$ to 40°C) have been obtained. A timing-wave recovery circuit consists of a phase-locked oscillator and has an equivalent $Q$ as high as $10^4$. By introducing reversible countersweep operation into the carrier recovery circuit, a pull-in range as wide as $\pm 15$ MHz has been obtained.

## 3. Field Trial

The field trial waveguide transmission line was constructed between terminals in the Ibaraki Electrical Communication Laboratory and the Mito Telegraph and Telephone Office with a 22.7-km span, as shown in the route map in Fig. 26 (Miyauchi *et al.*, 1975a). The Mukoyama repeater station is located 15.3 km from the Mito terminal station. A directly buried 4.2-km-long waveguide line was constructed in 1966. An 18.5-km-long waveguide line was newly constructed for the field trial. The waveguides were pulled into a 150-mm-bore, welded-steel conduit by attaching plastic roller rings to the waveguides. In Mito City, 40-mm-bore waveguides were laid for 600 m in an existing 75-mm-bore steel conduit line.

Waveguides 5 m long with a 51-mm bore are jointed to each other by specially designed connecting sleeves. Dielectric-lined and helix waveguides

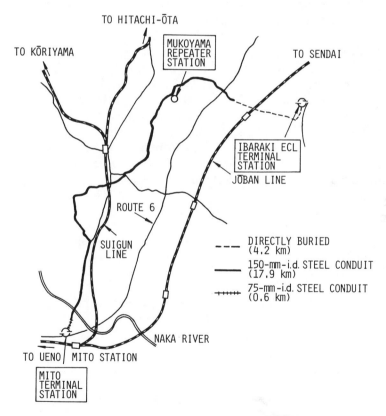

FIG. 26   Route map of the W-40G system field trial in Ibaraki Prefecture. (From Miyauchi *et al.*, 1975a.)

are laid in tandem in the ratio of 4 to 1 to eliminate undesired mode propagation. The minimum bending radius of the waveguide is as small as 35 m. At a sharper bend in the route, a single- or double-corner waveguide or a flexible helix, 20-mm-bore waveguide was installed in a manhole and a cable tunnel.

The undergound waveguide expands with temperature at a rate of 1.15 cm/km span. This thermal expansion was compensated for by expandable waveguide sections located in manholes. The waveguide was filled with nitrogen to avoid harmful absorption of the millimeter wave by oxygen molecules. Nitrogen-gas dams, which can bear pressures of up to 17 kg/cm², and which exhibit low reflection characteristics for millimeter waves, are installed at the entrance of each station. The welded-steel conduit is filled with dry air—the pressure is monitored to detect trouble in the transmission line.

The field trial link is equipped with 6-mm-wave signal channels, one each in the 50- and 80-GHz bands, and four in the 65-GHz band. Supervisory and control channels are also provided in the 65-GHz band.

The average attenuation constant of the waveguide line pulled into the 150-mm-i.d., welded-steel conduit is 3.1 dB/km at 43 GHz, 2.3 dB/km at 60 to 75 GHz, and 2.5 dB/km at 87 GHz. The standard deviation of the attenuation fluctuation is 0.35 dB in the 50-GHz band and 0.20 dB in the 70-GHz band. Figure 27 shows the attenuation characteristics of the experimental waveguide line. Figure 28 shows an example of the measured error rate performance. The C/N required to obtain an error rate of $10^{-9}$ is approximately 17 dB. Stable operation was confirmed for the results of continuously running tests in a chain of 10 repeaters.

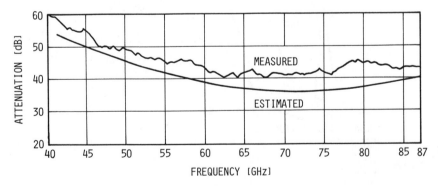

FIG. 27  Loss vs frequency of waveguide line installed in a 150-mm-i.d. steel conduit. (From Kuribayashi *et al.*, 1974.)

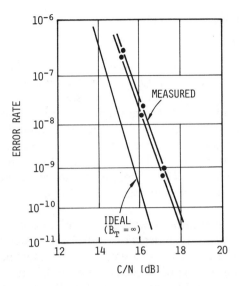

FIG. 28 Measured error rate vs carrier-to-noise ratio (C/N) of the W-40G system. (From Miyauchi *et al.*, 1975a.)

An intracity waveguide system (WT-40G) has also been developed (Shimada and Ishihara, 1978). The system is intended to bring the whole transmission capacity of the W-40G into the heart of large cities; the waveguide is installed in existing underground cable tunnels. In this case, because the waveguide line inevitably has a considerable number of curved sections and sharp bends and the frequency response of the installed line has many ripples and amplitude variations, low-loss flexible waveguides, mode filters, and amplitude equalizers are required.

To realize the low-loss characteristics of the flexible waveguides, the internal diameter and wall impedance are optimized by simulating the practical installation condition in cable tunnels. The measured loss of a 1.8-m-long, 14-mm-i.d. waveguide with one 90° arc was less than 0.6 dB over the frequency band of 40 to 90 GHz.

Two types of mode filters have been designed to absorb the $TE_{0n}$ spurious modes arising from curves and bends: a phase-inversion type and a circular-arc–polygonal type (see Section III.A.3.c). Through the field trial line installed in a cable tunnel in the Tokyo metropolitan area, sufficient mode suppression was confirmed. An amplitude equalizer was used to compensate for the loss peak that typically appears in the lower-frequency region of 40–60 GHz.

## C.  WT4 System

### 1.  *System Outline*

The WT4–WT4A system was developed by Bell Laboratories in the United States (Alsberg *et al.*, 1977). The regenerative repeater carries a digital signal stream of 274 Mbit/sec. The system error rate is less than $10^{-7}$ for a 6000-km transmission distance, and the service availability is better than 0.9998. Two-level, differentially coded PSK modulation (WT4) provides 238,000 two-way voice channels, and an upgraded version of four-level PSK modulation (WT4A) offers up to a total capacity of 476,000 two-way voice channels. This increase in capacity is achieved only by replacing some circuits in the repeaters without changing the repeater spacing.

The system uses a 60-mm-i.d. waveguide, and the operation bandwidth is 38–104.5 GHz,† which includes a total of 124 millimeter-wave channels, one-half of which carry east-to-west signals and the other half of which carry west-to-east signals. Of the 62 repeaters in each direction, 59 are used for signal transmission and 3 for protection. A band diplexer setup first splits the overall band into seven subbands. Each subband is then split into rf channels by a chain of tandem-connected channel diplexers. The 3-dB bandwidth of the channel-dropping filters is 475 MHz and the channel spacing is 500 MHz. The repeater spacing for the 60-mm waveguide is a maximum of approximately 60 km in relatively gentle or moderate terrain (typical of routes on the East Coast south of Washington, D.C.) and is reduced to 50 km in rugged terrain (such as the mountain area of Pennsylvania).

The band diplexers, high-pass–low-pass filters, and low-pass filters are made of a circular waveguide and operate in the $TE_{01}$ mode. The channel diplexer is made of a semicircular waveguide with two aperture-coupled rectangular cavities. The semicircular waveguide is made large enough to propagate the $TE_{01}$ mode but not large enough to propagate the $TE_{02}$ mode. The rectangular cavities operate in the dominant mode.

The multiplexing and demultiplexing network configuration and the waveguide parameters are selected so that the total loss between two repeaters matches the available power gain in the overall frequency band of 38–104.5 GHz. Figure 29 shows the overall loss budget for the waveguide loss, the multiplexer and demultiplexer network loss, and the excess margin for the WT4A system. Two cases are shown: for the waveguide installed in moderate routes (repeater spacing 60 km) and in rugged routes (repeater

---

† Based on the field evaluation test conducted between 1974 and 1976, the operation frequency bandwidth was shifted slightly downward from 40–110 to 38–104.5 GHz.

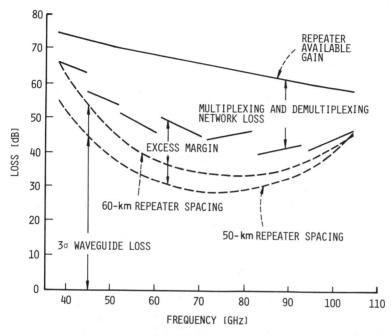

FIG. 29  WT4A system loss budget. (From Alsberg *et al.*, 1977. © 1969, AT&T. Reprinted from the Bell System Technical Journal by permission.)

TABLE V

WT4–WT4A SYSTEM FEATURES

| Feature | Dimension |
|---|---|
| Number of millimeter-wave channels | 62 (both E–W and W–E) |
| Overall transmission band | ~ 38–104.5 GHz |
| Transmission capacity of WT4 (of WT4A) | 16 Gbit/sec, two way (32); 274 Mbit/sec per millimeter-wave channel (548) |
| Transmission line | 99% dielectric lined, 1% helix |
| Conduit | 140 mm o.d. |
| Transmitter | Millimeter-wave modulation |
| Intermediate frequency | 1.37 GHz |
| Modulation and demodulation | Two-phase (four-phase for WT4A) PSK, differentially coherent detection |
| Maximum repeater spacing | 50 km (rugged route), 60 km (moderate route) |

spacing 50 km). The main parameters of the WT4–WT4A system are shown in Table V.

The 60-mm-i.d. circular waveguide, supported by spring rollers, is installed in a 140-mm-o.d. steel sheath, which is buried a minimum of 0.6 m below the ground surface. Individual sections of the waveguide and steel sheath are jointed by fusion welding, which provides good resistance to various kinds of mechanical damage. The installed waveguide is a 99% dielectric-lined and 1% helix waveguide. The helix waveguide is for spurious mode filtration. A waveguide loss of less than 1 dB/km for the overall frequency band of 40 to 120 GHz has been obtained, owing mainly to fully automated manufacture and computerized imperfection measurement, waveguide installation using spring roller supports and fusion jointing of the waveguide and steel sheath and minimization of the second-order mode conversion caused by the presence of tight horizontal and vertical bends and long mechanical periodicities in the installed waveguide (this was discovered in the course of the field evaluation test).

## 2. Repeaters

In the WT4–WT4A system, the repeater has four parts: the transmitter, receiver, line equalizer, and power supply (Barnes *et al.*, 1977). A repeater block diagram for two-level operation (WT4) is shown in Fig. 30.

In the transmitter, a cw millimeter-wave carrier generated from a DDR

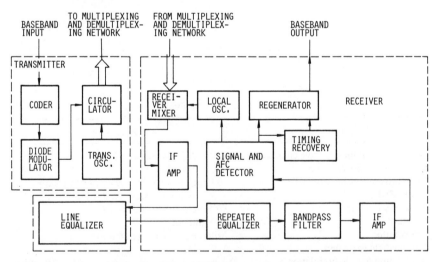

FIG. 30   WT4 system repeater block diagram. Trans. Osc., transmitter oscillator; Local Osc., local oscillator. (From Alsberg *et al.*, 1977. © 1977 AT&T. Reprinted from *The Bell System Technical Journal* by permission.)

IMPATT oscillator is PSK modulated (phase difference of 180°) by a $p-i-n$ diode modulator. This phase-modulated signal is sent via the multiplexing network and waveguide line to the receiver at the next station. The cavity loss of the IMPATT oscillator is less than 1 dB, which corresponds to a frequency drift of less than ±25 MHz. The $p-i-n$ phase modulator basically has the structure described in Section II.A.4. The insertion loss ranges from approximately 1.5 dB at 40 GHz to 3 dB at 110 GHz. The amplitude difference for two-phase stages is less than 0.25 dB. The power-handling capability is greater than 250 mW.

In the receiver, the millimeter-wave signal is converted by a double-balanced, beam-lead GaAs Schottky-barrier mixer to an intermediate frequency of 1.371 GHz. The IF signal is then amplified and sent to the line equalizer. The line equalizer compensates for delay distortion and undesirable loss frequency shaping arising in the waveguide line between the output of the transmitter and the input of the next receiver. The distortion arising in the electronic circuitry in the repeater is compensated for by a repeater equalizer. The local oscillator of the mixer is frequency controlled. A frequency shift of ±100 MHz or more is obtained by changing the diode (IMPATT) bias current.

All parts of the repeater (except the power supply) that carry rf and IF signals have aluminum housings and are pressurized with dry nitrogen to about 180 mm $H_2O$ to obtain complete isolation from the electromagnetic environment and improve component reliability. The repeater is water cooled at about 13°C by circulating water.

### 3. Field Evaluation Test

The field evaluation test was conducted in northern New Jersey between 1974 and 1976 (Cheng et al., 1977). A 14-km waveguide line was installed between the AT&T Metropolitan Junction Station in Netcong and a temporary experimental station in Long Valley (Fig. 31). The 14-km line includes various construction situations: route bends, road and stream crossings, steep grades, rocky terrain, and swamp areas.

A complete band diplexer setup, 12 repeaters, and 12-channel-dropping filters were installed. A total of 12 repeaters — 3 in the 40-GHz band, 3 in the 55-GHz band, 3 in the 80-GHz band, and 3 in the 110-GHz band — were made. The length of the field evaluation waveguide line being much shorter than the standard repeater spacing, pads were required to compensate for the excess margin of repeaters.

The principal items being measured during the test were loss and delay of the waveguide line and multiplexing and demultiplexing networks, repeater available gain, error rate characteristics, leakage and interference, baseline wander, and timing-recovery phase jitter. The tests were satisfactorily per-

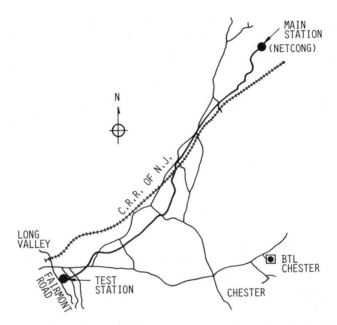

Fɪɢ. 31  WT4 system field evaluation test route. BTL: Bell Telephone Laboratory. (From Waters, 1976. © 1976 IEE.)

formed. They showed that the repeaters operated in the field essentially the same as in the laboratory. Figure 32 shows the measured loss-versus-frequency characteristics of the installed waveguide line. The loss is less than 1 dB/km over the entire frequency band of 40 to 110 GHz. Measurements made periodically for two years showed no changes in the accuracy of measurement.

The system error rate objective is $10^{-9}$ per repeater hop. The measured relative carrier power (RCP)† (or C/N ratio) for an error rate of $10^{-9}$ is shown in Table VI.

## D.  OTHER SYSTEMS

Guided millimeter-wave communication systems were also researched and developed in England, France, Italy, the USSR, and the Federal Republic of Germany. The main features of the systems are summarized in Table VII.

In England, a 14.2-km field trial route was installed between the Post

---

† The relative carrier power is defined as the ratio of the carrier power to the actual noise level in the signaling rate bandwidth.

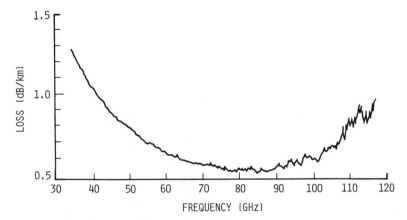

FIG. 32 Measured loss vs frequency of the WT4 system field evaluation test route (14 km). (From Alsberg *et al.*, 1977. © 1977, AT&T. Reprinted from *The Bell System Technical Journal* by permission.)

TABLE VI

WT4 RELATIVE CARRIER POWER (RCP) MEASUREMENT

| Frequency (dB) | Relative carrier power | | |
|---|---|---|---|
| | Back-to-back[a] (dB) | Line equalized[b] (dB) | ΔRCP[c] (dB) |
| 40.235 | 15.8 | 17.4 | 1.6 |
| 40.760 | 14.7 | 16.8 | 2.1 |
| 41.285 | 15.1 | 16.1 | 1.0 |
| 53.910 | 15.6 | 16.4 | 0.8 |
| 55.085 | 15.4 | 18.3 | 2.9 |
| 55.610 | 14.7 | 18.2 | 3.5 |
| 80.465 | 16.0 | 17.9 | 1.9 |
| 80.990 | 14.6 | 16.9 | 2.3 |
| 81.515 | 15.7 | 17.1 | 1.4 |
| 107.665 | 14.6 | 17.1 | 2.5 |
| 108.190 | 15.8 | 18.5 | 2.7 |
| 108.715 | 15.3 | 16.9 | 1.6 |

[a] Back-to-back test in the laboratory.
[b] Over the line with equalization.
[c] Increase in RCP above the back-to-back laboratory test.

TABLE VII

System Parameters of Guided Millimeter-Wave Transmission Systems

| Parameter | England | France | Federal Republic of Germany | Italy | USSR |
|---|---|---|---|---|---|
| | | | Dimension | | |
| Frequency band (GHz) | 30–110 | 30–60 | 35–115 | 40–90 | 25–40 |
| Signal format | 500 Mbit/sec, two-phase PSK, four-phase PSK (upconversion or direct modulation) | 580 Mbit/sec, four-phase PSK (upconversion) | 264 Mbit/sec, two-phase PSK (upconversion or direct modulation) | 800 Mbit/sec, four-phase PSK (upconversion) | |
| Transmission capacity (telephone channels) | 491,250[a] | 150,000 | | | |
| Waveguide inside diameter (mm⌀) | 50 | 50 | 70 | 50 | 60 |
| Waveguide type | Helix | Helix | Lined, helix | Lined, helix | Helix |
| Installation | 105 mm i.d., steel duct | Directly buried (iron cover) | 126 mm i.d., PVC duct | 114.3 mm i.d., steel duct | |
| Field trial (distance) | 1973, duct laid; 1975, trial on 14.2 km | 1973, trial on 15 km | 1968, 3 km × 3 | 1-km loop | |

[a] For future upgraded versions, the system has 64 two-way carriers each with 560 Mbit/sec of digital traffic.

Office Research Center at Martlesham Heath and Wickham Market. The field trial yielded the following conclusions (May, 1976).

(1) The mechanical stability of the duct is sufficient.

(2) The overall loss of the installed waveguide is less than 2.5 dB/km at 30 to 110 GHz. The repeater spacing is ~ 20–30 km.

(3) An error rate of $10^{-9}$ is obtained when the C/N at the demodulator input is 22 dB.

(4) Satisfactory transmission performance is achieved when the signal is relayed successively via different rf channels through the waveguide line. The waveguide system is the most promising system for meeting the future increased demand in digital communications.

In France, a 15-km-long waveguide line was installed near Lannion (Baptiste and Herlent, 1976; Dupuis and Joindot, 1974). A $10^{-8}$ error rate was obtained for a C/N of 20 dB. In the French project, a total error rate of $10^{-8}$ was assigned for a 1000-km link, and the error rate of one repeater spacing was $10^{-10}$. The repeater spacing was 20 or 30 km with the use of a 50- or 60-mm-i.d. waveguide.

Guided millimeter-wave communication systems were also researched and developed in Italy, the USSR, and the Federal Republic of Germany. The reader may refer to the digests of two international conferences: Conference on Trunk Telecommunications by Guided Waves, London, 1970; International Conference on Millimetric Waveguide Systems, London, 1976.

## IV. Millimeter-Wave Terrestrial Radio Communication Systems

### A. Fundamental Characteristics of Millimeter-Wave Terrestrial Radio Communication Systems

The radio transmission medium is "open" in the sense that it suffers from the effects of atmospheric environment and frequency interference between radio systems operating in the same space and frequency. In designing radio systems through this medium, the effects of a radio duct in the microwave band and precipitation in the millimeter-wave region should be considered. Figure 33 shows the relation between propagation characteristics and the principal deteriorations in system design. In the millimeter-wave region, the propagation characteristics are represented by frequency-nonselective attenuation and cross-polarization discrimination (XPD) degradation. The former is related mainly to thermal noise and interference. The latter affects the interference between polarizations. Attenuation caused by absorption

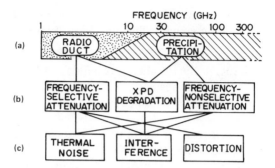

FIG. 33 Relation between (a) atmospheric phenomena, (b) propagation characteristics, and (c) main deteriorations in system design. The millimeter region ranges from 30 to 300 GHz.

and diffraction due to rainfall and by absorption due to water vapor and oxygen molecules is shown in Fig. 34 (CCIR, 1978).

Technologies of millimeter-wave equipment are rapidly progressing in many countries, because intensive research and development activities have been carried out on guided millimeter-wave transmission systems using frequency bands above 40 GHz. Radio systems using millimeter-wave bands below 100 GHz will become practical in the near future as a result of

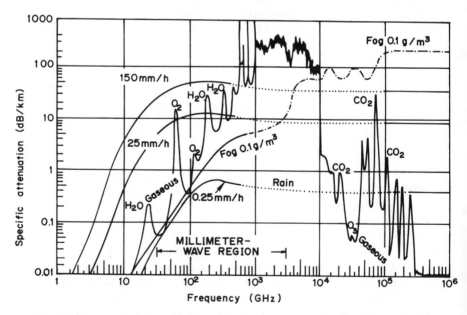

FIG. 34 Attenuation due to gaseous constituents and precipitation for transmission through the atmosphere. (From CCIR, 1978).

the application of these technologies. The present state of transmitting power and receiver noise figure are shown in Fig. 35.

Taking into account the conditions that we have described, we shall estimate the repeater station spacing of the following two systems, which represent the main technological applications:

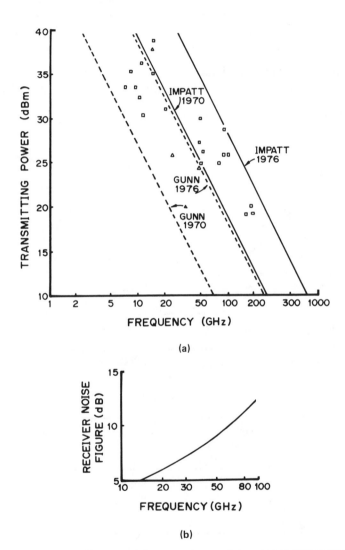

(a)

(b)

FIG. 35    Present state of (a) transmitting power (reported data: Δ IMPATT; □, GUNN) and (b) receiver noise figure.

(1)  a large-capacity, long-haul system using a wide bandwidth;
(2)  a small- or medium-capacity, short-haul system using the characteristics of smaller size and lighter weight of higher-frequency equipment.

The calculated results are given in Fig. 36. Using typical rainfall probability distribution and equipment performance in Japan (Morita and Higuchi, 1976), we have calculated the repeater station spacing as a function of the frequency and attenuation margin with respect to assumed unavailabilities in the 2500- and 20-km routes. The repeater station spacings of systems (1) and (2), for the frequency range 40–100 GHz, are as follows:

(1)  large-capacity, long-haul transmission system, $\sim 1.8-1.1$ km;
(2)  small-capacity, short-haul transmission system, $\sim 3.0-1.8$ km.

There are four radio windows (A–D) in the $\sim 40-300$-GHz band, in

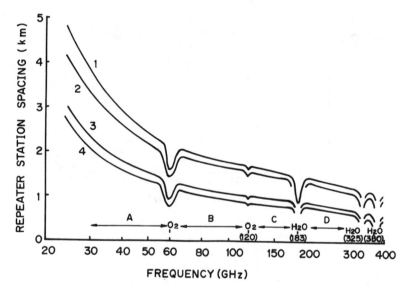

| Curve | Route length (km) | Transmission capacity (Mbit/sec) | Transmitting power[a] (dB m) | Antenna diameter (m$^\phi$) | Unavailability |
|---|---|---|---|---|---|
| 1 | 20 | 32 | 34 | 0.6 | 0.03%/20 km |
| 2 | 20 | 100 | 34 | 0.6 | 0.03%/20 km |
| 3 | 2500 | 400 | 37 | 1.0 | 0.3%/2500 km |
| 4 | 2500 | 400 | 34 | 1.0 | 0.3%/2500 km |

[a] 40 GHz.

FIG. 36  Repeater station spacing for 2500- and 20-km radio-relay systems.

which no absorption is caused by oxygen molecules and water vapor, as shown in Fig. 36. Absorption bands such as 60, 120, and 180–190 GHz can be effectively applied to high-density radio-relay systems because of less interference noise.

We shall investigate the ultimate transmission capacity of the millimeter-wave region by studying examples of large-capacity systems using total bandwidths of about 100 GHz in the four frequency regions. Considering that each large-capacity system is assumed to convey 20 radio channels using 16 Quadrature Amplification Modulations (QAM) with a capacity of ~ 1 to 4 Gbit/sec, the total transmission capacity might be equivalent to about 6,000,000 voice channels or 60,000 video channels.

Because a millimeter-wave antenna provides a sharp directional beam compared with that in the lower bands, the allowable minimum branching of the link angle can be made small, permitting installation of a large number of routes between two points. Thus efficiency in the use of the total spectrum, considering geographical factors, reaches and in some cases exceeds that of conventional systems.

However, when high-density networks of these systems are used to cope with immense traffic demands in large city areas, overreach interference will be unavoidable, especially under clear weather conditions. To solve this problem, studies concerning site selection, frequency selection, and adaptive transmitting power control are required.

## B. 20-GHz DIGITAL RADIO-RELAY SYSTEMS

Several countries have undertaken research and development of radio systems using a 20-GHz band (Stagg and Robinson, 1977; Volckman *et al.*, 1974). Here, the characteristics of two systems—the 20L-P1 and DR 18 systems developed in Japan (Nakamura *et al.*, 1977) and the United States (Longton, 1976), respectively—will be described.

### 1. *20L-P1 System*

NTT introduced the 20L-P1 system into its service in December 1976. This system can convey 400 Mbit/sec, which corresponds to 5760 telephone channels per radio channel for long-haul transmission. Eight working channels and one protection channel are arranged in the 3.5-GHz band between 17.7 and 21.2 GHz with the interleaved mode. The main 20L-P1 system parameters are given in Table VIII.

The 20-GHz digital radio-relay system consists of terminal radio stations, switching stations, and repeater stations. A terminal radio station is located at the end of a route and comprises circuit-connecting equipment, circuit-switching equipment, transmitter–receivers, and supervisory and control

56 K. MIYAUCHI

TABLE VIII

MAIN SYSTEM PARAMETERS OF THE 20L-P1 AND DR 18 SYSTEMS

| Parameter | Dimension | |
|---|---|---|
| | 20L-P1 (NTT) | DR 18 (Bell) |
| Frequency band (GHz) | ~ 17.7–21.2 | ~ 17.5–19.7 |
| Number of rf channels | 8 + 1 standby | 7 + 1 standby |
| Repeater spacing (km) | 3 | ~ 2.4–7.2ᵃ |
| Modem method | Four-phase PSK | Four-phase PSK |
| Transmission capacity (Mbit/sec per rf channel) | 400 | 274 |
| Transmitter output power (dB) | 22 | 22 |
| Noise Figure (dB) | 10 | 10 |
| Antenna (m$^\phi$) | 1.8 | 0.8 |
| Unavailability | 0.3%/2500 km | — |

ᵃ Anticipated value.

equipment. A circuit-switching station is located at every 30 repeater stations, composed of switching equipment in addition to transmitter–receivers. Both stations are usually installed in buildings. The repeater stations, located every 3 km, are installed in small housings, and mounted on top of 25-m-high steel pipes, as shown in Fig. 37.

The transmitter–receiver is solid state and makes use of high-power IMPATT diodes as a transmitting local oscillator, a low-noise Gunn diode as a receiving local oscillator, and microwave integrated circuits for most of the IF circuit. This configuration led to high reliability, miniaturization, and low-power consumption in the transmitter–receiver.

The repeater consists of four units—transmitter, receiver, demodulator, and dc power supply—as shown in Fig. 38. All units are the same size (350 × 91 × 400 mm) and employ a plug-in structure. The average weight of each unit is 6 kg, amounting to 24 kg total. Figure 39 shows a repeater block diagram.

The circuit-switching equipment has a function of automatic circuit switchover between eight working channels and one protection channel, transmission quality monitoring, and scrambling conversion to avoid undesirable spectrum concentration and ease clock signal extraction at each repeater. The circuit-connecting equipment converts a 397.2-Mbit/sec bipolar input signal to unipolar signal and then divides it into two synchronized series of 198.6-Mbit/sec unipolar signals.

Commercial tests of the 20L-P1 system were carried out on the Tokyo–Yokohama and Osaka–Kobe routes, two of the densest traffic routes in

FIG. 37    Pole-type repeater station of the 20L-P1 system.

Japan, between February 1976 and September 1977. The system was put into service in Tokyo and Osaka in December 1976 and March 1977, respectively. Many routes are planned, and it is expected that a long-haul route connecting major cities will be completed within 5 years.

## 2. DR 18 System

The DR 18 system was developed by Bell Laboratories. The expected applications of this system include metropolitan backbone and intercity trunks and eventual feeder routes for high-capacity, long-haul waveguide systems as an alternative to high-capacity coaxial cable systems. The system operates in the frequency band of 17.7–19.7 GHz, which is utilized to carry 274-Mbit/sec signals. This band has been channelized by the FCC for 16 wide-band channels. Thus seven working and one standby rf channels are

FIG. 38 Repeater of the 20L-P1 system, composed of four units: transmitter, receiver, demodulator, and dc power supply. Each unit is 350 mm in height, 91 mm in width, and 400 mm in depth.

set up in each direction, utilizing dual polarization of the radiated signals; half of the frequencies are for transmitters and the remaining half for receivers. A total route capacity of 28,244 voice circuits can be obtained using the standard Bell System digital hierarchy.

Among the main features of the DR 18 system are antenna canisters placed on top of tapered masts of various lengths depending upon site topography. As depicted in the cutaway view in Fig. 40, the canister contains the system's all-solid-state transmitters and receivers, antenna, and associated interconnecting equipment. The antenna is basically a shielded, inverted periscope consisting of a paraboloidal reflector, feed assembly, and mirror.

The main DR 18 system parameters are given in Table VIII. Anticipated typical repeater spacings are 2.4–7.2 km. The first field test of the DR 18 system was carried out on a 10-mile route in the White Plains–Nyack region in 1975.

## C. RADIO SYSTEMS IN FREQUENCY BANDS ABOVE 25 GHz

### 1. Small- and Medium-Capacity Radio Systems

a. *Radio Equipment Using 40-GHz Band for Video Conference System (VC-40G).* Radio equipment using the 40-GHz band to carry video confer-

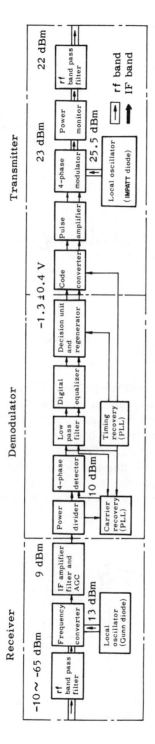

FIG. 39 Repeater block diagram of the 20L-P1 system. (From Kohiyama et al., 1976.)

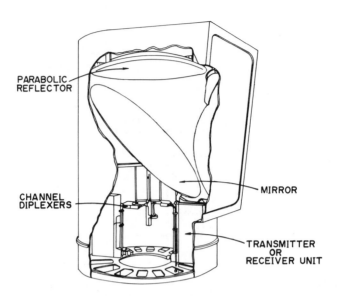

FIG. 40 Schematic diagram of a canister antenna showing salient features. (From Longton, 1976. © 1976 IEEE.)

ence system signals has been developed by NTT (Samejima and Kurokawa, 1979). Video conference systems, in which geographically dispersed meeting rooms are interconnected with pictures, sound, and other information associated with a conference, are becoming one of the most promising picture communication services. To make the video conference system efficient, it is preferable to install the video conference equipment on the user's premises. In this case, the transmission link connecting the meeting rooms will consist of a terminal link between the user's premises and a base

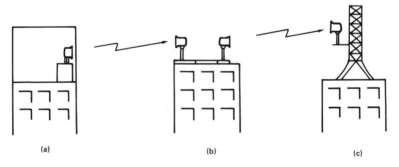

FIG. 41 Typical radio terminal link configuration: (a) user's premises, (b) intermediate relay station, (c) radio base station. (From Samejima and Kurokawa, 1979.)

station and a long-haul link between base stations. Whereas the long-haul link can be shared by numerous users, the terminal link is used exclusively by a single user, which means that this link must be economically constructed and maintained.

In designing a radio terminal link for a video conference system, the choice of standard repeater spacing should take account of the distribution of the potential users and the distance from each user to the nearest existing radio terminal station. Studies conducted in Tokyo show that a standard repeater spacing of 4 km can cover almost the entire area, with a maximum of two hops covering 8 km. From these considerations, a standard radio terminal link configuration has been decided upon, as shown in Fig. 41. This configuration is sufficient to cover almost any of the urban areas in Japan.

The following considerations were taken into account in selecting the carrier frequency.

(1) Demands for video conference service are concentrated mostly in urban areas, where radio spectrum congestion is severe. This has led to the use of a new frequency band.

(2) Short repeater spacing enables the use of higher frequencies.

(3) A wide usable frequency band is necessary to meet the growing demand for video conference service.

(4) It is preferable to use a higher frequency where equipment size can be reduced to facilitate terminal link construction for prompt service initiation.

Based on these conditions, the 40-GHz band was chosen. Because conference rooms are in buildings instead of private homes, the radio equipment installed on the user's premises will usually be mounted on the roof of a building. However, if the building is so tall that a line of sight path can be achieved through a window, indoor installation may be possible. The intermediate relay station will be located on the roof of a building, usually a telephone office. The equipment will be the same as that mounted on the user's roof. Radio equipment at the radio base station will be mounted on a tower. Equipment installed on a roof should have a self-supporting structure with a special design to withstand hard wind. To facilitate transportation and installation, the antenna, rf circuit, and FM modem are integrated in a single container.

Figure 42 shows a radio equipment block diagram. The main system parameters are given in Table IX. These pieces of equipment have been installed in Tokyo and Osaka to make up the video conference system radio terminal links shown in Fig. 43. They are presently being tested.

b. *40-GHz Digital Distribution Radio Equipment (MACT)*. A 40-GHz digital radio called MACT (millimeter-wave area coverage transmission

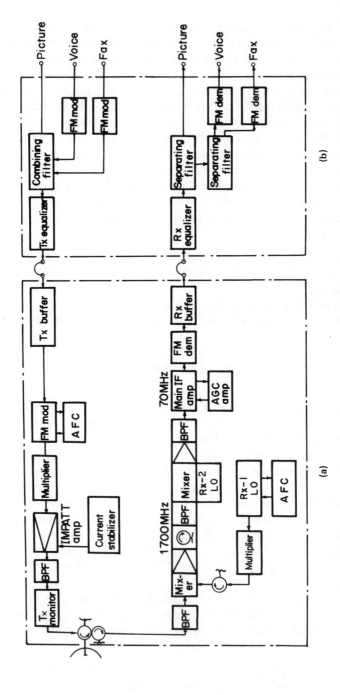

FIG. 42 Block diagram of the VC-40G system transportable radio equipment: (a) 40-GHz-band radio equipment, (b) picture and voice combining–separating equipment. Maximum distance between (a) and (b) is 200 m. Tx, transmitter; BPF, bandpass filter; FM mod, frequency modulation modulator; AFC, automatic frequency control; Rx, receiver; Rx-1 LO, first receiving local oscillator; Rx-2 LO, second receiving local oscillator; AGC, automatic gain control; Main IF amp, main intermediate frequency amplifier. (From Samejima and Kurokawa, 1979.)

TABLE IX

SMALL- AND MEDIUM-CAPACITY RADIO SYSTEMS USING THE FREQUENCY RANGE 25–40 GHz

| Parameter | Video conference system (VC-40G) | Digital distribution system (MACT) | Inter-telephone-office system (IT-27G) |
|---|---|---|---|
| Frequency band (GHz) | ~38.6–40 | ~38.6–40 | ~25.25–27.5 |
| Transmission capacity (per rf channel) | Color TV signal; 10 kHz voice and 2 control signal | 1.544 Mbit/sec (or 6.3 Mbit/sec) | 32 × 2 Mbit/sec |
| Repeater spacing (km) | 4 | 3 | 3 |
| Unavailability | 0.0125%/4 km | 0.01%/3 km | 0.00186%/3 km |
| Modulation | FM | Two-phase PSK | Four-phase PSK |
| Antenna (cm$^\phi$) | 60 | 45 | 60 |
| Transmitter output power (dB) | 20 | 7 | 30 |
| NF (dB) | 11.5 | <12 | 6 |
| Power consumption (W) | 50 | 65 | 50 |
| Weight (kg) | 40 | 35 | 20 |
| Present stage | Under test | Under test | Under development |

FIG. 43  Radio equipment installation of the VC-40G system: (a) roof installation, (b) indoor installation.

system) was developed by the OKI Electric Company (Hata, *et al.*, 1978) to meet the demand for business communication of voice and data in intracity and suburban areas. The equipment for such a purpose should be inexpensive because the number of loading channels is small, i.e., 1–24 channels, with a short radio span usually of less than 10 miles. Taking into account economy and ease of maintenance and installation, a new circuit configuration consisting of a single local IMPATT oscillator, for an upconverter and a receiver local source, is adopted.

Following an FCC regulation issued in late 1974, 14 two-way rf channels were assigned between 38.6 and 40 GHz; each channel has a 50-MHz bandwidth, and the channel separation for transmitter and receiver is just 700 MHz. The main specifications of the equipment are given in Table IX. In Fig. 44 is shown an outside view of the equipment, which can also house standby equipment. The dimensions of the cabinet are 20 × 20 × 23 in., and the weight of the cabinet is 94 lb. A block diagram of the single local transmitter–receiver is shown in Fig. 45. It is expected that this system can be effectively adapted as a low-density local distribution radio link in large cities.

c.  *A 27-GHz Inter-Telephone-Office Radio System (IT-27G).*  An inter-telephone-office system using the frequency range 25.25–27.5 GHz, called IT-27G, was proposed by Hashimoto *et al.* (1978). This system has the following features.

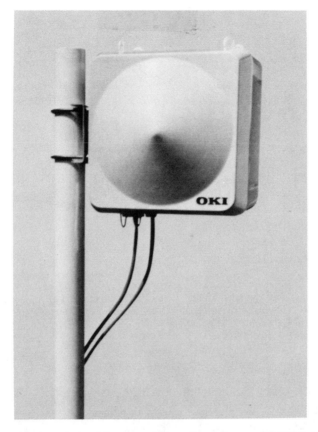

FIG. 44 Outside view of a MACT radio equipment. (From Itata *et al.*, 1980. © 1980 IEEE.)

(1)  It provides medium-capacity transmission facilities in large cities as a short-haul radio-relay link with high reliability.

(2)  It is useful for solving the problems of severe congest in manholes and conduits and of construction constraints encountered in large cities.

(3)  It can be easily transported and assembled and installed either outdoors or indoors, because equipment size can be reduced by using the 27-GHz band.

The main IT-27G system parameters are given in Table IX. Based on the distance distribution between the telephone offices in the central part of Tokyo, a 3-km repeater spacing was chosen, with a route length of up to 50 km. Figure 46 shows the outdoor installation. Transmitter–receiver circuits are integrated in a cylindrical container behind an antenna. By

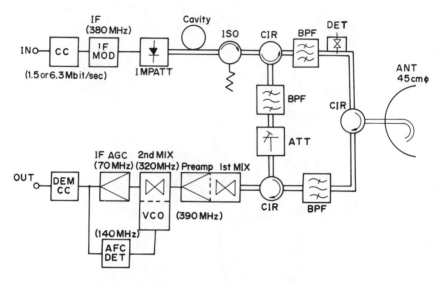

FIG. 45  Block diagram of the MACT system single local transmitter – receiver. IF, intermediate frequency; MOD, modulator; ISO, isolator; CIR, circulator, BPF, bandpass filter; DET,
detector, ANT, antenna; ATT, attenuator; MIX, mixer; VCO, voltage controlled oscillator;
AGC, automatic gain control; AFC, automatic frequency control; DEM, demodulator. (From
Itata *et al.*, 1980. © 1980 IEEE.)

adopting the offset parabolic antenna with low sidelobe level, this system
can be applied to high-density radio systems.

## 2.  Subscriber Radio Systems

a.  *A Subscriber Radio System for Distributing Broadband Digital Information* (*SR-27G*).   To answer the growing need for various new services,
broadband digital communications systems are required that transmit high-
speed digital data, facsimile, and video signals. Such systems should be
applicable to the simultaneous transmission of voice and broadband digital
signals (Menne, 1979; Mokhoff, 1979). The most important system that
meets these requirements is a subscriber system that distributes broadband
digital information.

The SR-27G system, now under investigation, is a broadband radio
communications system between a telephone office as base station and
office subscribers (Ohtomo *et al.*, 1979). The expected digital signal rate is
between 64 kbit/sec and several megabits per second, to meet the above-
mentioned high-speed services. The system has the following features.

(1)  It is capable of responding promptly to low-cost installation de-

FIG. 46 IT-27G system radio equipment model for outdoor installation. The antenna diameter is 60 cm.

mands and is especially suited for dispersed users with low traffic demands extending over a wide area.

(2)   It is adaptable to time-variable traffic, with effective use of the frequency band, because it employs time-division multiple access (TDMA) as a transmission mode.

(3)   Use of the 27-GHz band permits installation of the radio equipment either outdoors or indoors without special mounting engineering. In addition to permanent usage, this system is applicable to various temporary transmission links like a transportable radio system.

Figure 47 shows the concept of a millimeter-wave intracity radio distribution network, composed of a subscriber radio system (SR-27G) and an inter-telephone-office radio system (IT-27G). As shown in Fig. 48, the beam of the base station antenna is split into four 90° sectors with different radio frequencies. This configuration results in no interference between the subscriber systems. The radio equipment of the subscriber can easily be assembled and installed either on the roof or in a room of a building.

Each subscriber receives the TDMA signal from the base station and

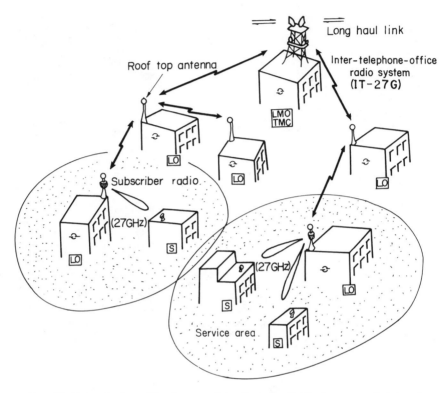

FIG. 47  Concept of a millimeter-wave intracity radio distribution network (subscriber radio system SR-27G). LO, local office; LMO, local tandem office; TMC, toll tandem office; S, subscriber. "Long haul link" includes the existing microwave link, the coaxial cable link, etc.

extracts the signal that is transferred to the subscriber at the terminal equipment. The data from the subscriber terminal equipment, on the other hand, are transmitted in a periodical burst at the time assigned by the control signal from the base station.

The expected parameters of this system are shown in Table X. The transmission rate is 12.3 Mbit/sec because the mean data burst is 512 kbit/sec and the number of required bursts for 100 subscribers is 24.

Because the transmitting power depends on the distance between the base station and the subscriber, it is effective to make the antenna diameter larger for remote subscribers, as shown in Fig. 49. An availability of 99.96% can be obtained when the transmitting power is 100 mW and the radius of the service area is 7 km. To handle the growing demands of the subscribers, additional radio systems can be installed.

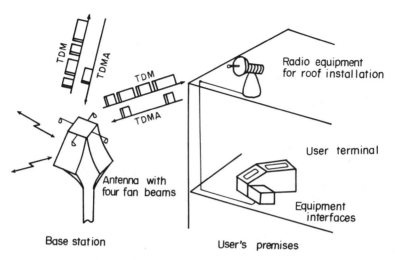

FIG. 48   Broadband digital subscriber radio system configuration. TDM, time division multiplex; TDMA, time division multiple access.

**b.** *A High-Density Subscriber System Using the 60-GHz Oxygen Absorption Band.* Absorption bands due to oxygen molecules in the atmosphere exist at around 60, 120, 180 (etc.) GHz. Data collected on the ground (Mitchell *et al.*, 1969) show that the maximum attenuation constant is 16.5 dB/km at a frequency of about 60.5 GHz. These bands are of considerable interest for terrestrial radio systems over short distances for the following reasons.

(1)   There is no mutual interference between terrestrial and space systems.

(2)   Interference between terrestrial systems is substantially reduced.

An example of possible applications of the 60-GHz absorption band may be high-density subscriber systems such as an intracity transmission system between an exchange plant and a subscriber in an office building (Morimoto *et al.*, 1978). To simplify the terminal equipment in these systems, the single subscriber per radio carrier method is useful. When the number of subscribers in an office building is large, a common transmitter and common receiver with many radio carriers are provided. The typical system parameters are as follows:

| | |
|---|---|
| route length | 7.5 km |
| repeater station spacing | 1 km |
| modulation | FM or two-phase PSK |

TABLE X

Main System Parameters of the SR-27G System

| Parameter | Dimension |
| --- | --- |
| Frequency band (GHz) | 27 |
| Transmission capacity | Broadband digital signals: TDMA, 12.3 Mbit/sec per rf channel, $\sim 0.064-1.544$ Mbit/sec per user; video signal, SCPC[a], 4 MHz per user |
| Service area radius (km) | 7 |
| Number of users | 100 per area or rf channel |
| Loss probability and traffic intensity per hour | 0.01 and 0.15 erlang (required burst number 24) |
| Availability | >99.96% |
| Transmitter output power (W) | 0.1 |
| Modulation | PSK |
| Antenna | |
|   Base station | Fan beam antenna with four 90° sectors |
|   User | 30 cm (service area radius 3 km), 60 cm$^\phi$ (5 km), 130 cm$^\phi$ (7 km) |

[a] SCPC: single channel per carrier.

| | |
| --- | --- |
| C/N (two-phase PSK) and S/N (FM) | 10.9 dB (normal) and 40 dB |
| transmitting power | 10 mW |
| antenna diameter | 0.3 m$^\phi$ |
| total subscribers | $N = 200$ |
| simultaneously operating carriers | $m = 20$ or $50$ |

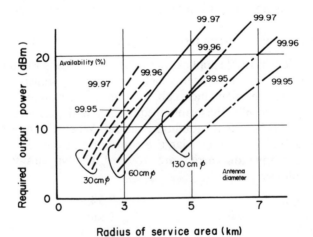

FIG. 49   Required transmitter output power vs radius of service area.

## 3. Large-Capacity Radio Systems

a. *An Intercity Highway Transmission System.* The feasibility of a large-capacity transmission system along a highway connecting large cities was proposed by Akiyama *et al.* (1977). A highway is an attractive radio transmission medium because the space above a highway is always open. Radio equipment can be installed near the top of highway street lights, because compact radio equipment can be realized by using millimeter-wave bands as shown in Fig. 50. Typical main system parameters of a highway transmission system using the 40-GHz band are as follows:

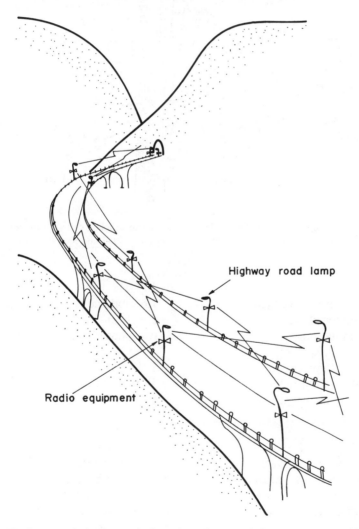

Highway road lamp

Radio equipment

FIG. 50 Concept of a large-capacity intercity radio transmission system along a highway.

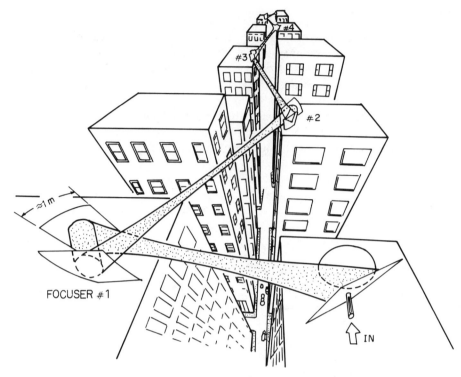

FIG. 51 Concept of a large-capacity intracity radio transmission system using frequency bands 50–160 GHz with 1.2 m × 1.2 m focuses. (From Arnaud and Ruscio, 1975. © 1975 IEEE.)

| | |
|---|---|
| frequency band | 40 GHz |
| transmission capacity | 1.6 Gbit/sec per rf channel |
| number of standby rf channels | 2 + 1 (s) |
| modulation | two-phase PSK |
| repeater spacing | 1 km |
| output power | 1 W |
| antenna diameter | 50 m$^\phi$ |
| anticipated repeater housing size | 50 cm × 90 cm$^\phi$ |

This system is promising as a very-large-capacity system between large cities using the millimeter-wave region.

b. *An Intracity Transmission System Using Hertzian Cable.* The feasibility of using large-capacity Hertzian cable to distribute and collect information within cities was proposed by Arnaud and Ruscio (1975). The Hertzian cable scheme makes use of millimeter-wave beams propagating through the atmosphere. There is therefore no need to dig trenches in the streets, as must be done for conventional cables. Millimeter-wave beams can be confined by sequences of pairs of cylindrical mirrors, each pair acting as a lens. A loss on the order of 2 dB/km was measured in clear weather at 100 GHz, with $1.2 \times 1.2$-m focusers spaced 80 m apart. This system is attractive mainly as a very-high-capacity system with many different paths crisscrossing in a densely populated city, as shown in Fig. 51.

### V. Satellite Communication in Millimeter Waves

#### A. GENERAL STATUS OF MILLIMETER-WAVE SATELLITE COMMUNICATION

The first millimeter-wave satellite (ATS-VI) (Wolff *et al.*, 1975) was launched by NASA in 1974. This satellite was used for communication and propagation experiments. The millimeter-wave transponder was used only for the latter purpose.

Recently, Japan has been vigorously developing millimeter-wave satellite systems. In December 1977, Japan launched a millimeter-wave satellite names CS, a medium-capacity communication satellite for experimental purposes. Experiments using the CS satellite began in May 1978 (Kaneda *et al.*, 1978). Furthermore, Japan plans to launch the first commerical milli-meter-wave satellite in 1983 (Fuguno *et al.*, 1979).

The European Space Agency (ESA) and the MIT Lincoln Laboratory have also conducted research and development on millimeter-wave satellite communications (Lopriore *et al.*, 1978; Solman *et al.*, 1979). In 1979, NASA announced that it would resume millimeter-wave satellite system research (Dement, 1979; Ward, 1979) after a 5-year interruption.

A list of millimeter-wave satellites is given in Table XI. Laboratories conducting millimeter-wave satellite research and development are listed in Table XII.

#### B. COMMUNICATION SATELLITE TECHNOLOGY

Spacecraft, transponders, multiplexers, and antennas will be described, with a special focus on Japan's CS satellite.

#### 1. *Spacecraft*

a. *Japan's CS.* Japan's CS spacecraft, shown in Fig. 52, is a spin-stabi-lized spacecraft weighing about 350 kg with a mechanical despun antenna. The features of the spacecraft are as follows.

(1) Frequency bands of 30 and 20 GHz are used. The spacecraft is equipped with six $K$-band transponders (27.5–31.0 GHz for uplink,

TABLE XI

MILLIMETER-WAVE SATELLITES

| Satellite | Frequency (GHz) | Purpose | Date launched | Country |
|-----------|-----------------|---------|---------------|---------|
| ATS-VI | ~29–31 and ~19–21 | Propagation experiment | 5/30/1974 | U.S. |
| COMSTAR | 19, 20 | Propagation experiment | 3/14/1976 | U.S. |
| LES-8 | ~36–38 | Communication experiment | 5/13/1976 | U.S. |
| LES-9 | ~36–38 | Communication experiment | 7/22/1976 | U.S. |
| ETS-II | 34.5 | Propagation experiment | 2/23/1977 | Japan |
| SIRIO | 18 | Propagation experiment | 8/17/1977 | Italy |
| CS | 30 and 20 | Communication experiment | 12/15/1977 | Japan |
| ECS | 35 and 22 | Communication experiment | 2/1980 | Japan |
| H-SAT | 30 and 20 | Propagation experiment | 1980, 3rd quarter | Europe |
| CS-II | 30 and 20 | Commercial use | 3/1983 | Japan |

## TABLE XII

### Laboratories Conducting Millimeter-Wave Satellite Research and Development

| Laboratory | Activities | Satellite | Reference |
|---|---|---|---|
| NASA (U.S.A.) | Propagation experiment using ATS-VI<br>Publication of future satellite system plan<br>Presentation of NASA programs using 30/20-GHz band<br>STS radar–communication subsystem | ATS-VI | (Wolff et al., 1975)<br>(Berkey and Mayer, 1976)<br>(Dement, 1979; Ward, 1979) |
| Bell Laboratories (U.S.A.) | Propagation experiment using COMSTAR beacon | STS<br>COMSTAR | (Cager, 1978)<br>(Cox, 1978)<br>(Arnold and Cox, 1978) |
| RRL[a] and NTT[b] (Japan) | Propagation experiment using ETS-II (34.5 GHz)<br>Communication experiment using CS<br>35/32-GHz experiments (ECS)<br>Commercial use (1983) | ETS-II<br>CS<br>ECS | (Fugono and Hayashi, 1978)<br>(Kaneda et al., 1978)<br>(Fugono et al., 1979) |
| ESA[c] (Europe) | H-SAT experiment (1980, 3d quarter)<br>Study on ECS (Europe) | H-SAT ECS (Europe) | (Herden, 1978)<br>(Lopriore, 1978) |
| Selenia SPA (Italy) | 18-GHz propagation experiment | SIRIO | (Perrotta, 1978) |
| MIT Lincoln Laboratory (U.S.A.) | Satellite-to-satellite data transfer and mobile communication experiment using LES-8, LES-9 | LES-8, LES-9 | (Solman et al., 1979)<br>(Collins et al., 1978) |

[a] Radio Research Laboratory of the Ministry of Posts and Telecommunications.
[b] Nippon Telegraph and Telephone Public Corporation
[c] European Space Agency.

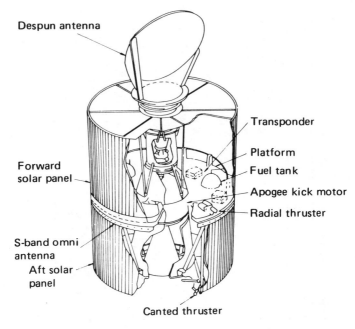

Despun antenna

Transponder

Platform

Forward
solar panel

Fuel tank

Apogee kick motor

Radial thruster

S-band omni
antenna

Aft solar
panel

Canted thruster

FIG. 52 The CS spacecraft. (From Fuketa *et al.*, 1980.)

17.7–21.2 GHz for downlink) and two *C*-band transponders (5.9–6.4 GHz for uplink, 3.7–4.2 GHz for downlink).

(2)   The antenna covers the 30- and 20-GHz and 6- and 4-GHz bands. The antenna radiation pattern is shown in Fig. 53.

The main performance features of the *K*-band equipment are as follows:

| | |
|---|---|
| transmitter output power | 34.0 dBm |
| receiver noise figure | 13 dB |
| antenna gain over mainland | 33 dB minimum |

b.   *COMSTAR.*   The spin-stablized COMSTAR satellite illustrated in Fig. 54 has millimeter-wave beacons with frequencies of 19 and 28 GHz. The COMSTAR can switch polarization at 19 GHz for propagation experiments. The beacons have an effective isotropically radiated power (EIRP) of 52 dBm at 19 GHz and 56 dBm at 28 GHz.

c.   *LES-8 and LES-9.*   The LES-9 spacecraft shown in Fig. 55 has two *K*-band (37-GHz) antennas. The dish-type antenna is a narrow-beam tracking antenna for cross-link or uplink–downlink communications. The horn-type antenna incorporates a fixed horn and provides wide-beam antenna coverage.

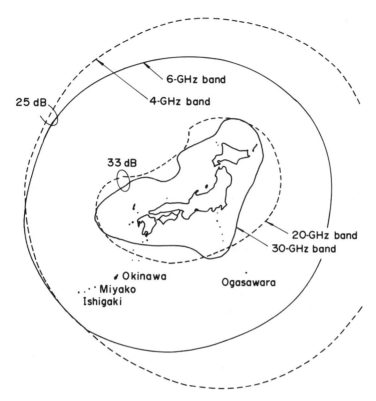

FIG. 53 CS antenna radiation pattern. (From Miyauchi, 1980. © 1980 IEEE.)

### 2. Antenna System

The CS antenna shown in Fig. 56 is composed of a shaped beam horn reflector, a reflector support, a conical horn, and a four-frequency band-splitting filter (Shinji et al., 1976). Its total weight is 14 kg. The reflector is made of a shallow sandwich shell with carbon fiber reinforced plastic (CFRP) skins and an aluminum honeycomb core. The reflector support and conical horn are made of CFRP and most of the filter parts are of aluminum. The measured surface roughness of this reflector is less than 0.2 mm rms.

### 3. Multiplexer

For the wide-band multiplexer, a periodic filter using a traveling-wave resonator is suitable (Kumazawa and Ohtomo, 1977). This is composed of a traveling-wave resonator, three directional couplers, and connecting wave-guides. A fabricated periodic filter for the CS receiver demultiplexer is

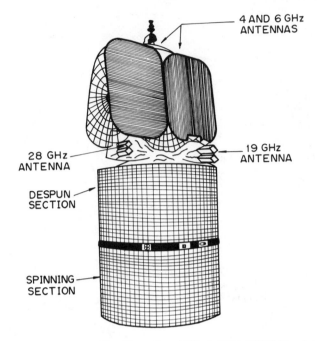

FIG. 54   The COMSTAR satellite. (From Cox, 1978. © 1978, AT&T. Reprinted from *The Bell System Technical Journal* by permission.)

shown in Fig. 57. It is made of aluminum to reduce its weight of 93 g, and its dimensions are about $150 \times 90 \times 40$ mm$^3$. The 3-dB bandwidth is 900 MHz and the insertion loss at the center frequency is less than 0.42 dB.

For the narrow-band multiplexer, ring-type filters are utilized. The fabricated filter for the CS transmitter multiplexer is shown in Fig. 58. The 3-dB bandwidth of this filter is 180 MHz. It has a two-pole maximally flat frequency response and weighs less than 140 g.

## 4.   Receiver and Transmitter

The GaAs Schottky-barrier mixer is usually used as a low-noise mixer for the millimeter-wave receiver. The GaAs field-effect transistor (FET) low-noise amplifier is rapidly being developed and is expected to be put into practical use soon.

The traveling-wave tube (TWT) or IMPATT diode amplifier is most suitable for the final power amplifier in a millimeter-wave transponder transmitter. Currently, the CS uses the TWT and the LES-8 and COMSTAR use the IMPATT.

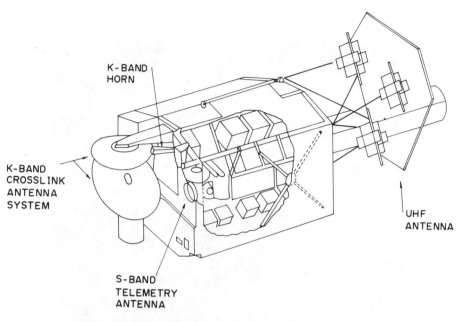

K-BAND
HORN

K-BAND
CROSSLINK
ANTENNA
SYSTEM

UHF
ANTENNA

S-BAND
TELEMETRY
ANTENNA

FIG. 55   The LES-9 spacecraft. (From Solman *et al.*, 1978.)

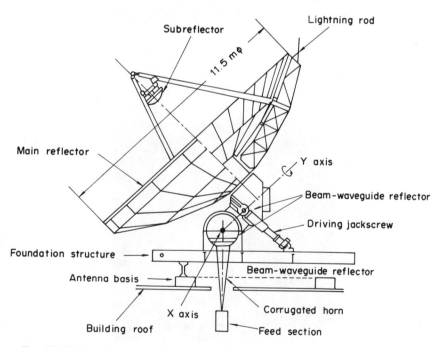

Subreflector

Lightning rod

11.5 mφ

Main reflector

Y axis

Beam-waveguide reflector

Driving jackscrew

Foundation structure

Beam-waveguide reflector

Antenna basis

Corrugated horn

X axis

Building roof

Feed section

FIG. 56   Fabricated antenna of the CS earth station. (From Yokosuka Electrical Communication Laboratory, 1980.)

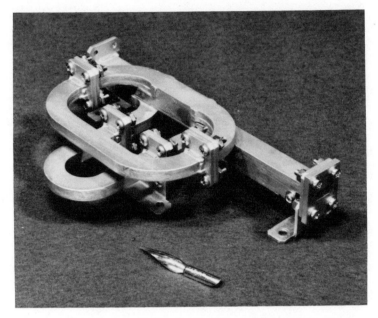

FIG. 57 Fabricated periodic filter for the CS on-board receiver multiplexer.

FIG. 58 Fabricated ring-type filter for the CS on-board transmitter multiplexer.

## C. EARTH STATION FACILITIES

In this section we describe the earth station antenna, high-power amplifer (HPA), and low-noise amplifier (LNA), which are used exclusively in satellite communications.

### 1. Earth Station Antenna

The following type of millimeter-wave earth station antennas have been developed:

(1) axisymmetric Cassegrain antenna (Fig. 59);
(2) offset Cassegrain antenna (Fig. 60).

FIG. 59 An axisymmetric Cassegrain antenna.

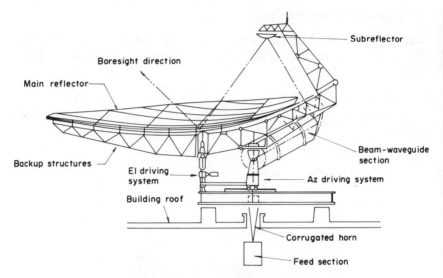

FIG. 60   An offset Cassegrain antenna. El, elevation; Az, azimuth. (From Yokosuka Electrical Communication Laboratory, 1980.)

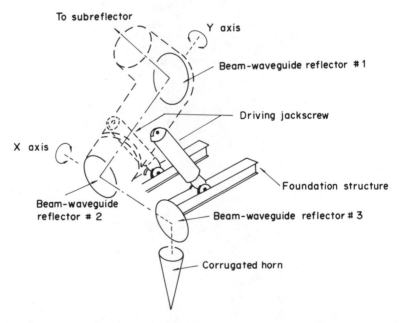

FIG. 61   A beam-waveguide section and drive system for an earth station antenna. (From Yokosuka Electrical Communication Laboratory, 1980.)

Offset antennas are rather expensive owing to their asymmetric reflectors; they feature low sidelobe and wind load.

Earth station antennas provide tracking mechanisms that make the feeder system more intricate than that of terrestrial systems. An example of this feeder system is shown in Fig. 61.

### 2. HPA and LNA

A klystron or TWT is utilized for the HPA for millimeter-wave satellite systems. A list of HPA performance specifications is given in Table XIII.

For the LNA, paramps (parametric amplifiers) are usually used (Egami, 1975). A 19-GHz paramp with a new type of in-line signal circuit is shown in Fig. 62. The recent remarkable progress in the development of FETs indicates that in the near future an FET LNA suitable for many satellite earth stations will be realized. A list of LNA performance specifications is given in Table XIV.

### D. SYSTEM EXAMPLES

### 1. COMSTAR

Beacons on the COMSTAR satellites are used for propagation experiments as well as for their original purpose. Two satellites with 19- and 28-GHz beacons were launched in 1976. The 19-GHz signals are switched between two orthogonal linear polarizations. The 28-GHz signal is linearly polarized and phase modulated to produce a coherent sideband of 264 MHz for the carrier.

### 2. LES-8 and LES-9

The Lincoln experimental satellites LES-8 and LES-9 are a pair of experimental communication satellites designed and built by the MIT

TABLE XIII

HPA PERFORMANCE SPECIFICATIONS

| Frequency (GHz) | Power (dBW) | Tube type | Country |
|---|---|---|---|
| 27.5–31 | 27 | klystron | Japan |
| | 29 | TWT | Japan |
| | 30 (duty 30%) | klystron | Japan |
| | 33 | klystron | U.S.A. |
| | 37 | TWT | Federal Republic of Germany |
| 36–38 | 27 | TWT | Japan |
| | 30 | TWT | Federal Republic of Germany |
| 40–50 | 23 | TWT | U.S.A. |
| | 26 | TWT | Federal Republic of Germany |

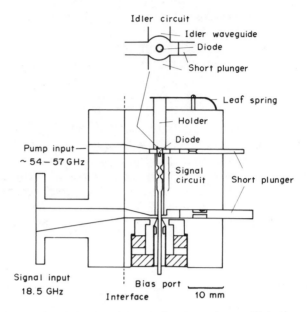

FIG. 62   An 18.5-GHz double-tuned paramp for a low-noise amplifier. (From Egami, 1975. © 1975 IEEE.)

Lincoln Laboratory. The LES-8 and LES-9 are designed to operate in a synchronous ecliptic orbit and to communicate on a cross link between them as well as with earth stations.

A geosynchronous satellite has a ground visibility area about 13,000 km in diameter. With cross-link communication between two satellites spaced thousands of kilometers apart, a pair of satellites can cover more than three-fourths of the earth's surface. The arrangement of the LES-8 and LES-9 in orbit is shown in Fig. 63, which indicates some of the possible communication links at uhf and $K$ band.

TABLE XIV

LNA PERFORMANCE SPECIFICATIONS

| Frequency (GHz) | Noise temperature (K) | Type | Country |
|---|---|---|---|
| 17.7–20.2 | 50–60 | Cryogenic paramp | U.S., Japan |
| | 200–400 | Uncooled paramp | Japan, U.S. |
| | 500 | Cooled FET amp | Japan |
| | 600–1000 | Schottky-barrier mixer | Japan, U.S. |
| | 600–1000 | FET amp | Japan, U.S. |
| 34–37 | 400–600 | Uncooled paramp | Japan, U.S. |

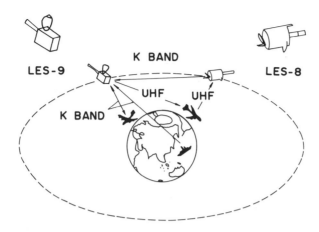

FIG. 63   LES-8 and LES-9 communication links. (From Collins *et al.*, 1978.)

## 3. *NASA Advanced Communication Systems*

At the International Conference of Communications in 1979, NASA announced its satellite communications program (Derment, 1979), which focuses on the development of frequency reuse technology, including multibeam antennas and on-board switching for spacecraft of the future. NASA has been making a study of wide-band (30- and 20-GHz bands), narrow-band (uhf), and large-capacity aggregate systems. Wide-band studies are being directed forward fixed trunking and direct-to-user services in the 30- and 20-GHz bands. The studies of large-capacity aggregate systems address a more advanced stage than the other studies.

## 4. *Japan's Domestic Satellite Systems*

NTT has conducted research on domestic satellite communication systems since 1967 (Miyauchi *et al.*, 1978). The 30- and 20-GHz bands were selected for the following reasons.

(1)   Interference between the satellite communication systems and terrestrial radio links, which are widespread in Japan, is less severe than with the 6- and 4-GHz bands.

(2)   A 3.5-GHz bandwidth can be used in the 30- and 20-GHz band system, but only the 500-MHz bandwidth is available in the 6- and 4-GHz band system.

The NTT medium-capacity domestic satellite communication system is described in Table XV. This system can be used for communication links between important regional centers on the mainland in case of emergency.

TABLE XV

Proposed Medium-Capacity Domestic Satellite Communication Systems

| Dimension | Fixed earth station system | Remote island system | Transportable earth station system | | |
|---|---|---|---|---|---|
| Frequency band (GHz) | 30 and 20 | 6 and 4 | 6 and 4 | 6 and 4 | 30 and 20 |
| Transmission mode | TDMA, time preassignment | TDMA, fixed preassignment | FM, point–point | FM, point–point | FM, point–point |
| Transmission capacity | 2880 two-way voice circuits | 192 two-way voice circuits, 2 color TV | 60 two-way voice circuits | 1 color TV, 1 voice order wire channel | 132 two-way voice circuits |
| Transponders | 6 | 1 | 1 | 1 | 1 |
| Earth Station | Small earth station with 11.5-m antenna installed on telephone office | Earth station with 10-m antenna | Vehicle-mounted transportable 3-m antenna, two small containers | Vehicle-mounted transportable 3-m antenna, two small containers | Vehicle-mounted transportable 2.7-m antenna, two small containers |

TABLE XVI

TDMA-60M System Specifications

| Specification | Dimension |
|---|---|
| Bit rate | 64.136 Mbit/sec |
| System capacity | 960 channels |
| Number of accessing stations | Six stations per transponder |
| Modulation | Two-phase PSK without differential encoding |
| Demodulation | Coherent detection |
| Acquisition | Combination of high-level preestimation method and low-level PN sequence transmission method |
| Terrestrial interface | Direct digital interface at 1.544-Mbit/sec PCM signal stage |

It can also be used between the mainland and isolated islands in Japan as well as between distant isolated islands.

a. *Fixed Earth Station Satellite Communication Systems.* The time-division multiple access (TDMA) system was developed for the 30- and 20-GHz fixed earth station satellite communication system. It has the following advantages.

(1) Instantaneous change in channel capacities between two earth stations is easily accomplished.

(2) The satellite transponder can be operated at its saturation point. There is no intermodulation problem.

The TDMA system specifications are given in Table XVI. The main features of the TDMA system are as follows.

(1) Network clock synchronization is used to attain high frame utilization efficiency.

(2) The communication network configuration is readily changed on a time preassigned basis by assigning the necessary transmission capacity to the appropriate earth station by the 24-telephone-channel unit.

(3) Direct digital interface at 1.544 Mbit/sec is realized by using pulse stuffing techniques in cooperation with a multiplexed digital echo suppressor.

The 30- and 20-GHz system is planned to replace temporarily damaged transmission lines in case of natural disasters or similar emergencies. The network configuration is designed to be changed easily and quickly, with communication circuit control and channel allotment organized by a network control station. Examples of network configurations are shown in Fig. 64. The network shown in Fig. 64a is the normal configuration for fixed

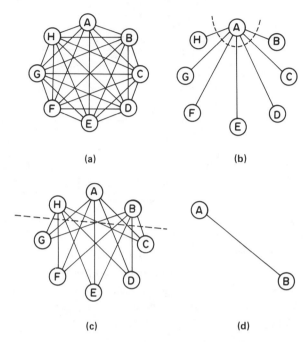

(a)                                      (b)

(c)                                      (d)

FIG. 64   Examples of satellite communication network configurations: (a) mesh; (b), broadcast mode; (c), partial mesh; (d) point-to-point. (From Miyauchi *et al.*, 1978.)

FIG. 65   A 30- and 20-GHz-band transportable earth station.

TABLE XVII

Specifications of a 30- and 20-GHz-Band
Transportable Earth Station

| Specification | Dimension |
| --- | --- |
| Frequency | 30- and 20-GHz bands |
| Antenna | 2.7-m beam-waveguide-type feeder system |
| HPA | 20 dBW (klystron) |
| LNA | 300 K (uncooled paramp) |
| Transmission capacity | 132 telephone channels |
| Transport | Exclusively by use of truck or helicopter |
| Setup time | $\sim 1$ h |

earth stations in large cities. When a terrestrial transmission line fails, the network configuration is changed as shown in Figs. 64b and 64d.

Channel allotment is made in units of 24 channels—in other words, a 1.544-Mbit/sec PCM signal. Each earth station has a maximum transmission capacity of 120 $\times$ 1.544 Mbit/sec signals (2800 two-way channels).

b. *Transportable Earth Station Communication System.* When telephone service is damaged in emergencies, recovery of the service is of the utmost urgency. Recovery in a local area will be quickly achieved by employing a transportable earth station that connects telephone sets or a local switching plant directly to the undamaged telephone network through a satellite link.

For the purpose of experiments with such a communication service, two kinds of transportable earth stations were manufactured: 30- and 20-GHz and 6- and 4-GHz-band transportable earth stations. An outside view of a 30- and 20-GHz-band transportable earth station is shown in Fig. 65. The specifications are given in Table XVII.

ACKNOWLEDGMENTS

The author is greatly indebted to Dr. S. Seki, Dr. M. Akaike, Dr. I. Ohtomo, and Dr. K. Kohiyama of Electrical Communication Laboratories, Nippon Telegraph and Telephone Public Corporation, for their collaboration in preparing this chapter.

REFERENCES

Akaike, M., and Ohnishi, K. (1977). *IEEE Trans. Microwave Theory Tech.* **MTT-25**(12), 1059–1064.
Akaike, M., Kanmuri, N., and Kato, H. (1975). *Rev. Elec. Commun. Lab.* **23**(7–8), 904–918.

Akaike, M., Kato, H., and Kanmuri, N. (1976a). *IEEE Trans. Microwave Theory Tech.* **MTT-24**(11), 693–701.

Akaike, M., Kato, H., and Yuki, S. (1976b). *IEEE Trans. Microwave Theory Tech.* **MTT-24**(3), 147–151.

Akaike, M., Yuki, S., Kanmuri, N., and Kawasaki, R. (1977). *Rev. Elec. Commun. Lab.* **25**(9–10), 997–1014.

Akiyama, T., Hashimoto, S., and Kurita, O. (1977). "A Consideration on Millimeter-Wave Radio Systems on Highways." Paper Tech. Group on Commun. Syst. IECE Jpn. CS76–179.

Allanic, J. (1976). *Int. Conf. Millimetric Waveguide Syst.*, pp. 42–45. IEE, London.

Alsberg, D. A., Bankert, J. C., and Hutchison, P. T. (1977). *Bell Syst. Tech. J.* **56**(10), 1829–1848.

Ando, M., Haga, I., Kaneko, K., and Kanmuri, N. (1978). *IEEE MTT-S Int. Microwave Symp. Dig.*, p. 312–314.

Arnaud, J. A., and Ruscio, J. T. (1975). *IEEE Trans. Microwave Theory Tech.* **MIT-23**(4), 377–379.

Arnold, H. W., and Cox, D. C. (1978). *AIAA Conf. Rec.* San Diego, California, pp. 616–621.

Baptiste, C., and Herlent, Y. (1976). *Int. Conf. Millimetric Waveguide Syst.*, pp. 1–4. IEE, London.

Barnes, C. E., Brostrup-Jensen, P., Harkless, E. T., Muise, R. W., and Nardi, A. J. (1977). *Bell Syst. Tech. J.* **56**(10), 2055–2075.

Bendayan, J., Comte, G., and Trezeguet, J. P. (1974). *Annales Télécommunication,* **29**(9–10), 387–401.

Berkey, I., and Mayer, H. (1976). *Astron. Aerona.* **14**(7/8), 34–63.

Bernardi, P., Falciasecca, G., Ferrentino, A., Grasso, G., and Occhini, E. (1976). *Int. Conf. Millimetric Waveguide Syst.*, pp. 34–37. IEE, London.

Boyd, R. J., Cohen, W. E., Doran, W. P., and Tuminaro, R. D. (1977). *Bell Syst. Tech. J.* **56**(10), 1873–1897.

Cager, R. H., (1978). *IEEE Trans. Comm.* **26**(11), pp. 1604–1619.

Carlson, E. R., Schneider, M. V., and McMaster, T. F. (1978). *IEEE Trans. Microwave Theory Tech.* **MTT-26**(10), 706–715.

Carratt, M. (1976). *Int. Conf. Millimetric Waveguide Syst.*, p. 38. IEE, London.

CCIR (1976). Rept. 233: Influence of the non-ionized atmosphere on wave propagation. Conclusions of the interim meeting on SG-5.

CCIR (1978). Rept. 719: Attenuation by Atmospheric Gases, pp. 97–102.

Cheng, S. S., Harkless, E. T., Muise, R. W., Studdiford, W. E., and Zukermann, D. N. (1977). *Bell Syst. Tech. J.* **56**(10), 2147–2156.

Clemetson, W. J., Kenyon, N. D., Kurokawa, K., Mohr, T. W., and Schlosser, W. O. (1972). *Proc. IEEE* **60**(7), 912–913.

Clemetson, W. J., Kenyon, N. D., Kurokawa, K., Owen, B., and Schlosser, W. O. (1971). *Bell Syst. Tech. J.* **50**(9), 2917–2945.

Collins, L. J., Jones, L. R., McElroy, D. R., Siegel, D. A., Ward, W. W., and Willim, D. K. (1978). *AIAA Conf. Rec.,* San Diego, California, pp. 471–478.

Corazza, G. C., Koch, R., Mastellari, G. A., Straccs, G. B., and Valdoni, F. (1974). *Alta Freq.* **43**(9), 628–633.

Cox, D. C., (1978). *Bell Syst. Tech. J.* **57**(5), 1231–1255.

Dement, D. K. (1979). *ICC 1979 Conf. Rec.,* Boston, Massachusetts, pp. 15.1.1–15.1.4.

Dunn, C. N., Petersen, O. G., and Redline, D. C. (1977). *Bell Syst. Tech. J.* **56**(10), 2119–2134.

Dupuis, P., and Joindot, M. (1974). *1974 Eur. Microwave Conf.*, pp. 639–643.

Egami, S. (1975). *IEEE Trans. Microwave Theory Tech.* **MTT-23**(3), 282–287.

Fugono, N., and Hayashi, R. (1978). *AIAA Conf. Rec,* San Diego, California, pp. 638–646.

Fugono, N., Yoshimura, K., and Hayashi, R. (1979). *IEEE Trans. Comm.* **27**(10), 1381–1391.

Fukatsu, Y., Akaike, M., and Kato, H. (1971). *IEEE Int. Solid-State Circuit Conf. Dig.*, pp. 172–173.

Fuketa, H., Iso, A., and Nakatani, I. (1980). *Rev. Elec. Commun. Lab.* **28**(7–8), 528–550.

Harkless, E. T., Nardi, A. J., and Wand, H. C. (1977). *Bell Syst. Tech. J.* **56**(10), 2089–2101.

Hashimoto, K. (1974). *Trans. IECE Japan* **57-B**(1), 37–44.

Hashimoto, K. (1976). *IEEE Trans. Microwave Theory Tech.* **MTT-24**(1), 25–31.

Hashimoto, R., Kurita, O., and Nakamura, Y. (1978). *Nat. Conf. Rec. IECE Japan*, Tokyo, p. 1640.

Hata, M., Fukasawa, M., Bessho, M., Makino, S., and Higuchi, M. (1978). *1978 IEEE MTT-S Int. Microwave Symp.*, Ottawa, Canada, pp. 236–238.

Hata, M., Fukasawa, A., Bessho, M., Makino, S., and Higuchi, M. (1980). *IEEE Trans. Microwave Theory Tech.* **MTT-28**(9), 951–962.

Herden, B. L. (1978). *AIAA Conf. Rec.* San Diego, California, pp. 698–708.

IECE Japan (1979). "Handbook for Electronics and Communication Engineers," p. 775.

Inada, K., Akimoto, T., and Hayakawa, T. (1975). *1975 IEEE MTT-S Int. Microwave Symp.*, pp. 235–238.

Ishii, S., and Oki, K. (1975). *Int. Microwave Symp. Dig.*, Palo Alto, California, pp. 232–234.

Johnston, R. L., DeLoach, B. C., and Cohen, B. G. (1965). *Bell Syst. Tech. J.* **44**(2), 369–372.

Kaneda, H., Tsukamoto, K., and Fuketa, H. (1978). *AIAA Conf. Rec.*, San Diego, California, pp. 601–607.

Kanmuri, N., and Kawasaki, R. (1975). *Trans. IECE Japan* **58-B**(5), 247–253.

Kanmuri, N., and Kawasaki, R. (1976). *IEEE Trans. Microwave Theory Tech.* **MTT-24**(5), 259–261.

Kawasaki, R., Ishihara, H., Horiguchi, Y., and Kitazume, S. (1977). *Trans. IECE Japan.* **60-B**(11), 901–903.

Kita, S., Kanmuri, N., and Fukatsu, Y. (1973). *Rev. Elec. Commun. Lab.* **21**(1–2), 77–86.

Kitazume, S., and Ishihara, H. (1975). *IEEE Int. Microwave Symp. Dig.* Palo Alto, California, pp. 224–228.

Kohiyama, K., Horikawa, I., Monma, K., Yamamoto, H., and Morita, K. (1976). *Rev. Elec. Commun. Lab.* **24**(7–8), 552–572.

Krahn, F. (1976). *Int. Conf. Millimetric Waveguide Syst.*, pp. 46–49. IEE, London.

Kumazawa, H., and Ohtomo, I. (1977). *IEEE Trans. Microwave Theory Tech.* **MTT-25**(8), 683–687.

Kuno, H. J., and English, D. L. (1973). *IEEE Trans. Microwave Theory Tech.* **MTT-21**(11), 703–706.

Kuribayashi, M., Yamaguchi, K., Nihei, F., Nagahama, N., and Tamura, Y. (1974). *Rev. Elec. Commun. Lab.* **22**(1–2), 51–62.

Longton, A. C. (1976). *ICC 1976 Conf. Rec.*, Philadelphia, Pennsylvania, pp. 18-14–18-17.

Lopriore, M., Maccoll, D., and Woode, A. (1978). *AIAA Conf. Rec,* San Diego, California, pp. 250–257.

Mahien, J. R. (1973). *1973 Eur. Microwave Conf. Dig.*, Brussels, Session No. B.13, Paper No. 6.

Marcatili, E. A. (1964). *Proc. Symp. Quasi-Optics*, pp. 535–543. Polytechnic Press, New York.

May, C. A. (1976). *Int. Conf. Millimetric Waveguide Syst.*, pp. 22–25. IEE, London.

Menne, D. (1979). *IEEE Spectrum* **16**(1), 38–42.

Miller, S. E. (1954). *Bell Syst. Tech. J.* **33**(9), 1209–1265.

Mills, B. A., Bodonji, J., and Watson, B. K. (1976). *1976 Int. Conf. Millimetric Waveguide Syst.*, pp. 128–131. IEE, London.

Mitchell, R. L., Reber, E. E., and Carter, C. J. (1969). *Microwave J.* **12**(11), 75–81.

Miyauchi, K. (1975). *IEEE MTT-S Int. Microwave Symp.*, Palo Alto, California, pp. 208–211.

Miyauchi, K. (1980). *IEEE Trans. Comm.* **18**(5), 25–34.

Miyauchi, K., Seki, S., and Ishida, N. (1975a). *Japan Telecommun. Rev.* **17**(3), 189–196.

Miyauchi, K., Seki, S., Ishida, N., and Izumi, K. (1975b). *Rev. Elec. Commun. Lab.* **23**(7–8), 707–741.

Miyauchi, K., Fuketa, H., and Watanabe, Y. (1978). *AIAA Conf. Rec.*, San Diego, California, pp. 488–496.

Mokhoff, N. (1979). *IEEE Spectrum* **16**(10), 28–50.

Moore, A. J., Taylor, D. A., and Greene, D. J. (1976). *Int. Conf. Millimetric Waveguide Syst.*, pp. 59–63. IEE, London.

Morimoto, S., Kurita, O., Komaki, S., and Nakamura, Y. (1978). *1978 Nat. Conf. Rec. IECE Japan*, pp. 7–210.

Morita, K., and Higuchi, I. (1976). *Rev. Elec. Commun. Lab.* **24**(7–8), 651–668.

Nagao, H., Hasumi, H., Katayama, S., and Ohmori, M. (1978). *IEEE MTT-S Int. Microwave Symp. Dig.*, Ottawa, Canada, pp. 366–368.

Nakagami, T., and Takenaka, S. (1975). *IEEE MTT-S Int. Microwave Symp. Dig.*, Palo Alto, California, pp. 229–231.

Nakagami, T., Takenaka, S., and Ishizaki, M. (1974). *1974 Nat. Conf. Rec. IECE Japan*, p. 950.

Nakajima, N. (1978). *Trans. IECE Japan* **E61**(9), 750–751.

Nakamura, Y., Yoshikawa, T., Hosoda, A., and Mukai, T. (1977). *ICC 1977 Conf. Rec.* vol. III, Chicago, Illinois, pp. 88–92.

Nicholson, B. F., and Watson, B. K. (1976). *Int. Conf. Millimetric Waveguide Syst.*, p. 121.

Nippon Electric Company Ltd. (1979). Data sheet V588.

Oguchi, B., *Rev. Elec. Commun. Lab.* **16**(3–4), 169–200.

Ohara, T., Shibata, S., and Suzuki, T. (1976). *Elec. Commun. Lab. Tech. J.* **25**(11), 1745–1752.

Ohtomo, I., Suzuki, N., and Nakajima, N. (1975). *Rev. Elec. Commun. Lab.* **23**(7–8), 783–798.

Ohtomo, I., Kurita, O., and Aikawa, M. (1979). "Systems Consideration of 28 GHz Band Intra-City Radio Communication Systems." Paper Tech. Group on Commun. Syst. IECE Japan, CS79-29.

Perrotta, G. (1978). *AIAA Conf. Rec.*, San Diego, California, pp. 736–745.

Read, W. T. (1958). *Bell Sys. Tech. J.* **37**(2), 401–446.

Ritchie, W. K., and Childs, G. H. L. (1976). *Int. Conf. Millimetric Waveguide Syst.*, pp. 50–54. IEE, London.

Samejima, S., and Kurokawa, S. (1979). *Japan Telecommun. Rev.* **21**(8), 364–369.

Scharfetter, D. L., Evans, W. J., and Johnston, R. L. (1970). *Proc. IEEE* **58**(7), 1131–1133.

Schnitger, H., and Unger, H. G. (1974). *IEEE Trans. Comm.* **22**(1), 1374–1377.

Seidel, T. E., Dawis, R. E., and Igresias, D. E. (1971). *Proc. IEEE* **59**(8), 1222–1228.

Seki, S., Akaike, M., and Kanmuri, N. (1974). *1974 Eur. Microwave Conf. Dig.*, pp. 624–628.

Seki, S., Ishida, N., Akaike, M., Kanmuri, M., Ishio, H., and Yamamoto K. (1975). *Rev. Elec. Commun. Lab.* **23**(7–8), 755–781.

Shimada, S., and Ishihara, F. (1978). *Rev. Elec. Commun. Lab.* **26**(3–4), 345–351.

Shinji, M., Shimada, S., Koyama, M., and Kumazawa, H. (1976). *Rev. Elec. Commun. Lab.* **24**(3–4), 274–300.

Solman, F. J., Berglund, C. D., Chick, R. W., and Clifton, B. J. (1978). A Collection of Technical Papers, AIAA 7th C.S.S.C., 78-562, pp. 208–215.

Solman, F. J., Berglund, C. D., Chick, R. W., and Clifton, B. J. (1979). *J. Spacecraft,* **16**(3), 181–186.

South, C. R. (1976). *Int. Conf. Millimetric Waveguide Syst.*, pp. 64–66. IEE, London.

Stagg, L. J., and Robinson, J. N. (1977). *1977 Eur. Microwave Conf.*, Copenhagen, Denmark, pp. 305–311.

Suchkov, V. F., Kerzhenseva, N. P., and Aron, V. A. (1973). *1973 Eur. Microwave Conf.*, Brussels, Session No. B.11, Paper No. 4.

Suzuki, K. (1975). *Rev. Elec. Commun. Lab.* **23**(7–8), 939–947.

Suzuki, T., and Ohara, T. (1975). *Tech. Group Microwaves IECE Japan*, MW75-37, pp. 71–79.

Suzuki, N., Ohtomo, I., and Shimada, S. (1972). *Elec. Comm. Lab. Tech. J.* **21**(3), 475–503.

Unger, H. G. (1957). *Bell. Syst. Tech. J.* **36**(5), 1253–1278.

Unger, H. G. (1958). *Bell. Syst. Tech. J.* **37**(6), 1599–1647.

Volckman, P. A., Gibbs, S. E., and Denniss, P. (1974). *ICC 1974 Conf. Rec.*, Minneapolis, Minnesota, pp. SF-1–SF-2.

Ward, J. J. (1979). *ICC 1979 Conf. Rec.*, Boston, Massachusetts, pp. 15.2.1–15.2.5.

Watanabe, R., and Nakajima, N. (1978). *Electronics Letters* **14**(4), 81–82.

Waters, W. D. (1976). *Int. Conf. Millimetric Waveguide Syst.*, pp. 18–21. IEE, London.

Wolff, E. A., *et al.* (1975). *IEEE Trans. AES* **11**(6), 984–993.

Yamaguchi, K., Nihei, F., and Nagahama, N. (1974). *Rev. Elec. Commun. Lab.* **22**(1–2), 20–38.

Yokosuka Electrical Communication Laboratory (1980). Tech. Publication No. 187, pp. 1–16.

Young, D. T., and Irvin, J. C. (1965). *Proc. IEEE* **53**(12), 2131.

CHAPTER 2

# A Comparative Study of Millimeter-Wave Transmission Lines†

*Tatsuo Itoh and Juan Rivera**

*Department of Electrical Engineering*
*The University of Texas at Austin*
*Austin, Texas*

## I. Introduction

Because of the advances in microwave and millimeter-wave technologies, there exists a large number of relatively new transmission lines that may be used in the upper frequency bands. The state of the art in the areas of material science, solid-state devices, and photoetching techniques has advanced rapidly and created a need for innovative and efficient methods of guiding electromagnetic energy. In fact, new types of waveguides are added to the list from time to time. With the advent of high-speed computers, it is possible to analyze new structures using numerical techniques that yield accurate and reliable results. Hence, a brief discussion of various transmission lines for use in the millimeter-wave frequency bands is useful.

In this chapter, a detailed analysis of each transmission line is not provided. Rather, a discussion of the properties that characterize the guiding structure is given. The information was gathered from journals, publications, and discussions with people whose expertise in this area is recognized. The purpose of this work, therefore, is not to ascertain which guiding method is best, but to illustrate the wide range of guiding media and provide

† Supported by a grant from Texas Instruments, Equipment Group, the Army Research Office under Grant No. DAAG29-81-K-0053, and the Joint Services Electronics Program under Grant No. F49620-79-C-0101.
* Present Address: TRW Inc., One Space Park-133/3645, Redondo Beach, CA 90278.

a familiarity with the most common designs. The most "suitable" design will depend on the particular application for which it is intended.

Because the primary motivation for developing new transmission lines is to form integrated circuits at millimeter-wave frequencies, we shall not discuss the conventional rectangular waveguides, circular waveguides, and ridge waveguides. Characteristics of these waveguides are well known among researchers who work in this area.

## II. List of Transmission Lines

Owing to similarities in their design, the transmission lines of interest may be grouped into four categories:

(1) *dielectric waveguides,* image guide and its variants, inverted strip dielectric waveguide;

(2) *semiopen structures,* H guide, groove guide;

(3) *non-TEM lines,* slot line, fin line;

(4) *quasi-TEM lines,* coplanar waveguide, microstrip, suspended microstrip, inverted microstrip, trapped inverted microstrip.

The lines will be treated in the order indicated. A brief description of each structure is provided, followed by a discussion of electrical properties such as impedance, phase velocity, dispersion, and loss. In several instances, two transmission lines are grouped together owing to similarities in their performance characteristics. In each case, the second transmission line is slightly modified from the basic structure, but unless specifically stated, the discussion applies to both lines.

### A. DIELECTRIC WAVEGUIDES

Dielectric waveguides have been used extensively in optical integrated circuits and have become increasingly popular in the millimeter-wave region. Owing to differences in operating frequencies, actual structures of millimeter dielectric waveguides are different from their optical counterparts. Nevertheless, the operating principles are identical. Although some versions of dielectric waveguide contain metal conductors, the wave guidance does not rely upon the existence of conductors. Rather, the wave is guided by means of total internal reflection at the dielectric interface.

The fundamental form of dielectric waveguides is the rectangular dielectic rod shown in Fig. 1a. However, in integrated circuit applications, this is often modified into the forms of an image guide or insulated image guide shown in Figs. 1b and 1c. The ground plane can be used as a structural support and, when solid-state devices are implemented, as a heat sink and dc bias return. Various analytical methods for these waveguides are summar-

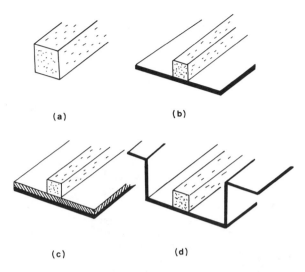

(a)       (b)

(c)       (d)

FIG. 1 Dielectric waveguides: (a) dielectric rod, (b) image guide, (c) insulated image guide, (d) trapped image guide.

ized in Volume 4 of "Infrared and Millimeter Waves" (Itoh, 1981). The simplest of them is that developed by Marcatili (1969). When compared with microstrip lines at millimeter frequencies, the size of the dielectric waveguides is much larger. The thickness and width of the guides are on the order of the operating wavelength, whereas the microstrip line must be less than one-tenth the wavelength. This size advantage allows easier fabrication and more economical production. However, fabrication of the dielectric waveguide does not lend itself to photoetching techniques.

1. *Image Guides and Insulated Image Guides*

In rectangular rod dielectric waveguides, the dominant modes of the two orthogonal polarizations are called $E_{11}^y$ and $E_{11}^x$. For all practical purposes, these modes are degenerate. Imperfections, discontinuities, or bends will cause coupling of these modes (Marcatili, 1969). The degeneracy is most easily resolved by using the image guide arrangement shown in Fig. 1b, because the $E_x$ field component that would be maximum at the center of the dielectric waveguide for the $E_{11}^x$ mode is now shorted by the conductor. Dispersion characteristics based on the effective dielectric constant approach are presented in Fig. 2 for the dominant $E_{11}^y$ mode and two higher order modes, $E_{21}^y$ and $E_{31}^y$. (Knox and Toulios, 1970).

The loss characteristics of the image guide have been studied by a number of workers (Knox and Toulios, 1970; Solbach, 1978). In a straight wave-

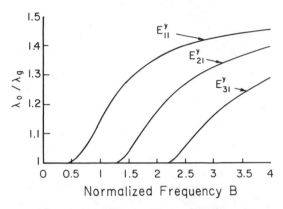

FIG. 2  Dispersion characteristics of image guide. $\varepsilon_r$, relative dielectric constant of the material; $\lambda_0$, free-space wavelength. $B = (4b/\lambda_0)\sqrt{\varepsilon_r - 1}$. (From Knox and Toulios, 1970. Reprinted from *Proc. of the Symp. on Submillimeter Waves*, pp. 497–516. © Polytechnic Press of the Polytechnic Institute of Brooklyn.)

guide section, the loss is made up of contributions from imperfect dielectrics and the ground plane. In practice, however, any imperfections in shape and material cause radiation. Radiation is also caused by bent or curved sections of guide. The fundamental loss mechanism in a curved system is discussed by Lewin *et al.* (1977) and that at junctions by Rozzi (1978).

Actually, interest in the image guide has appeared not only in recent years but also as far back as the 1950s. For instance, Wiltse (1959) reported various characteristics of image guides with a semicircular cross section. At that time, however, these waveguides were not viewed as components for integrated circuits.

For many applications, use of high dielectric materials such as alumina is advantageous because the field is more strongly attached to the guide. Although the physical dimensions become smaller for single-mode operation as the dielectric constant increases, they are still much larger than those required for microstrip lines.

The insulated image guide (Knox, 1976) has an additional layer between the guiding rod and the ground plane. The primary purpose of this layer is to reduce conductor losses and to minimize the effect of air gaps between the rod and the ground plane. Table I shows typical values for the insulated image guide.

As with the image guide, losses in a straight uniform section of insulated image guide are of two types: conductor loss on the image plane and dielectric loss. Relationships have been developed that allow determination of the parameters that should be changed to reduce the attenuation. By using metals with high conductivity, the conductor loss may be kept relatively

TABLE I

TYPICAL PROPERTIES OF DIELECTRIC WAVEGUIDES[a]

| Property | Specifications |
|---|---|
| Dielectric Materials | |
| $\epsilon_1$ | Alumina, semiconductors |
| $\epsilon_2$ | Plastic |
| Dielectric constant | |
| $\epsilon_1$ | 10–15 or higher |
| $\epsilon_2$ | 2.5 |
| Index ratio $\sqrt{\epsilon_1/\epsilon_2}$ | 2.0 |
| Waveguide width (in $\lambda_g$) | 0.5 |
| Radius of curvature (in $\lambda_g$) | 2–5 |
| Degenerate $E^y_{11}$ mode suppressed? | yes |

[a] From Knox (1976). Reprinted from *IEEE Trans. Microwave Theory Tech.* **24**, 806–814. © 1976 IEEE.

small. Low-loss plastics such as polystyrene or polyethylene have been used in insulated image lines and their sizes adjusted to give the desired attenuation.

A comparison of attenuation characteristics is made for three types of waveguides in Table II. Dispersion characteristics are shown in Fig. 3. For the dispersion curves, a cross section in which the guide width equals the guide height is used to obtain optimum bandwidth and suppression of higher-order modes.

Although the attenuation due to the material loss is small, that due to leakage can be significant. Radiation occurs if conditions on the guide deviate in any way from the perfectly straight ideal. Specifically, leakage exists when there is a bending of the guide, an obstacle or device inserted into the guide or the field surrounding it, a change of cross-sectional dimensions, a change in dielectric material, or surface roughness on the walls of the guide.

The greatest advantage of the dielectric waveguides is their low propagation loss at high frequencies. Theoretically, they are very attractive and would seem ideal as guiding media. However, in the actual design, radiation occurs easily and cannot be completely eliminated except by the use of shielding. One way to reduce the radiation loss at a bend is to place the image guide in a trough structure (Fig. 1d). This structure, called the trapped image guide, has been analyzed and tested at Ka band. It was reported that the radiation loss can be reduced significantly, especially at the lower end of the frequency range (Itoh and Adelseck, 1980).

Up to around 100 GHz, the image guide family is not very competitive

TABLE II

THEORETICAL ATTENUATION CHARACTERISTICS FOR
INSULATED IMAGE, MICROSTRIP, AND METAL WAVEGUIDES[a]

| Description | $f$ (GHz) | $\lambda_g$ (cm) | $2a$ (cm) | $t$ (cm) | $\alpha$ (dB/cm) | $\alpha$ (dB/$\lambda_g$) | $Q_u$ |
|---|---|---|---|---|---|---|---|
| Insulated | 15 | 0.955 | 0.534 | 0.053 | 0.0093 | 0.0089 | 3063 |
| image | 30 | 0.476 | 0.268 | 0.027 | 0.0224 | 0.0107 | 2551 |
| | 60 | 0.237 | 0.134 | 0.013 | 0.0554 | 0.0131 | 2072 |
| | 90 | 0.158 | 0.090 | 0.009 | 0.0955 | 0.0151 | 1802 |
| Microstrip, | 15 | 1.210 | 0.108 | 0.054 | 0.0200 | 0.0242 | 1124 |
| gold on | 30 | 0.605 | 0.054 | 0.027 | 0.0562 | 0.0340 | 800 |
| fused quartz | 60 | 0.302 | 0.027 | 0.014 | 0.1542 | 0.0466 | 583 |
| | 90 | 0.201 | 0.018 | 0.009 | 0.2802 | 0.0563 | 483 |
| Rectangular | 15 | 2.58 | 1.580 | | 0.0019 | 0.0049 | 5551 |
| metal wave | 30 | 1.40 | 1.067 | | 0.0066 | 0.0092 | 2956 |
| guide, silver | 60 | 0.699 | 0.376 | | 0.0156 | 0.0104 | 2615 |
| plated | 90 | 0.441 | 0.254 | | 0.0300 | 0.0132 | 2060 |

[a] From Knox (1976). Reprinted from *IEEE Trans. Microwave Theory Tech.* **24**, 806–814. © 1976 IEEE.

with other forms of transmission lines such as suspended microstrips. The structure should be more attractive with increasing frequencies. To take advantage of the image guide, interfacing problems with solid-state devices must be solved or new types of devices invented that reduce the junction problems. It is also conceivable to combine the dielectric waveguide structure with quasi-optical structures as attempted by Yen *et al.* (1981).

## 2. Strip and Inverted Strip Dielectric Waveguide

The waveguide called the strip dielectric guide (SDG), in which the dielectric constant of the guiding layer, $\varepsilon_2$, is chosen to be larger than those of the dielectric strip and substrate, $\varepsilon_1$ and $\varepsilon_3$, respectively, is shown in Fig. 4a. Because of this distribution of dielectric constants, the electromagnetic energy is concentrated in the guiding layer immediately below the strip (Itoh, 1976).

The inverted strip dielectric guide (ISDG) is a natural extension of the SDG and operates on the same principle. The structure, shown in Fig. 4b, consists of a guiding layer with dielectric constant $\varepsilon_2$ placed on dielectric strip $\varepsilon_1$, which in turn rests on a ground plane. Because $\varepsilon_1 < \varepsilon_2$, most of the energy is carried in the guiding layer directly above the strip.

For both waveguides, the conductor loss is small because most of the

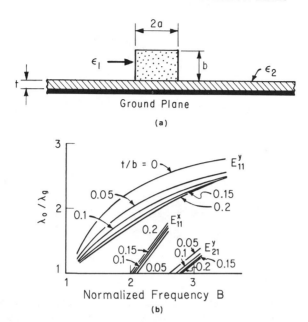

Fig. 3  Insulated image guide: (a) cross-sectional view, (b) dispersion curves for $B = (4b/\lambda_0)\sqrt{\varepsilon_1 - 1}$. $a/b = 0.5$; $\varepsilon_1 = 9.8$, $\varepsilon_2 = 2.25$. (From Knox, 1976. Reprinted from *IEEE Trans. Microwave Theory Tech.* **24**, 806–884. © 1976 IEEE.)

energy is in the guiding layer away from the ground plane. In addition, the ground plane provides a heat sink and is convenient for dc biasing in solid-state device applications. In the ISDG, only two dielectrics are involved, which may reduce the dielectric loss caused in the substrate of the SDG. The ISDG can be held together by mechanical pressure between the guiding layer and the ground plane, and thus bonding material that causes additional propagation loss can be eliminated.

Theoretical and experimental results have been obtained for the field distributions in the cross sections of the ISDGs. The decay in the transverse direction is on the order of 10 dB one wavelength away from the edge of the strip (Itoh, 1976).

The normalized propagation constant $(k_z/k_0)$ is found to vary significantly for small ratios of strip to guiding layer thickness. However, as this ratio increases, and hence the effect of the ground plane is reduced, $k_z/k_0$ approaches some asymptote. Hence, a moderately large ratio (greater than 0.8 for a guiding layer thickness $\simeq 1.5$ mm) might be chosen if one desires little effect on $k_z/k_0$ due to temperature variation (Itoh, 1976).

Compared with image guides and other dielectric rod waveguides, the radiation loss due to surface roughness is reduced in this structure. The

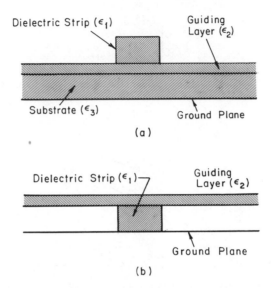

FIG. 4 Cross-sectional view of waveguide structures for millimeter-wave integrated circuits: (a) strip dielectric waveguide, (b) inverted strip dielectric waveguide. (From Itoh, 1976. Reprinted from *IEEE Trans. Microwave Theory Tech.* **24,** 821–827. © 1976 IEEE.)

dielectric strip creates a "lens" effect in the structure and propagation occurs in the planar guiding layer. Energy in dielectric rod waveguides propagates such that surface roughness of the side walls enhances radiation loss.

The disadvantages present in this structure are similar to those in image guides. Radiation problems may still occur and transitions are difficult to achieve. In fact, the radiation at bends and irregularities may be greater in the ISDG than in the image guide, because the wave is more weakly guided in the ISDG.

A recent theoretical development has revealed an interesting physical phenomenon of wave leakage in the form of a surface wave. This phenomenon can take place only in the subclass of dielectric waveguides that reduces to planar layered waveguides in the limit where the size of rod or strip is reduced to zero. Therefore, the phenomenon can occur in the SDG, ISDG, or insulated image guide but not in the image guide or dielectric rod guide. The leakage from the central "core" region to the sideward "layered" region can occur owing to interaction of the fields at the junction between the respective constituent regions. A detailed explanation and the condition for such leakage phenomena are given by Oliner *et al.* (1981) and Peng and Oliner (1981); here, we only indicate the potential existence of such phenomena to draw the attention of waveguide designers.

In the past few years, significant advances have been made on integrated circuits made of dielectric waveguides, especially in the area of passive devices and antennas. Implementation of active devices still presents a great challange, as discontinuities are automatically encountered. Perhaps the monolithic technique may enhance the usefulness of dielectric waveguides because solid-state devices may be created in a GaAs waveguide without any significant discontinuity.

## B. SEMI-OPEN WAVEGUIDES

The semi-open structures are characterized by the absence of side walls and resemble a parallel-plate waveguide. By inserting a dielectric slab between the plates or by creating a grooved region, the electromagnetic energy is guided along this type of transmission line. These structures are believed to be especially useful in the near millimeter-wave frequency band because of their large overall dimensions and very low attenuation factors.

### 1. H Guide

The H Guide consists of two parallel conducting plates and a dielectric slab between them. The basic guide form is shown in Fig. 5 along with the field distributions. The H guide is open along one of the transverse directions. The dielectric slab concentrates the transport of electromagnetic energy in its vicinity between the conducting strips. Only a small fraction of the energy is transported outside the guide (Tischer, 1956).

Attenuation is mainly due to energy dissipation in the walls and dielectric losses in the slab, but it is rather low compared to that of a rectangular waveguide. Wall and dielectric attenuation of the H guide calculated for different wavelengths is shown in Figs. 6 and 7. An effective metal conductivity of $10^7$ S/m and a loss tangent of $10^{-3}$ were assumed for the calculation of these curves. As Fig. 6 shows, the attenuation is very small and decreases with increasing frequency.

A practical design is illustrated in Fig. 8. By choosing an appropriate

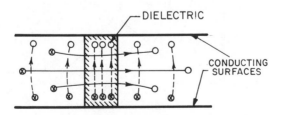

FIG. 5   H-guide cross section and transverse field distribution: ——, E; ---, H. (From Harris *et al.*, 1978. Reprinted from *IEEE Trans. on Microwave Theory Tech.* **26**, 998–1001. © 1978 IEEE.)

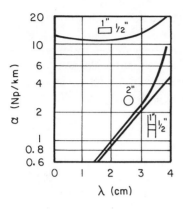

Fig. 6 Attenuation of H guide compared to that of rectangular and circular waveguides at $X$-band frequencies. (Tischer, 1956).

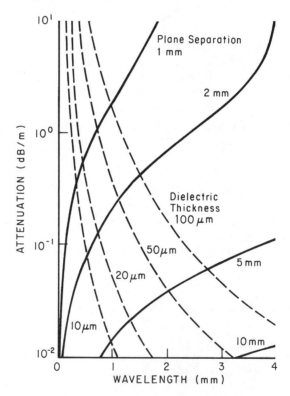

Fig. 7 H-guide attenuation:——, conductor loss ($\sigma_m = 5.8 \times 10^7$ S/m);——, dielectric loss ($\varepsilon_r = 2.1$, tan $\delta = 0.001$). (From Harris *et al.,* 1978. Reprinted from *IEEE Trans. Microwave Theory Tech.* **26,** 998–1001. © 1978 IEEE.)

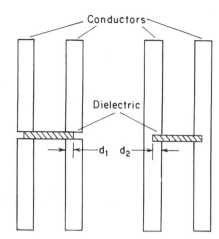

FIG. 8 Configuration for H-guide dielectric support and slot adjustment. (From Harris *et al.*, 1978. Reprinted from *IEEE Trans. Microwave Theory Tech.* **26,** 998–1001. © 1978 IEEE.)

thickness of the thin dielectric film and adjustment of the dimension *d*, it is possible to excite only the fundamental mode. However, excessive reduction of the guide width for mode purity would not be feasible because of the high wall attenuation for small plane separations. Yet, a decrease in attenuation by increasing plane separation would introduce higher-order modes (Doswell and Harris, 1973).

H-guide construction requires the use of thin dielectric film, and the design would have to be modified to include some means for locating and holding the dielectric between the conducting plates. Schematic suggestions made by Doswell and Harris (1973) are shown in Figs. 9a and 9b. The inclusion of the wall parallel to the dielectric alters the surface wave exponential damping term outside the dielectric into a hyperbolic function (Doswell and Harris, 1973). In fact, the extra wall has little effect if most of the energy is located close to the dielectric. Suggestions made by Tischer

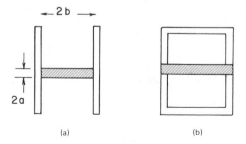

(a)                                (b)

FIG. 9 H-guide transmission lines. (From Doswell and Harris, 1973. Reprinted from *IEEE Trans. Microwave Theory Tech.* **21,** 587–589. © 1973 IEEE.)

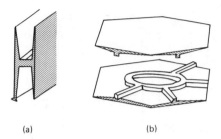

(a)                                    (b)

Fig. 10  (a) H-guide and (b) ring circuit design for easy fabrication (Tischer, 1956).

(1956) are shown in Figs. 10a and 10b. Fabrication of these structures could be accomplished by plastic molding techniques; the two outer surfaces are metalized by electroplating and the two parts glued together.

## 2. Groove Guide

The groove guide shown in Fig. 11 belongs to the group of H-guide structures and has similar properties. The grooves in the walls cause the field distribution to decrease exponentially from the center, and, because no dielectrics are used in the guide, the dielectric losses are eliminated with a corresponding reduction of attenuation (Tischer, 1963). Analyses of groove-guide propagation characteristics have been performed (Harris and Lee, 1977; Nakahara and Kurauchi, 1965).

Overall dimensions are large compared with the wavelength. Although theoretical wall losses are greater than conductor losses for the H guide, practical losses could still be an order of magnitude less than those of a rectangular waveguide. The transverse-resonance technique employed by Harris and Lee (1977) showed that energy can propagate out in the transverse direction and "leak" from the groove region if the proper groove depth is exceeded.

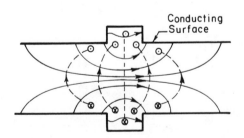

Fig. 11  Groove-guide cross section and transverse field distribution:——, E; –––, H. (From Harris et al., 1978. Reprinted from IEEE Trans. Microwave Theory Tech. **26**, 998–1001. © 1978 IEEE.)

The H and groove guides are simple to construct, and the capability exists for suppression of higher-order modes. The dimensions are reasonable and attenuation figures are very low, so that this class of waveguide is very promising for the short millimeter-wave region. However, radiation loss problems exist. Because these waveguides are open in construction along one of the dimensions, coupling to the radiation mode occurs at bends and discontinuities just as in dielectric waveguides. Therefore, problems of mounting devices should be resolved if these semi-open waveguides are to be used for circuit components.

A recent addition to this class of waveguides is the nonradiative dielectric (NRD) waveguide (Yoneyama and Nishida, 1981). Its cross-sectional structure is essentially identical to that of the H guide. However, the separation of the top and bottom conducting plates is such that the field is below cutoff in the air region and above cutoff in the dielectric when the guide is excited with its electric field parallel to the conducting walls. Because of this operation below cutoff, no radiation should occur at a bend.

## C. NON-TEM LINES

In this section, we shall treat printed transmission lines in which substantial axial field components exist. Two of the most widely used examples are slot lines and fin lines. Their field distribution and propagation characteristics resemble those of a closed metal waveguide. On the other hand, the quasi-TEM transmission lines described later have properties similar to those found in a coaxial cable or a parallel wire transmission line.

### 1. Slot Line

The slot line, introduced by Cohn in 1969 for microwave applications, consists of a slot or gap in a conductive coating on a dielectric substrate as shown in Fig. 12. It is particularly useful for applications requiring regions of circularly polarized magnetic field and/or shunt mounted elements.

The slot mode configuration is such that the electric field extends across the slot, whereas the magnetic field is in a plane perpendicular to the slot and forms closed loops at half-wave intervals. Hence, the slot mode has a region of elliptical polarization that should have application for nonreciprocal ferrite devices (Mariani et al., 1969).

The slot line uses a dielectric substrate of sufficiently high permittivity ($\varepsilon_r = 16$) that the fields are closely confined to the slot region and the guide wavelength is substantially less than the free-space wavelength. The typical width of the slot varies from 21 to about 30 mils for measurement in the 1–10-GHz range.

Calculations by Cohn (1969) show that the impedance variation with frequency is relatively small—generally less than $\pm 10\%$ over a two-octave

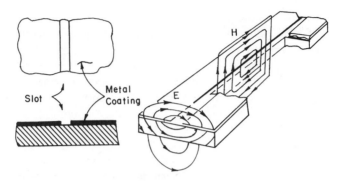

FIG. 12　Schematic representation of slot line. (From Robinson and Allen, 1969. Reprinted from *IEEE Trans. Microwave Theory Tech.* **17**, 1097–1101. © 1969 IEEE.)

bandwidth. For different ratios of slot width to substrate thickness, the characteristic impedance varies from 40 up to 200 Ω. The guide wavelength and characteristic impedance of typical slot-line structures are shown in Fig. 13.

Transmission losses for slot lines are about the same as those of conventional microstrips (Robinson and Allen, 1969). Although measured slot-line losses show no strong dependence on substrate thickness, they tend to increase when the gap width becomes very narrow (5 mils or less). The conductor thickness, so long as it is several skin depths, has little effect.

As a rule, if the guide wavelength is roughly 30–40% of the free-space value, the fields will be adequately contained. For example, at 8.5 GHz, a 21-mil slot on a 55-mil substrate with a dielectric constant of 16 will have a guide wavelength of about 40% of the free-space value (Robinson and Allen, 1969). Based on Cohn's (1969) analysis, the voltage measured along a semicircular path $\frac{1}{2}$ in. from the slot is 40 dB below the voltage across the slot, indicating that the fields are tightly bound. This was confirmed by experiments indicating that conductors placed $\frac{1}{4}$ in. above the substrate had a negligible effect (Robinson and Allen, 1969).

Tolerances at some specified slot widths and frequencies have been observed. A 1% increase in dielectric constant produces a 0.5% decrease for both the wavelength ratio $\lambda_g/\lambda_0$ and the characteristic impedance $Z_0$ with the substrate thickness fixed. A 3% increase in the substrate thickness leads to approximately a 0.5% decrease in $\lambda_g/\lambda_0$ and $Z_0$. Moreover, for a given dielectric substrate of a certain thickness and at a specified frequency, a 20% increase in slot width leads to only a 1% increase in $\lambda_g/\lambda_0$ and a 6% increase in $Z_0$ (Mariani *et al.*, 1969).

If microstrips and slot lines of proper characteristic impedance cross each

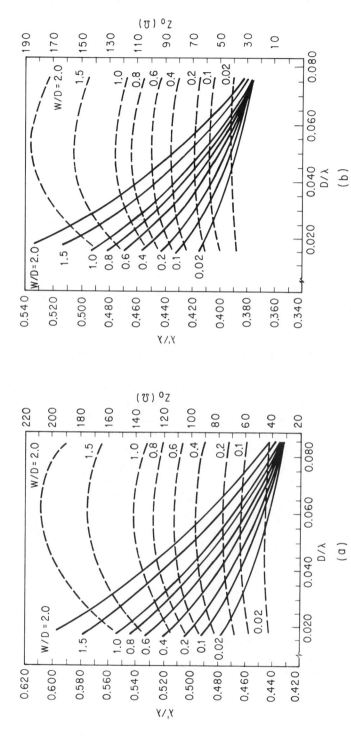

FIG. 13 Slot-line characteristics for (a) $\varepsilon_r = 9.6$ and (b) $\varepsilon_r = 13.0$: ——, $\lambda'/\lambda$; — — —, $Z_0$. (From Mariani *et al.*, 1969. Reprinted from *IEEE Trans. Microwave Theory Tech.* **17**, 1091–1096. © 1969 IEEE.

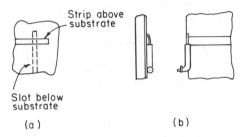

FIG. 14    (a) Transition between slot line and microstrip. (b) Broadband transition between slot line and miniature semirigid coaxial cable. (From Cohn, 1969. Reprinted from *IEEE Trans. Microwave Theory Tech.* **17**, 768–778. © 1969 IEEE.)

other at right angles and extend $\frac{1}{4}\lambda$ beyond the crossing point as shown in Fig. 14a, coupling will be strong and a transition covering approximately 30% of the bandwidth can be achieved (Robinson and Allen, 1969). A wide-band transition is shown in Fig. 14b; a miniature, semirigid coaxial line is placed at the end of an open-circuited line with both the inner and outer conductors of the coaxial line soldered to the conductive plating of the slot line. A low-frequency test (3 GHz) of such a configuration yielded acceptable results [voltage standing-wave ratio (VSWR) of 1.2 into 50 Ω].

Slot line and microstrip are often assumed to be complementary. This is not exactly true, because there is no longitudinal path of conduction currents in the slot line that would support quasi-TEM waves as in a microstrip. Fabrication of the slot line lends itself to known photoetching techniques, which would help reduce costs. In addition, incorporation of solid-state devices is simple, but attenuation figures may be excessively high.

At millimeter-wave frequencies such as the Ka band and beyond, the slot line is rarely used in its original form. Slot-line modes often appear in the fin-line structure or the coplanar waveguide, discussed later. For instance, the coupled slot-line mode can be used in a balanced mixer in which both balanced and unbalanced fields must exist.

## 2. Fin Lines

The fin line was introduced by Meier (1974) as a low-loss transmission line compatible with batch-processing techniques and superior to microstrip lines in several respects at millimeter-wave frequencies. Some configurations are compatible with hybrid devices and can easily be interfaced with waveguide instrumentation.

There are several forms of fin lines, as is shown in Fig. 15. They are in effect printed ridge waveguides, and hence can be designed to have a wider useful bandwidth than a conventional rectangular waveguide. It is interest-

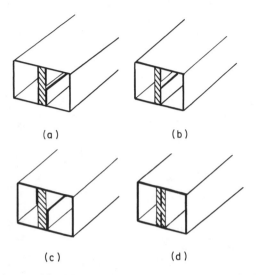

(a)          (b)

(c)          (d)

FIG. 15 General types of fin lines: (a) bilateral, (b) unilateral, (c) antipodal, (d) insulated.

ing to note that when the slot line is used in a channel—that is, when the slot-line substrate bridges the broad walls of a rectangular waveguide—a unilateral fin line is realized.

In most millimeter-wave applications, the fin lines are constructed from a soft substrate such as RT Duroid, which has a relatively low dielectric constant of 2.2. These soft substrates are inexpensive and are widely used with fin-line components for frequencies of operation as high as 94 GHz.

The propagation constant and characteristic impedance have been calculated by a number of authors (Hoefer, 1978; Jansen, 1979; Saad and Schünemann, 1978; Scmidt and Itoh, 1980). Figure 16 shows typical frequency characteristics of the guide wavelength and characteristic impedance computed by Knorr and Shayda (1980). The impedance definition is based on the transmitted power and slot voltage. It is seen from the figure that a fin line with a characteristic impedance of less than 100 Ω is not easy to realize in a unilateral form. This difficulty is equally applicable to bilateral fin lines. The antipodal fin line described by Hofmann (1977a,b) is advantageous in realizing a low-impedance fin line (Fig. 17). This structure, however, does not seem to be well suited for mounting shunt devices.

Attenuation of fin lines is about two or three times as high as that of an unloaded waveguide of the same dimensions. Tests show that attenuation increases with increasing frequency and dielectric constant (Saad and Begemann, 1977). Attenuation curves are shown in Fig. 18. The structural parameters $D$, $W$, $a$, $b$, and $h$ are defined as in Fig. 16a, except that the

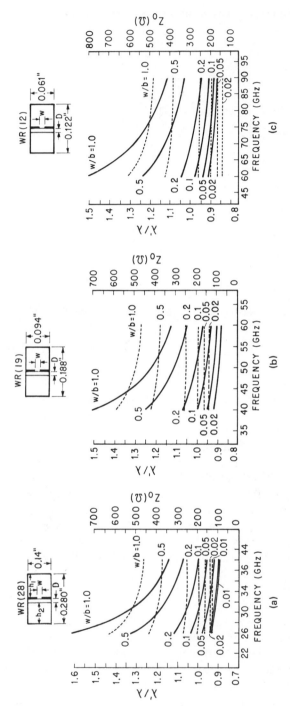

Fig. 16  Guided wavelength and characteristic impedance for fin lines with (a) WR(28), (b) WR(19), and (c) WR(12) shields:——, $\lambda'/\lambda$; —––, $Z_0$. $D = 0.005$ in., $\varepsilon_r = 2.2$. (a) $a = 0.280$ in., $b = 0.140$ in., $b/D = h_1/D = 28$, $h_2/D = 27$; (b) $a = 0.188$ in., $b = 0.094$ in., $b/D = h_1/D = 18.8$, $h_2/D = 17.7$; (c) $a = 0.122$ in., $b = 0.061$ in., $b/D = h_1/D = 12.2$, $h_2/D = 11.2$. The fins are centered. (From Knorr and Shayda, 1980. Reprinted from IEEE Trans. Microwave Theory Tech. **28**, 737–743. © 1980 IEEE.)

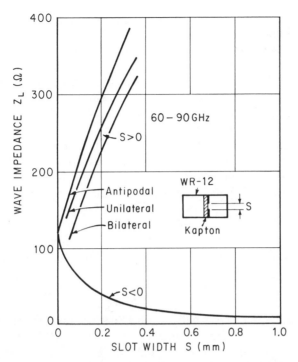

FIG. 17 Wave impedance $Z_L$ vs slot width $S$ for unilateral fin line: $\varepsilon_r = 3.0$, $d = 0.05$ mm.) (From Hofmann, 1977b. Reprinted from *IEEE MTT-S Int. Microwave Symp.*, pp. 381–384. © 1977 IEEE.)

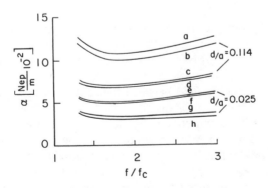

FIG. 18 Fin-line attenuation constant $\alpha$ vs $f/f_c$ ($f_c$: cutoff frequency). $b/a = 0.5$, $\varepsilon_r = 2.22$. $h_1 = 0.3a - 0.5D$ for curves a, d, e, and h, and $h_1 = 0.5a - 0.5D$ for curves b, c, f, and g. $W/a = 0.1$ for curves a, b, e, and f, and $W/a = 0.3$ for curves c, d, g, and h. (From Saad and Begemann, 1977. Reprinted from *IEE J. Microwaves, Opt. Acoust.* **1**, 81–88. © 1977 IEE.)

structure here has another fin inserted on the other side of the substrate (bilateral fin line).

The advantages of the fin line are its wide bandwidth, ease of fabrication by photoetching processes, and easy design and installation of shunt elements. Also, transitions to the waveguide are easily achieved by way of a tapered section. Mixers, couplers, $p-i-n$ switches, and other millimeter-wave components have been successfully developed and are being used in the frequency range up to at least 94 GHz.

A structure similar to fin lines has been independently developed by Konishi et al. (1974). The structure is called a planar circuit mounted in a waveguide. It consists of a metal fin placed between two broad walls of a rectangular waveguide and is realized if the dielectric substrate is removed from a unilateral fin line. An inexpensive but high-performance receiver at 12 GHz and above has been built with this structure. Both this structure and fin-line structures may be summarily called the $E$-plane circuit, because they are inserted in parallel with the $E$ plane of the rectangular $TE_{10}$ waveguide.

D.  QUASI-TEM LINES

These structures are essentially two conductor transmission lines that can be designed to operate in a mode similar to the TEM mode. Extensive information is available on most of them, and they are popular because fabrication can be accomplished with known photoetching techniques. The useful operating frequencies have been steadily extended owing to advances in fabrication techniques, materials, and design, and operations at the millimeter-wave bands are now possible for a number of quasi-TEM lines.

1.  *Coplanar Waveguide*

A coplanar waveguide consists of a strip of thin metallic film deposited on the surface of a dielectric slab with two ground electrodes running adjacent and parrallel to the strip on the same surface, as is shown in Fig. 19. There is no low-frequency cutoff because of the quasi-TEM mode of propagation. However, the rf electric field between the center conducting strip and the ground electrodes tangential to the air–dielectric boundary produces a discontinuity in displacement current density at the interface, giving rise to an axial as well as transverse component of the rf magnetic field (Wen, 1969). As in the slot line, an elliptical polarization is provided. Characteristic impedances $Z_0$, of coplanar waveguides on dielectric half planes with relative dielectric constants $\varepsilon_r$ have been calculated as a function of strip-to-slot ratio (Wen, 1969); the results are plotted in Fig. 20, in which three experimentally obtained points are also shown.

The thickness of the dielectric substrate becomes less critical with higher

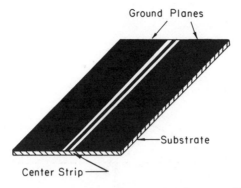

FIG. 19   General configuration of a coplanar waveguide.

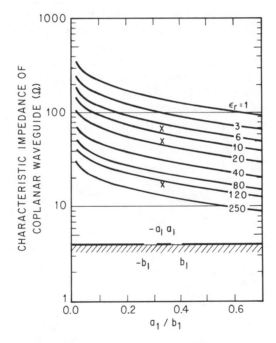

FIG. 20   Characteristic impedance $Z_0$ of coplanar waveguide as a function of the ratio $a_1/b_1$ with the relative dielectric constant as a parameter. Experimentally obtained points are indicated by ✕. (From Wen, 1969. Reprinted from *IEEE Trans. Microwave Theory Tech.* **17,** 1087–1090. © 1969 IEEE.)

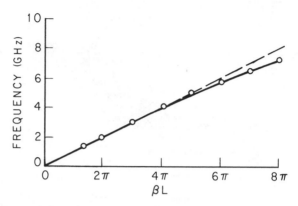

FIG. 21  Measured dispersion properties of a coplanar waveguide on a TiO$_2$ substrate. (From Wen, 1969. Reprinted from *IEEE Trans. Microwave Theory Tech.* **17,** 1087–1090. © 1969 IEEE.)

relative dielectric constants. For infinitely large $\varepsilon_r$, the characteristic impedance increases by less than 10% when the thickness of the substrate is reduced from infinity to $b - a$, the width of the slots. The finite thickness does affect the dispersion characteristics, however, as shown in Fig. 21. The experimentally measured results indicated with ○ are for a coplanar waveguide fabricated on a single-crystal rutile substance. Dispersion is seen as deviation of measured data from the dashed line.

At microwave frequencies up to $X$ band, the attenuation is due mainly to the copper loss of the conductors if the loss tangent of the dielectric is 0.001

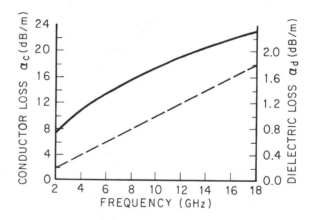

FIG. 22  Conductor and dielectric loss coefficient for coplanar waveguide configured with $Z_0 = 50\Omega$:———, $\alpha_c$;— —, $\alpha_d$. (From Spielman, 1976.)

or smaller. Work done by Spielman (1976) indicates that the losses in a coplanar waveguide are rather high (greater than for a microstrip). His results for a 50-$\Omega$ line with a metalization thickness of 6.35 $\mu$m and a dc conductivity of 4.10 $\times$ 10$^7$ S/m are presented in Fig. 22.

Ease of fabrication (by photoetching techniques) and accessibility to external elements are the best features of the coplanar waveguide. Values of less than 10 $\Omega$ can be obtained for the characteristic impedance but only at the expense of high dissipation losses. The losses also increase rapidly with increasing frequency.

The coplanar waveguide is also well suited for connecting various elements in monolithic microwave and millimeter-wave integrated circuits. Gopinath (1979) compared the coplanar waveguide with the microstrip line as applied to monolithic circuits. Conductor and dielectric losses are lower in a microstrip than in coplanar waveguides when the substrate height is equal to the ground plane spacing in the coplanar waveguide. When radiation loss is included and the ground plane spacing is allowed to increase in the coplanar waveguide, the guides are comparable.

The coplanar configuration allows easy installation of shunt elements such as active devices. However, care must be taken so that the structural symmetry is not destroyed. Otherwise, unwanted slot modes may be generated. Jansen (1979) studied the guide wavelength and the power–voltage characteristic impedance of the even and odd modes in the coplanar

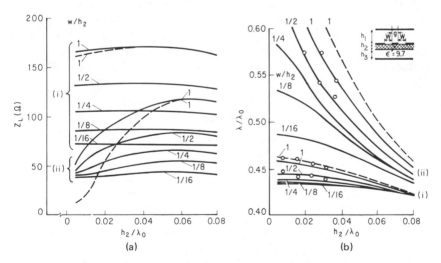

FIG. 23   (a) Characteristic impedances for coupled slots of different widths and (b) dispersion characteristics of even and odd modes.——, $h_1 = 20h_2$;— —, $h_1 = h_2$. (i), even mode; (ii), odd mode. $g/h_2 = \frac{1}{2}$, $h_3 = 20h_2$. In (b), $\circ$ represents a test measurement. (From Jansen, 1979. Reprinted from *IEE J. Microwaves Opt. Acous.* **3** (1), 14–22. © 1979 IEE.)

waveguide (Fig. 23). They are the dominant (coplanar waveguide) mode and the coupled slot mode, respectively. In other applications, the coplanar waveguide is often used as a transition between the fin line and suspended strip line for millimeter-wave balanced mixers operating up to $W$ band (Bates and Coleman, 1979; Bui and Ball, 1982).

## 2. Microstrip Lines

A microstrip line is a broadband transmission medium formed by a narrow strip conductor placed on a dielectric substrate that is in turn backed by a conducting ground plane (Fig. 24). Both the strip width and the substrate thickness are usually small fractions of the operating wavelength to ensure single-mode propagation. This transmission line is one of the most widely studied, and some monographs have appeared (Edwards, 1981; Mittra and Itoh, 1974). A list of many of the published works on microstrip lines is available (Wolff, 1979).

Most of the early work on microstrip lines assumed that the field guided by the microstrip line is nearly TEM or quasi-TEM and reduced the analysis to finding the capacitance per unit length of the line (Bryant and Weiss, 1968; Stinehelfer, 1968; Wheeler, 1965; Yamashita and Mittra, 1968). The characteristic impedance versus the width-to-thickness ratio computed by Schneider (1969a) is plotted in Fig. 25 for several dielectric substances. The guide wavelength that is shorter than the free-space value and larger than that in the bulk dielectric material is also shown in Fig. 26, as computed by Schneider (1969a). In Fig. 26, $(\varepsilon_{eff})^{1/2}$ is equal to $\lambda/\lambda_g$, where $\lambda$ and $\lambda_g$ are the free-space and guide wavelengths, respectively. From Fig. 25 it is readily seen that a wide range of impedance can be realized. This is one of the features that make the microstrip lines very popular for integrated circuit applications.

At lower frequencies, the quasi-TEM treatment is quite satisfactory.

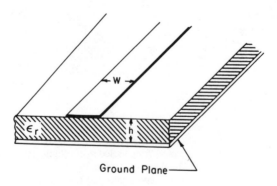

FIG. 24   Conventional microstrip.

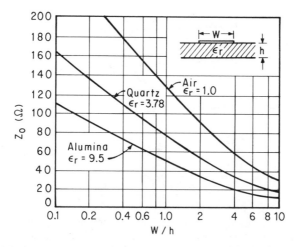

FIG. 25  Characteristic impedance of microstrip for $\varepsilon_r = 1.0$, 3.78, and 9.5 as a function of $W/h$. (From Schneider, 1969a. Reprinted from *The Bell System Technical Journal* by permission. © 1969 AT&T.)

However, as the frequency is increased, various "wave" phenomena such as dispersion and radiation appear. Dispersion characteristics can be obtained by formulations based on the full-wave analysis. Accurate numerical solutions for the dispersion characteristics of microstrip lines are available in a number of recent publications (Denlinger, 1971; Hornsby and Gopinath, 1969; Itoh, 1981; Kowalski and Pregla, 1971). All of these analyses agree that the effective dielectric constant $\varepsilon_{\text{eff}} = (\lambda/\lambda_g)^2$ increases with frequency and

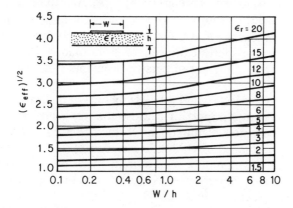

FIG. 26  Square root of the effective dielectric constant for the standard microstrip as a function of $W/h$. (From Schneider, 1969a. Reprinted from *The Bell System Technical Journal* by permission. © 1969 AT&T.)

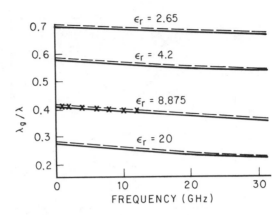

FIG. 27 Normalized guide wavelength $\lambda_g/\lambda$ vs frequency for microstrip:———, zero order;——, closed microstrip; ×, open microstrip experiment. $W = 1.27$ mm; $d = 1.27$ mm. (From Itoh and Mittra, 1973. Reprinted from *IEEE Trans. Microwave Theory Tech.* **21**, 496–499. © 1973 IEEE.)

approaches that of the substrate, $\varepsilon_r$, at higher frequencies. Some numerical results of $\lambda_g/\lambda$ versus frequency are presented in Fig. 27. It is clear that microstrip lines are not very dispersive and are therefore convenient for circuit design.

In addition to these numerical approaches, some work has been done on the development of a simple formula to predict the dispersion characteristics of microstrip lines (Getsinger, 1973; Schneider, 1972). The formula developed by Edwards and Owens (1976) is believed to be accurate within ±0.8% for the dielectric substrate, $10 < \varepsilon_r < 12$. The dispersive nature of the characteristic impedance is still not well understood, mainly because the definition of the characteristic impedance is not unique. Depending on the choice of definition, different dispersive behaviors can be predicted (Edwards, 1981).

At higher frequencies beyond 30 GHz, choice of dielectric materials for the substrate becomes important because the loss tends to increase with frequency. At extremely high frequencies, the preferred choice seems to be Z-cut quartz (Scarman *et al.*, 1981). In fact, contrary to the common belief that microstrip lines cannot be used at frequencies much beyond 30 GHz, microstrip mixer structures have been developed that operate successfully up to 110 GHz (Scarman *et al.*, 1981).

The attenuation in the microstrip line is caused by conductor loss, dielectric loss, radiation, and surface wave excitation. Because the last two are mainly associated with junctions and discontinuities and can be isolated,

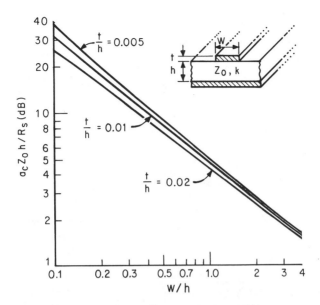

FIG. 28 Theoretical conductor attenuation factor of microstrip as a function of aspect ratio. (From Pucel *et al.*, 1968. Reprinted from *IEEE Trans. Microwave Theory Tech.* **16**, 342–350. © 1968 IEEE.)

only the first two are significant in straight sections of microstrip lines and are mentioned in this chapter. A number of studies have performed on the conductor attenuation $\alpha_c$ and dielectric attenuation $\alpha_d$ (Schneider, 1969b). Figure 28 shows the conductor attenuation calculated by Pucel *et al.* (1968). In general, conductor losses are larger than dielectric losses for microstrip lines on alumina substrates. This is not necessarily the case when other substrate materials are used. It is worth mentioning that in monolithic circuit configurations where microstrip lines are created on GaAs or Si substrates, the dielectric loss is much larger than the conductor loss (Edwards, 1981).

The overall attenuation constant of a microstrip line on a typical nonsemiconductor substrate is on the order of 0.1 dB per wavelength. Because this value is relatively large, most engineers try to use alternative forms of microstrip lines such as suspended or inverted microstrip lines to reduce the attenuation. Some exceptions do exist, as mentioned earlier (Scarma *et al.*, 1981); nevertheless, the upper limit of the frequency for the microstrip line would not be much beyond 100 GHz.

The power-handling capacity of the microstrip line, like any other dielectric field transmission line, is limited by breakdown and dielectric heating

TABLE III

COMPARISON OF AVERAGE POWER-HANDLING CAPACITY FOR
VARIOUS SUBSTRATES AT 2.0, 10, AND 20 GHz[a]

| Substrate | Maximum average power (kW) | | |
|---|---|---|---|
| | 2.0 GHz | 10 GHz | 20 GHz |
| Polystyrene | 0.325 | 0.126 | 0.0759 |
| Quartz | 1.22 | 0.531 | 0.363 |
| Si | 3.33 | 2.33 | 1.72 |
| GaAs | 3.61 | 1.50 | 0.962 |
| Sapphire | 11.80 | 5.19 | 3.53 |
| Alumina | 12.30 | 5.26 | 3.46 |
| BeO | 176.0 | 76.80 | 52.4 |

[a] From Bahl and Gupta (1979). Reprinted from *IEE J. Microwaves, Opt. Acoust.* 3(1), 1–4. © 1979 IEE.

(Bahl and Gupta, 1979). For cw operation, dielectric heating is usually the limiting factor. Breakdown will be more serious for high-power pulse operation.

For a microstrip with a line $\frac{7}{32}$ in. wide on a base of Teflon-impregnated fiberglass $\frac{1}{16}$ in. thick at 3 GHz with a cw power of 300 W, the rise in temperature under the strip line is about 50°C above the ambient temperature (20°C) (Arditi, 1955).

Under pulse conditions, corona effects appear at the edge of the strip conductor for a pulse power of 15 kW at 9 GHz (Arditi, 1955). However, gradients are higher at transducers, and there is a greater possibility of breakdown at these points. Special precautions should be taken to round off all sharp edges for pulse operation, and it may be desirable to paint the edges of the strip with a dielectric paint to increase the breakdown voltage. Table III compares various dielectrics for maximum power at a maximum operating temperature of 100°C. It is clear that as the frequency increases, the power-handling capability decreases in keeping with the smaller size of the structure.

In the low-frequency range, the microstrip line is considered a convenient transmission medium. Its low cost and ease of fabrication make it an attractive transmission line that is now quite commonly used in industry. The usefulness of the microstrip line, however, deteriorates gradually with frequency owing to increased losses. Extreme care in the choice of material and fabrication process is needed for applications beyond 30 GHz. A number of the attractive features of microstrip lines are preserved at higher frequencies in modified versions of microstrip lines, as will be seen shortly.

### 3. Suspended and Inverted Microstrip Lines

When the ground plane of a microstrip line is detached from the substrate, the attenuation can be reduced. This arrangement is called the suspended microstrip line (Fig. 29a). The inverted microstrip line also has a detached ground plane, but the center strip is located on the underside of the substrate as shown in Fig. 29b. For millimeter-wave integrated circuit applications, these transmission lines are placed in a shielding environment such as a split block to reduce interference and excitation of higher-order modes. In such a shield, the distinction between suspended and inverted microstrip lines diminishes.

The field configurations in the suspended and inverted microstrip lines (either shielded or open) are closer to TEM than those in the conventional microstrip line because a substantial portion of the field exists in the air region below the substrate and the dielectric constant of the substrate is usually low: 3.8 for fused quartz and 2.2 for RT Duroid.

Because the substrate is suspended from the ground plane, both sides of the substrate can be used for circuit layouts. This is particularly convenient in complex structures such as balanced mixers in which several different frequencies are involved and possibly several different transmission lines (e.g., fin lines, coplanar waveguides, and suspended microstrip lines) are used.

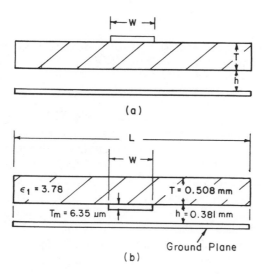

FIG. 29 General configuration of (a) suspended and (b) inverted microstrip. [(b) From Spielman, 1977. Reprinted from *IEEE Trans. Microwave Theory Techniques* **25**, 648–656. © 1977 IEEE.]

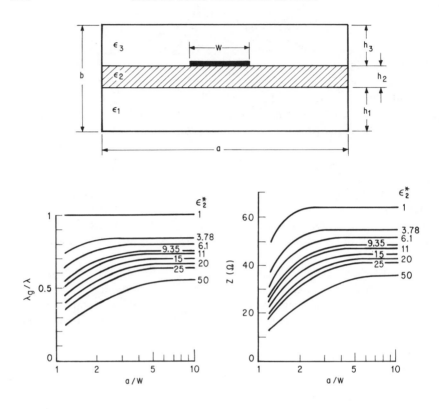

FIG. 30 Normalized guide wavelength $\lambda_g/\lambda$ and characteristic impedance $Z$ of suspended microstrip: $\varepsilon_1 = \varepsilon_3 = \varepsilon_0$, $h_1 = h_3 = 0.4b$, $h_2 = 0.2b$, $W = b$. (From Yamashita and Atsuki, 1970. Reprinted from *IEEE Trans. Microwave Theory Tech.* **18**, 238–244. © 1970 IEEE.)

Because dispersions are relatively small, quasi-TEM analyses yield accurate results for many applications even at millimeter-wave frequencies. An example of characteristic impedance and guide wavelength of a suspended microstrip line in a shield is presented in Fig. 30, which shows results obtained by Yamashita and Atsuki (1970) using a Green's function formulation. The attenuation characteristics are given in Fig. 31.

The results of Spielman's (1976) static analysis are presented for a structure using fused silica as the substrate. This material is suitable for the inverted microstrip at millimeter-wave frequencies (Spielman, 1977). Figure 32 shows the range of characteristic impedance and phase velocity as a function of the aspect ratio. It is evident that a relatively wide range of impedances is available and that the increasing phase velocity for larger strip widths causes the fields to be distributed mostly in the region under the strip.

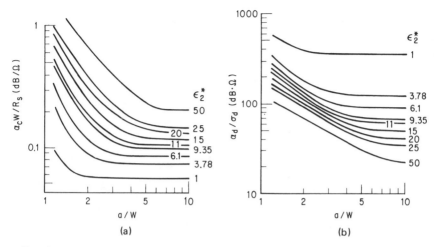

FIG. 31   (a) Conductor loss and (b) dielectric loss constants for suspended microstrip: $\varepsilon_1 = \varepsilon_3 = \varepsilon_0$, $h_1 = h_3 = 0.4b$, $h_2 = 0.2b$, $W = b$. (From Yamashita and Atsuki, 1970. Reprinted from *IEEE Trans. Microwave Theory Tech.* **18**, 238–244. © 1970 IEEE.)

The information in Fig. 33 correctly indicates that conductor and dielectric losses would be approximately the same for narrow strips, but eventually the conductor losses would dominate for larger widths as less energy is distributed in the dielectric slab. Note that this argument is valid only for the inverted microstrip line without top and side shields.

The dielectric slab serves to reduce the loss of energy to radiation by

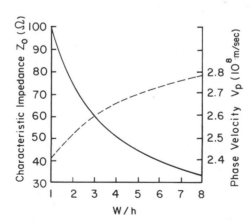

FIG. 32   Characteristic impedance and phase velocity vs $W/h$ for inverted microstrip:——, $Z_0$; – – –, $V_p$. (From Speilman, 1977. Reprinted from *IEEE Trans. Microwave Theory Tech.* **25**, 648–656. © 1977 IEEE.)

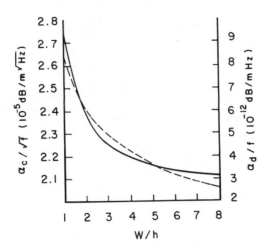

FIG. 33 Loss constants vs aspect ratio for inverted microstrip: $\tan \delta_1 = 2.0 \times 10^{-4}$, $\sigma = 4.10 \times 10^7 \, S/m$; ——, $\alpha_c/\sqrt{f}$; – – –, $\alpha_d/f$. (From Spielman, 1977. Reprinted from *IEEE Trans. Microwave Theory Tech.* **25**, 648–656. © 1977 IEEE.)

trapping the fields. However, for small aspect ratios and thick slabs, the surface wave modes may be sufficiently excited that energy will leak from the edges of the slab.

With proper design, this transmission line can be quite useful because the attenuation figures are quite low and it has been shown to have very little dispersion in the microwave frequency bands.

### 4. Trapped Inverted Microstrip

The trapped inverted microstrip (TIM) line is a modified version of the standard inverted microstrip line. The trapped inverted microstrip offers smaller losses and better isolation than the conventional alumina substrate microstrip, but has a smaller physical size than the air–dielectric strip line.

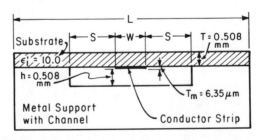

FIG. 34 Generic cross section for trapped inverted microstrip. (From Spielman, 1977. Reprinted from *IEEE Trans. Microwave Theory Tech.* **25**, 648–656. © 1977 IEEE.)

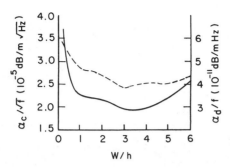

FIG. 35 Loss constants vs aspect ratio for trapped inverted microstrip: $\tan \delta_1 = 6.0 \times 10^{-4}$; $\sigma = 4.10 \times 10^7$ $S/m$;——, $\alpha_c/\sqrt{f}$; ---, $\alpha_d/f$. (From Spielman, 1977. Reprinted from *IEEE Trans. Microwave Theory Tech.* **25,** 648–656. © 1977 IEEE.)

The TIM line structure shown in Fig. 34 is basically a trough line with a dielectric overlay. The dominant mode of propagation is quasi-TEM, with much of the energy concentrated under the printed conductor in the channel. Fringing fields tend to be trapped in the dielectric layer and are confined close to the channel. The effective dielectric constant is about 3 for typical cases, and the line is slightly dispersive (Bera and Wallace, 1979).

Calculated values of the conductor and dielectric loss coefficients are plotted in Fig. 35. The irregular appearance of these characteristics is attributed to the effects of channel side-wall interaction with the conducting strip. A similar situation is apparent in Fig. 36, where the characteristic impedance and phase velocity are shown.

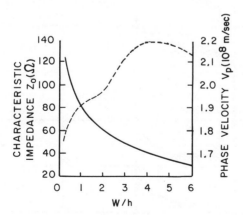

FIG. 36 Characteristic impedance and phase velocity vs $W/h$ for trapped inverted microstrip:——, $Z_0$; ---, $V_p$. (From Spielman, 1977. Reprinted from *IEEE Trans. Microwave Theory Tech.* **25, 648**–656. © 1977 IEEE.)

TABLE IV

| TYPE OF LINE | CONFIGURATION | TRANSITION TO WAVEGUIDE | FABRICATION | COMPAT. ACTIVE DEVICES |
|---|---|---|---|---|
| DIELECTRIC SURFACE WAVEGUIDE | | Moderate | Injection Molding Machining | Difficult |
| IMAGE GUIDE | | Moderate | Injection Molding Machining | Difficult |
| INSULATED IMAGE GUIDE | | Moderate | Injection Molding Machining | Difficult |
| INVERTED STRIP DIELECTRIC GUIDE | | Moderate | Machining | Difficult |
| H GUIDE | | Difficult | Machining Molding | Difficult |
| GROOVE GUIDE | | Difficult | Machining | Difficult |
| FIN LINE | | Moderate | Etching | Yes |
| SLOT LINE | | Moderate | Etching | Yes |
| COPLANAR WAVEGUIDE | | Moderate | Etching | Yes |
| MICROSTRIP | | Moderate | Etching | Yes |
| SUSPENDED MICROSTRIP | | Easy | Etching | Yes |
| INVERTED MICROSTRIP | | Easy | Etching | Yes |
| TRAPPED INVERTED MICROSTRIP | | Easy | Etching | Yes |

OVERALL COMPARISON OF MILLIMETER-WAVE TRANSMISSION LINES

| COMPAT. PASSIVE DEVICES | ISOLATION | PROPAG. LOSS | RADIATION LOSS | COMPAT. WITH RADIATORS |
|---|---|---|---|---|
| Requires Develop. | Poor to Moderate at Bends | Low | High | Yes – Dielectric Antennas |
| Requires Develop. | Good at 3" – 35 GHz 3/4" – 70 GHz | Low | High | Yes – Dielectric Antennas |
| Requires Develop. | Poor to Moderate at Bends | Low | High | Yes – Dielectric Antennas |
| Requires Develop. | −10 dB at 1 Wavelength Away | Low | High | Yes – Dielectric Antennas |
| Difficult | Poor to Moderate | Low | Moderate to High | Yes – Dielectric Antennas |
| Difficult | Poor to Moderate | Low | Moderate to High | Yes – Dielectric Antennas |
| Yes | Moderate to Good | High | None | Need Transition to Waveguide |
| Yes | Moderate to Poor | Same or > Than $\mu$-Strip | Moderate | Horns or Slot Antennas |
| Yes | Poor at 1/2 Wavelength | > Than $\mu$-Strip | Moderate | Coplanar Antennas |
| Yes | Moderate | High | Moderate | $\mu$-Strip Antennas or |
| Yes | Good | Same as Inverted Microstrip | Low | Transition to Waveguide Horn |
| Yes | Moderate to Good | 75 % Less Than $\mu$-Strip | Low | $\mu$-Strip or |
| Yes | Moderate to Good | 70 % Less Than $\mu$-Strip | Low | Transition to Waveguide Horn |

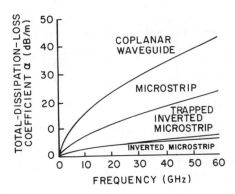

FIG. 37 Comparison of total dissipative losses for four 50-Ω transmission lines. (From Spielman, 1977. Reprinted from *IEEE Trans. Microwave Theory Tech.* 25, 648–656. © 1977 IEEE.)

The total dissipation loss characteristics are plotted in Fig. 37 for a nominally 50-Ω configuration of trapped inverted microstrip, coplanar waveguide, and inverted microstrip. It is interesting to note that the losses incurred in inverted and trapped inverted microstrips at frequencies as high as 60 GHz are comparable to those incurred in a microstrip at frequencies in the 5–10-GHz range. It can also be seen from Fig. 37 that the coplanar waveguide appears to be considerably more lossy than the microstrip.

The conducting strip widths required for microstrip and coplanar waveguide are approximately 2.5–3 times narrower than those required for inverted and trapped inverted microstrips. This feature enhances the attractiveness of the inverted and trapped inverted microstrip lines by virtue of the mitigation of fabrication difficulties. By concentrating more of the field energies in air (with correspondingly lower $\varepsilon_{eff}$), wider strips are possible for a prescribed impedance level (compared to the other two lines). These advantages should be realized even if the dimensions must be contracted to suppress higher-order modes at higher frequencies. It should be noted that these transmission lines may have to be shielded for some practical circuit applications. When these shields are provided, however, the width of the strip is restricted and the lowest impedance level attainable is often higher than desired. This is one drawback of this type of line.

## III.  Conclusion

It should be stated once more that the purpose of this chapter is neither to provide complete analytical results for various transmission lines nor to argue the superiority of one particular type over the others. We have attempted to present well-known and commonly accepted features of these transmission lines so that the user will have an idea of their fundamental properties.

Choice of transmission lines is dependent on various factors and is often dictated by the objective of the user. For instance, the dielectric waveguide family is quite appropriate if one intends to build a leaky wave antenna. On the other hand, active components are difficult to develop with dielectric waveguides at present.

In lieu of further concluding remarks, we present in Table IV an overall assessment of the various transmission lines included in this chapter.

## REFERENCES

Arditi, M. (1955). *IEEE Trans. Microwave Theory Tech.* **3**, 31–56.
Bahl, I. J., and Gupta, D. C. (1979). *IEE J. Microwaves, Opt. Acoust.* **3**(1), 1–4.
Bates, R. N., and Coleman, M. D. (1979). *9th Eur. Microwave Conf.* Brighton, England, pp. 721–725. Microwave Exhibitions and Publishers Ltd., Kent, England.
Bera, R. F., and Wallace, R. N. (1979). *IEEE MTT-S Int. Microwave Symp.,* Orlando, Florida, pp. 306–308.
Bryant, T. G., and Weiss, J. A. (1968). *IEEE Trans. Microwave Theory Tech.* **16**, 1021–1027.
Bui, L., and Ball, D. (1982). *IEEE MTT-S Int. Microwave Symp.,* Dallas, Texas, pp. 204–205.
Cohn, S. B. (1969). *IEEE Trans. Microwave Theory Tech.* **17**, 768–778.
Denlinger, E. J. (1971). *IEEE Trans. Microwave Theory Tech.* **19**, 30–39.
Doswell, A., and Harris, D. J. (1973). *IEEE Trans. Microwave Theory Tech.* **21**, 587–589.
Edwards, T. C. (1981) "Foundations for Microstrip Circuit Design." Wiley, New York.
Edwards, T. C., and Owens, R. P. (1976) *IEEE Trans. Microwave Theory Tech.* **24**, 506–513.
Getsinger, W. J. (1973). *IEEE Trans. Microwave Theory Tech.* **21**, 34–39.
Gopinath, A. (1979). *IEEE MTT-S Int. Microwave Symp.,* Orlando, Florida, pp. 109–111.
Harris, D. J., and Lee, K. W. (1977). *Electronics Lett.* **2**, 775–776.
Harris, D. J., Lee, K. W., and Reeves, J. M. (1978). *IEEE Trans. Microwave Theory Tech.* **26**, 998–1001.
Hoefer, W. J. R. (1978). *IEEE MTT-S Int. Microwave Symp.,* Ottawa, Canada.
Hofmann, H. (1977a). *Arch. Elektron. Ubertragungstechn* **31**, 40–44.
Hofmann, H. (1977b). *IEEE MTT-S Int. Microwave Symp.,* San Diego, California, pp. 381–384.
Hornsby, J. S., and Gopinath, A. (1969). *IEEE Trans. Microwave Theory Tech.* **17**, 684–690.
Itoh, T. (1976). *IEEE Trans. Microwave Theory Tech.* **24**, 821–827.
Itoh, T. (1981). *In* "Infrared and Millimeter Waves" (J. Wiltse and K. Button, eds.), Vol. 4, pp. 199–273. Academic Press, New York.
Itoh, T., and Adelseck, B. (1980). *IEEE Trans. Microwave Theory Tech.* **28**, 1433–1437.
Itoh, T., and Mittra, R. (1973). *IEEE Trans. Microwave Theory Tech.* **21**, 496–499.
Jansen, R. H. (1979). *IEE J. Microwaves Opt. Acoust.* **3**(1), 14–22.
Knorr, J. B., and Shayda, P. M. (1980). *IEEE Trans. Microwave Theory Tech.* **28**, 737–743.
Knox, R. M. (1976). *IEEE Trans. Microwave Theory Tech.* **24**, 806–814.
Knox, R. M., and Toulios, P. P. (1970). *Proc. Symp. Submillimeter Waves,* 497–516. Polytechnic Institute of Brooklyn, New York.
Konishi, Y., Uenakada, K., Yazawa, N., Hoshino, N., and Takahashi, T. (1974). *IEEE Trans. Microwave Theory Tech.* **22**, 451–454.
Kowalski, G., and Pregla, R. (1971). *Arch. Elektron. Uebertragungstechn* **25**, 193–196.

Lewin, L., Chang, D. C., and Kuester, E. F. (1977). "Electromagnetic Waves and Curved Structures." Peter Peregrinus Ltd., England.

Marcatili, E. A. J. (1969). *Bell Syst. Tech. J.* **48**, 2071–2102.

Mariani, E. A., Heinzman, C. P., Figrios, J. P., and Cohn, S. B. (1969). *IEEE Trans. Microwave Theory Tech.* **17**, 1091–1096.

Meier, P. J. (1974). *IEEE Trans. Microwave Theory Tech.* **22**, 1209–1216.

Mittra, R., and Itoh, T. (1974). *In* "Advances in Microwaves" (L. Young and H. Sobol, eds.), Vol. 8, pp. 67–141. Academic Press, New York.

Nakahara, J., and Kurauchi, N. (1965). *Sumitomo Elec. Tech. Rev.* **5**, 65–71.

Oliner, A. A., Peng, S.-T., Hsu, T.-I., Sanchez, A. (1981). *IEEE Trans. Microwave Theory Tech.* **29**, 855–870.

Peng, S.-T., and Oliner, A. A. (1981). *IEEE Trans. Microwave Theory Tech.* **29**, 843–855.

Pucel, R. A., Masse, D. J., and Hartwig, C. P. (1968). *IEEE Trans. Microwave Theory Tech.* **16**, 342–350.

Robinson, G. H., and Allen, J. L. (1969). *IEEE Trans. Microwave Theory Tech.* **17**, 1097–1101.

Rozzi, T. E. (1978). *IEEE Trans. Microwave Theory and Tech.* **26**, 738–746.

Saad, A. M. K., and Begemann, G. (1977). *IEE J. Microwaves Opt. Acoust.* **1**, 81–88.

Saad, A. M. K., and Schünemann, K. (1978). *IEEE Trans. Microwave Theory Tech.* **26**, 1002–1007.

Scarman, R. E., Lowbridge, P. L., and Oxley, T. H. (1981). *11th Eur. Conf. Dig.*, Amsterdam, The Netherlands, pp. 327–332. Microwave Exhibitions and Publishers, Ltd., Kent, England.

Schmidt, L.-P., and Itoh, T. (1980). *IEEE Trans. Microwave Theory Tech.* **28**, 981–985.

Schneider, M. V. (1969a). *Bell Syst. Tech. J.* **48**, 1421–1444.

Schneider, M. V. (1969b). *Bell Syst. Tech. J.* **48**, 2325–2332.

Schneider, M. V. (1972). *Proc. IEEE* **60**, 144–146.

Solbach, K. (1978). *IEEE Trans. Microwave Theory Tech.* **26**, 755–758.

Spielman, B. E. (1976). "Computer Aided Analysis of Dissipation Losses in Isolated and Coupled Transmission Lines for Microwave and Millimeter Wave Integrated Circuit Applications." Naval Res. Lab. Report 8009, Washington, D.C.

Spielman, B. E. (1977). *IEEE Trans. Microwave Theory Tech.* **25**, 648–656.

Stinehelfer, H. E. (1968). *IEEE Trans. Microwave Theory Tech.* **16**, 439–444.

Tischer, F. J. (1956). *IRE Conv. Rec.*, New York, Pt. 5.

Tischer, F. J. (1963). *IEEE Trans. Microwave Theory Tech.* **11**, 291–296.

Wen, C. P. (1969). *IEEE Trans. Microwave Theory Tech.* **17**, 1089–1090.

Wheeler, H. A. (1965). *IEEE Trans. Microwave Theory Tech.* **13**, 172–185.

Wiltse, J. C. (1959). *IRE Trans. Microwave Theory Tech.* **7**, 65–70.

Wolff, I. (1979). "Microstrip-Bibliography 1948–1978." Verlag Henning Wolff, Aachen, Germany.

Yamashita, E., and Atsuki, K. (1970). *IEEE Trans. Microwave Theory Tech.* **18**, 238–244.

Yamashita, E., and Mittra, R. (1968). *IEEE Trans. Microwave Theory Tech.* **16**, 251–256.

Yen, P., Paul, J. A., and Itoh, T. (1981). *IEEE MTT-S Int. Microwave Symp.*, Los Angeles, California, pp. 114–116.

Yoneyama, T. and Nishida, S. (1981). *IEEE Trans. Microwave Theory Tech.* **29**, 1188–1192.

INFRARED AND MILLIMETER WAVES, VOL. 9

CHAPTER 3

# Dielectric Waveguide Electrooptic Devices

*Marvin B. Klein*

Hughes Research Laboratories
Malibu, California

## I. Introduction

In recent years, integrated circuit techniques have become increasingly important in implementing components and subsystems at millimeter wavelengths. A number of passive and active devices in the dielectric waveguide geometry have been demonstrated, including directional couplers, filters, phase shifters, scanning antennas, detectors, and sources. In most cases, the active devices studied required careful mounting of a discrete

133

electrical device in close proximity to the waveguide structure in a way that optimizes coupling to the waveguide. In this chapter we describe active devices using dielectric waveguides that can be realized by direct electrical control of the *bulk* dielectric properties of the transmission medium. These devices have intrinsic high speed and are directly compatible with common waveguide structures.

Two general classes of materials have bulk dielectric properties that can be electrically controlled: semiconductors and electrooptic crystals. In semiconductors, the generation of carriers leads to changes in both the real and the imaginary parts of the complex permittivity. These effects have been used to construct phase shifters (De Fonzo *et al.,* 1979; Jacobs and Chrepta, 1974) and switches (Lee *et al.,* 1978) in dielectric waveguide using optical or electrical generation of carriers in silicon. Large modulation amplitudes have been observed, but the switching speed in most cases is limited by the carrier recombination time to values in the range $0.1 - 1.0$ $\mu$sec. Gallium arsenide is the preferred material when high speed is required, because the recombination time in high-resistivity chromium-doped samples is on the order of 100 psec (Lee *et al.,* 1977). Response times in this range have been demonstrated in the optoelectronic modulation of 95-GHz radiation in a GaAs dielectric waveguide (Li *et al.,* 1982); however, electrical switching has not yet been reported.

In electrooptic materials (or nonlinear dielectrics) an applied voltage produces a direct change in the real part of the dielectric constant, owing to distortions of the crystal lattice. Materials of interest in this category include ferroelectric crystals commonly used for modulation in the visible region, such as $LiNbO_3$ (lithium niobate) and $LiTaO_3$ (lithium tantalate). In these materials the medium response time is instantaneous at frequencies up to $\sim 1000$ GHz, and the modulation rate is limited only by circuit conditions. Because of their intrinsic high-speed response, nonlinear materials are also promising for harmonic generation and frequency mixing applications at frequencies throughout the millimeter band.

The usefulness of nonlinear materials for frequency mixing in the millimeter band was first demonstrated by Boyd *et al.* (1971). In these and later experiments (Boyd and Pollack, 1973), sideband generation on a 56-GHz carrier was observed at an applied frequency of 100 MHz, and nonlinear coefficients for eight materials were determined from the sideband-to-carrier power ratio. The measured values were found to agree within a factor of two with values obtained from a phenomenological model based on the motion of the electrons and ions in an anharmonic potential. The nonlinearity in $BaTiO_3$ was found to be particularly large. At the same time, electrooptic modulation was demonstrated in $LiNbO_3$ at a carrier frequency of 126 GHz (Vinogradov *et al.*, 1970), and an approximate measurement of the electrooptic coefficient $r_{222}$ was obtained. Subsequent to this work, Fetterman *et al.* (1975) demonstrated electrooptic modulation in $LiTaO_3$ at 4 K (with a carrier frequency of 890 GHz) and measured the electrooptic coefficient $r_{333}$. Cooling of the sample was necessary to reduce the absorption loss present at the high carrier frequency. More recently, we have measured, with improved accuracy, the electrooptic coefficients for several orientations in $LiNbO_3$ and $LiTaO_3$ at a carrier frequency of 94 GHz (Klein, 1979, 1981). The resulting values agree closely with values obtained using the nonlinear coefficients reported by Boyd and Pollack (1973). Ho *et al.* (1981) have also observed a large electrooptic effect at 58 GHz in SBN $(Sr_{0.61}Ba_{0.39}Nb_2O_6)$.

In all of these experiments the sample was mounted either in free space or in a conventional metal waveguide. Free-space mounting requires large samples to minimize diffraction loss; mounting in a conventional waveguide introduces severe dimensional tolerances and complicates the application of modulating voltages. Overmoded circular waveguide ($TE_{01}$ mode) is another candidate for these experiments, but sample mounting is still difficult and the circular mode polarization is awkward for most modulation schemes. If, instead, the electrooptic material is prepared in the form of a dielectric waveguide, then the requirement for large samples or critical dimensional tolerances is removed, modulating voltages can easily be applied, and compatibility with other integrated circuit components is facilitated.

In this chapter we shall present the results of analysis and experiments on dielectric waveguide electrooptic devices. In Section II we describe the electrooptic properties of ferroelectric materials, and in Section III candidate waveguide structures for phase shifting are considered. In Sections IV and V we describe the properties of the $TE_{10}$ mode in a $LiNbO_3$ H guide. In Section VI we describe experiments on $LiNbO_3$ H-guide phase shifters at a carrier frequency of 95 GHz, and in Section VII our analysis of other control devices is presented. In Section VIII we discuss the prospects for improved materials, and our conclusions are presented in Section IX.

## II. Electrooptic Properties of Ferroelectric Materials

The linear electrooptic effect in crystals is the change in the refractive indices that is proportional to the applied electric field. This effect exists only in crystals that do not possess inversion symmetry (as is true for all polar crystals). In centrosymmetric crystals, a higher-order (quadratic) electrooptic effect exists, but this will not be considered here. The properties of linear electrooptic materials and the characteristics of electrooptic modulation have been discussed by Kaminow (1974) and Yariv (1975). Our review here will be brief; the reader is referred to the cited references for more details.

## A.  ELECTROOPTIC COEFFICIENT AND INDUCED PHASE CHANGE

The electrooptic coefficient $r_{ijk}$ is defined in terms of the impermeability tensor $(1/n^2)_{ij}$ by

$$\Delta(1/n^2)_{ij} = \sum_{k=1}^{3} r_{ijk}E_k, \tag{1}$$

where $n_{ij}$ is the refractive index and $E_k$ is the component of the applied electric field along the $k$ direction in the crystal. The indices $i, j, k$ refer to the principal crystalline axes and can take on the values $i, j, k = 1, 2, 3$, where $1 = x, 2 = y$, and $3 = z$. The terms with $i = j$ correspond to a change in the refractive index along a principal crystalline axis, and the terms with $i \neq j$ correspond to a rotation of the principal axes.† For each noncentrosymmetric crystal class, only certain components of $r_{ijk}$ are nonzero. For example, in LiNbO$_3$ and LiTaO$_3$ (crystal class $3m$), the only nonzero components are $r_{333}, r_{223} = r_{113}, r_{222} = -r_{112} = -r_{121}$, and $r_{131} = r_{232}$.

In most applications the modulating field is directed along a principal crystalline axis. Furthermore, for most applications we are interested only in modulation geometries with $i = j$. With these assumptions Eq. (1) becomes

$$\Delta(1/n^2)_{ii} = 2\,\Delta n_{ii}/n_{ii}^3 = r_{iik}E_k \tag{2}$$

or

$$\Delta n_{ii} = \tfrac{1}{2}n_{ii}^3 r_{iik}E_k. \tag{3}$$

For example, if the modulating field is directed along the $z$ axis and the carrier is polarized along the $x$ axis, then

$$\Delta n_{11} = \tfrac{1}{2}n_{11}^3 r_{113}E_3. \tag{4}$$

Most materials of interest are uniaxial and are characterized by an ordinary refractive index $n_o = n_{11} = n_{22}$ and an extraordinary refractive index $n_e = n_{33}$.

For an index change $\Delta n_{ii}$, the induced phase shift is

$$\Delta\phi_{ii} = (2\pi\,\Delta n_{ii}/\lambda_0)L = (\pi L/\lambda_0)n_{ii}^3 r_{iik}E_k, \tag{5}$$

where $\lambda_0$ is the free-space wavelength and $L$ is the interaction length. Thus, a

---

† The electrooptic coefficient can also be written in contracted tensor form as $r_{mk}$ ($m = 1, ..., 6, k = 1, 2, 3$), with $1 = x, 2 = y, 3 = z, 4 = yz, 5 = xz$, and $6 = xy$.

carrier wave polarized along any principal crystalline direction will undergo pure phase modulation, as described by Eq. (5). If the input carrier polarization has components along two principal directions, differential phase retardation can result, thus causing amplitude modulation of the output signal transmitted through an analyzer. The waveguide devices described in this chapter have a geometry more favorable for phase modulation than for amplitude modulation. However, as will be shown in Section IV, the output from a phase modulator can always be converted to amplitude modulation through the use of a phase bridge.

With reference to Eq. (5), we can define a material figure of merit given by

$$R_{iik} = n_{ii}^3 r_{iik}. \tag{6}$$

One important reason for utilizing the electrooptic effect for microwave frequency carriers is that the refractive indices and electrooptic coefficients (and thus the figures of merit) for certain materials in the microwave region are much larger than those in the visible region (Boyd *et. al.*, 1971). This enhancement is an important requirement, because the expected phase shift varies inversely with wavelength [see Eq. (5)] and would otherwise be small in the far-infrared and millimeter spectral regions. We shall show in the following section that crystals with the largest microwave electrooptic (and nonlinear) coefficients are those with the largest values of *linear* susceptibility at the same frequencies (Boyd *et al.*, 1971). The linear susceptibility in the microwave region can be much larger than that in the visible region owing to the contribution from lattice polarization—this leads to the required enhancement in the figure of merit.

## B. MODELS FOR ELECTROOPTIC AND NONLINEAR COEFFICIENTS

The dependence of the nonlinear susceptibility on the corresponding linear susceptibilities was first noted by Miller (1964) for the case of second-harmonic generation in the visible region. Miller derived a specific relationship by first writing the free energy of the crystal as a power series in the polarization (Devonshire, 1954):

$$U = \tfrac{1}{2}\alpha P^2 + \tfrac{1}{3}\delta P^3 + \tfrac{1}{4}\gamma P^4 + \cdots. \tag{7}$$

The electric field is then given by

$$E = dU/dP = \alpha P + \delta P^2 + \gamma P^3 + \cdots. \tag{8}$$

The term in Eq. (8) that is quadratic in $P$ is

$$E(2\omega) = \delta[P(\omega)]^2. \tag{9}$$

The second-harmonic coefficient or susceptibility $d(2\omega)$ relates the electric field at the fundamental frequency to the induced polarization at the second-harmonic frequency:

$$P(2\omega) = d(2\omega)[E(\omega)]^2. \tag{10}$$

Combining Eqs. (9) and (10) and using $P = \chi E$ yields

$$d(2\omega) = \chi(2\omega)[\chi(\omega)]^2\delta. \tag{11}$$

Using measured values of nonlinear coefficients and linear susceptibilities for a variety of materials, Miller (1964) determined that the parameter $\delta$ varied much less from material to material than did the nonlinear coefficient $d(2\omega)$. Miller also pointed out that $\delta$ has the full tensor properties of $d(2\omega)$, so that in tensor form Eq. (11) becomes

$$d_{ijk}(2\omega) = \chi_{ii}(2\omega)\chi_{jj}(\omega)\chi_{kk}(\omega)\delta_{ijk}. \tag{12}$$

Although the utility of expressing nonlinear coefficients in the form of Eq. (11) or (12) was first recognized by Miller, he considered only frequency doubling in the visible region, where electron motion alone contributes to the linear and nonlinear susceptibility. Furthermore, no physical argument was presented as to why $\delta$ should be invariant from material to material. To avoid these limitations it is useful to consider a coupled anharmonic oscillator model (Garrett, 1968; Garrett and Robinson, 1966). A single electronic and a single ionic oscillator are considered whose (one-dimensional) equations of motion include terms that are quadratic in displacement and that account for cross coupling. The linear susceptibility is also divided into separate electronic and ionic contributions:

$$\chi = \epsilon_r - 1 = \chi_e + \chi_i, \tag{13}$$

with

$$\chi_e = n_v^2 - 1, \tag{14}$$

where $n_v$ is the refractive index in the visible region and $\epsilon_r$ is the relative dielectric constant. The nonlinear equations of motion may be solved for the nonlinear coefficient as a function of linear susceptibilities. The particular terms in the solution depend on the frequency regime of interest. If all applied frequencies are above the lattice frequencies, the ionic motion can be neglected and the solution is a generalized form of Eq. (11):

$$d(\omega_3 = \omega_1 + \omega_2) = \chi_e(\omega_3)\chi_e(\omega_2)\chi_e(\omega_1)\delta. \tag{15}$$

The factor $\delta$ is given by

$$\delta = 3\epsilon_0 D/e^3 N^2, \tag{16}$$

where $D$ is the strength of the anharmonic term and $N$ is the electron density. Garrett and Robinson (1966) and Yariv (1975) have been able to estimate $D$ from physical arguments regarding the shape of the electrostatic potential and the magnitude of the electron displacement at a lattice site. The resulting values of $\delta$ agree with measured values within a factor of $\sim 4$. This suggests that the invariance of $\delta$ is due to common factors in lattice geometry, in spite of the wide range of materials involved.

In the second case of interest, one applied frequency is below the ionic resonance frequency and the other two frequencies are above it. Examples of this interaction include the electrooptic effect in the visible region and difference frequency generation of far-infrared radiation. The nonlinear coefficient in this case contains a term similar to Eq. (15) plus a second term reflecting electron–ion coupling.

In the third case, all frequencies are below the ionic resonance frequencies; this is the regime of interest here. The nonlinear coefficient in this case (Boyd and Pollack, 1973) contains terms accounting for the separate and coupled motion of both electrons and ions:

$$d(\omega_3 = \omega_1 + \omega_2) = [\chi_i(\omega_1)\chi_i(\omega_2)\chi_i(\omega_3) + \chi_i(\omega_1)\chi_i(\omega_2)\chi_e(\omega_3)]\delta_{AB}$$
$$+ \chi_i(\omega_1)\chi_e(\omega_2)\chi_e(\omega_3)\delta_c + \chi_e(\omega_1)\chi_e(\omega_2)\chi_e(\omega_3)\delta_D. \quad (17)$$

For materials with $\chi_i \gg \chi_e$, Eq. (17) reduces to

$$d(\omega_3 = \omega_1 + \omega_2) \simeq \chi_i(\omega_1)\chi_i(\omega_2)\chi_i(\omega_3)\delta_{AB}, \quad (18)$$

so that the primary contribution to $d$ is from ion motion alone. The enhancement of $d$ in going from the visible region to microwave frequencies is

$$d_m/d_v \simeq \chi_i^3/\chi_e^3, \quad (19)$$

which can be on the order of $10^3$–$10^4$ in ferroelectrics with large values of dielectric constant.

The linear electrooptic coefficient $r$ is related to $d$ by (Yariv, 1975)

$$r = 4d/\chi_i(\omega_1)\chi_i(\omega_2) \simeq 4\chi_i(\omega_3)\delta_{AB}. \quad (20)$$

The electrooptic figure of merit at microwave frequencies [defined in Eq. (6)] can then be written as

$$R \simeq 4[\chi_i(\omega_3)]^{5/2}\delta_{AB} \simeq 4[n(\omega_3)]^5\delta_{AB}, \quad (21)$$

where $\chi_i(\omega_3) \simeq [n(\omega_3)]^2$.

C. MEASUREMENT OF ELECTROOPTIC AND NONLINEAR
COEFFICIENTS AT MICROWAVE CARRIER FREQUENCIES

Past measurements of electrooptic and nonlinear coefficients at microwave frequencies were reviewed briefly in the Introduction. Here we expanded on two of these studies that are of particular relevance in later sections.

Nonlinear interactions in the microwave region were first studied in depth by Boyd et al. (1971). In these experiments, sideband generation on a 56-GHz carrier was observed at an applied frequency of 100 MHz, and nonlinear coefficients for eight materials were determined from the sideband-to-carrier ratio. Value of $\delta_{AB}$ were then obtained using expressions similar to Eq. (17). Table I (Boyd et al., 1971) presents values of the measured microwave nonlinear coefficient $d_m$, the equivalent coefficient in the visible $d_v$, the Miller coefficient $\delta_{AB}$, the ionic susceptibility $\chi_i$, the electronic susceptibility $\chi_e$, and the microwave intensity absorption coefficient $\alpha_I$ for the most promising materials studied. The intensity absorption coefficient is related to the loss tangent by

$$\alpha_I = (2\pi/\lambda_0)\sqrt{\epsilon_r} \tan \delta \simeq (2\pi/\lambda_0)\sqrt{\chi_i} \tan \delta, \qquad (22)$$

where we assume that $\tan \delta \ll 1$.

As expected, the data in Table I indicate that the materials with the largest values of $\chi_i$ also give the largest values of $d_m$. The most notable example is BaTiO$_3$, whose large microwave nonlinear (and electrooptic) coefficients offer great promise for a variety of device applications. The major challenge in the design of devices is the absorption loss, as characterized by the large values of $\alpha_I$. The selection and optimization of materials for specific device applications will be discussed in Section VI. Note also that the measured values of $\delta_{AB}$ lie entirely in the range $0.02-0.64 \times 10^{-12}$ m/V, but the variation among materials in a given class (e.g., LiNbO$_3$ and LiTaO$_3$) is less. Later data for other orientations in LiNbO$_3$ and LiTaO$_3$ (Boyd and Pollack, 1973) yielded values of $\delta_{AB}$ that also fall within the range indicated. This suggests a common physical origin for the microwave nonlinearity in ferroelectric materials, in direct analogy with the observation by Miller (1964) for nonlinearities in the visible region.

Although BaTiO$_3$ has the largest nonlinear coefficient among the materials measured by Boyd et al., available samples of this material are relatively impure and limited in size and have significant absorption losses at microwave frequencies. On the other hand, LiNbO$_3$ and LiTaO$_3$ samples are commercially available in large sizes and with high purity levels, and the measured absorption coefficients (see Table II as well as Table I) are small enough for certain device applications. We have recently measured both

## TABLE I

### LINEAR AND NONLINEAR PARAMETERS FOR MATERIALS AT 56 GHz[a]

| Material | Orientation | Electronic susceptibility[b] $\chi_e$ | Ionic susceptibility $\chi_i$ | Intensity absorption coefficient $\alpha$ (cm$^{-1}$) | Visible nonlinear coefficient d ($10^{-12}$ m/V) | Microwave nonlinear coefficient $d_m$ ($10^{-12}$ m/V) | Delta coefficient $\delta_{AB}$ ($10^{-12}$ m/V) |
|---|---|---|---|---|---|---|---|
| BaTiO$_3$ | 333 | 4.25 | 52 | 1.7 | 6.8 | 97,000 | 0.64 |
| LiTaO$_3$ | 333 | 3.58 | 35 | 1.0 | 20 | 16,000 | 0.33 |
| LiNbO$_3$ | 333 | 3.66 | 21 | 0.5 | 40 | 6,700 | 0.63 |
| KDP | 123 | $\left\{\begin{array}{l}1.24\\1.13\end{array}\right.$ | $\left\{\begin{array}{l}42\\19\end{array}\right.$ | 0.9 | 0.48 | 1,850 | 0.05 |
| ADP | 123 | $\left\{\begin{array}{l}1.28\\1.16\end{array}\right.$ | $\left\{\begin{array}{l}55\\12\end{array}\right.$ | 1.2 | 0.59 | 970 | 0.02 |

[a] From Boyd et al. (1971).

[b] The upper and lower values of $\chi_e$ and $\chi_i$ correspond to the 1 and 3 axes, respectively.

linear parameters (refractive index and absorption coefficient) and electrooptic coefficients for several orientations in LiNbO$_3$ and LiTaO$_3$ at a
carrier frequency of 94 GHz (Klein, 1981). Our samples were mounted in
free space between two horns to avoid intricate mounting in a metal
waveguide. The sample faces were oriented normal to the principal crystalline axes, and typical sample dimensions were $\sim 1.5$ cm on a side. To
measure the refractive index and absorption coefficient the sample transmission was recorded as a function of frequency, using a 75–110-GHz
swept frequency source. Characteristic Fabry–Perot fringes were observed
in transmission, owing to Fresnel reflections (caused by impedance mismatch) at the parallel entrance and exit faces. The refractive index was
obtained from the fringe spacing, and the absorption coefficient was obtained from the fringe contrast. Measurements were performed for both
ordinary and extraordinary polarization in the crystal. Measured values of
the linear parameters are given in Table II.

In the same experiment, the electrooptic coefficients for several sample
orientations were determined by measurements of the phase shift induced
by a modulating transverse electric field. With this technique, the interpretation of the data is much simpler if the Fresnel reflections from the sample
faces can be eliminated. To accomplish this, we pressure-contacted quarter-
wave matching plates on both faces. The matching plates are fabricated
from materials whose microwave refractive index is as close as possible to
the square root of the crystal index. Table II lists the matching plate
materials and refractive indices for both ordinary and extraordinary polarization in the LiNbO$_3$ and LiTaO$_3$ samples.

In our experiment, the electrooptic phase shift was measured using a
phase bridge, as shown in Fig. 1. For each sample and orientation, the bridge

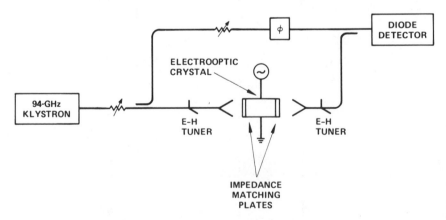

FIG. 1   Phase bridge for measurement of induced phase shift.

## TABLE II

LINEAR DIELECTRIC PARAMETERS FOR LiNbO$_3$ AND LiTaO$_3$ AT 94 GHz AND MICROWAVE REFRACTIVE INDICES OF MATCHING PLATE MATERIALS

| | Nonlinear material | | | | Matching plate material | |
|---|---|---|---|---|---|---|
| Symbol | Refractive index n | Dielectric constant $\epsilon_r$ | Intensity absorption coefficient $\alpha_I$ (cm$^{-1}$) | Loss tangent $\tan \delta$ (10$^{-3}$) | Symbol | Refractive index n |
| LiNbO$_3$ | 5.2 ($n_e$) | 27 ($\epsilon_e$) | ~0.2 | ~2 | NaF | 2.3 |
| | 6.7 ($n_o$) | 45 ($\epsilon_o$) | ~0.2 | ~1.5 | CaF$_2$ | 2.6 |
| LiTaO$_3$ | 6.5 ($n_e$) | 42 ($\epsilon_e$) | ~0.9 | ~7 | SrF$_2$ | 2.5 |
| | 6.5 ($n_o$) | 42 ($\epsilon_o$) | ~0.4 | ~3 | SrF$_2$ | 2.5 |

## TABLE III

ELECTROOPTIC FIGURES OF MERIT AND DELTA COEFFICIENTS FOR SEVERAL ORIENTATIONS IN LiNbO$_3$ AND LiTaO$_3$

| Material | Orientation | Figure of merit $R = n^3 r$ | | Delta coefficient $\delta_{AB}$ | |
|---|---|---|---|---|---|
| | | Calculated (cm/V) | Measured ($10^{-10}$ cm/V) | Measured ($10^{-10}$ cm/V) | Previous work[a] ($10^{-10}$ cm/V) |
| LiNbO$_3$ | 333 | $(1.4 \times 10^4)\delta_{AB}$ | $5.0 \times 10^3$ | 0.36 | 0.30 |
| | 113 | $(2.9 \times 10^4)\delta_{AB}$ | $4.8 \times 10^3$ | 0.16 | 0.14 |
| | 222 | $(9.3 \times 10^4)\delta_{AB}$ | $5.7 \times 10^3$ | 0.06 | 0.03 |
| LiTaO$_3$ | 333 | $(4.6 \times 10^4)\delta_{AB}$ | $1.5 \times 10^4$ | 0.33 | 0.23 |
| | 113 | $(4.6 \times 10^4)\delta_{AB}$ | $5.2 \times 10^3$ | 0.11 | 0.11 |

[a] From Boyd and Pollack, 1973.

was first balanced by adjusting the variable attenuator for maximum fringe contrast. The adjustable phase shifter was then set at the midpoint of the fringe pattern, i.e., in the linear region. For this bias condition, an ac voltage applied to the crystal causes an amplitude variation on the detector output at the applied frequency of 1 kHz. The fractional modulation can then be related directly to the peak phase shift, which in turn is related to the figure of merit $R$ through Eqs. (5) and (6).

Our measured values of $R$ for several orientations in $LiNbO_3$ and $LiTaO_3$ are given in Table III. The calculated values were obtained from the linear susceptibilities, using Eq. (21). We also list the values of $\delta_{AB}$ obtained from our measurements and from the frequency mixing experiments of Boyd and Pollack (1973). The agreement in the values of $\delta_{AB}$ is very good, considering the difference in the two measurement techniques.

### III. Candidate Waveguide Structures for Phase Shifting

We shall now consider possible dielectric waveguide structures for the particular case of electrooptic phase shifting. A candidate structure should be judged against three major requirements.

(1) The structure must contain two (or more) metallic conductors for application of modulating voltages.

(2) The microwave and modulating fields must be confined in the dielectric for optimum interaction.

(3) The electric field lines for the carrier and modulating fields should be straight and parallel to a principal crystal direction to exploit a single tensor component of the electrooptic coefficient.

Requirements (1) and (2) are straightforward. Requirement (3) is necessary to avoid phase interference effects or unwanted amplitude modulation that can occur when more than one tensor component of the electrooptic coefficient contributes to the carrier phase. The characteristics of a large number of dielectric guiding structures are described in detail in Chapter 2 by Itoh and Rivera. In Fig. 2 we list four candidate structures for phase shifting and the particular requirements satisfied by each. Image guide is an attractive structure for millimeter-wave integrated circuits because of its low wall loss, ease in construction, and ready compatibility with a variety of components. Furthermore, the metal image plane can mechanically support the dielectric material and provide a heat sink and electrical ground for active devices. However, modulating voltages cannot be applied to the dielectric, because only one metallic conductor is used. Note that this is not a disadvantage for frequency-doubling applications.

| STRUCTURE | CONFIGURATION | REQUIREMENTS SATISFIED |
|---|---|---|
| IMAGE GUIDE | 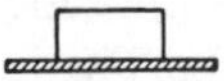 | (2), (3) |
| MICROSTRIP | 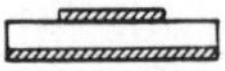 | (1), (2) |
| SLOT LINE | 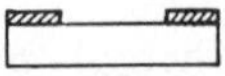 | (1), (2) |
| H GUIDE (PM$_{11}$ MODE) |  | (1), (2) |
| H GUIDE (TE$_{10}$ MODE) |  | (1), (2), (3) |

FIG. 2   Dielectric waveguide structures for electrooptic phase shifting.

Microstrip is a commonly used structure for printed circuit applications. However, the field lines in this structure are not straight, and thus requirement (3) is not satisfied. Furthermore, microstrip is unsuitable for use above $\sim 100$ GHz, because of the increasing wall loss and the small physical dimensions required. Slot line (the "inverse" of microstrip) also suffers from the same limitation.

The fourth structure in Fig. 2 is H guide, a partially dielectric-loaded parallel-plate waveguide consisting of two infinite conducting planes enclosing a rectangular dielectric slab (Cohn, 1959; Tischer, 1959). Energy is confined in the vertical direction by the metal plate and in the lateral direction by the dielectric–air interface. H guide shares some of the advantages of image guide (heat sinking and support by metal plates) and has the important addition of the second metallic conductor, which allows the application of dc bias or modulating voltages [requirement (1)].

Prior analyses (Cohn, 1959; Cohn and Eikenberg, 1962; Tischer, 1959) have shown that two classes of modes (PE$_{mn}$ and PM$_{mn}$) can propagate in an H guide (see Fig. 3). The designation PE, for *parallel electric*, refers to the fact that the electric field lines of these modes lie entirely in planes parallel to the dielectric–air interfaces ($E_x = 0$). These modes can be viewed as resulting from TE surface waves propagating in an infinite slab of the same material and thickness as the strip and bouncing back and forth between the metal walls. Similarly, the designation PM, for *parallel magnetic*, refers to the fact that the magnetic field lines of these modes lie entirely in planes parallel to the dielectric–air interfaces ($H_x = 0$). These modes in turn can be viewed as resulting from TM surface waves propagating in an infinite slab and bouncing back and forth between the sidewalls. The index $m$ gives the rank of the mode and is indicative of the manner in which the fields within the dielectric strip vary as a function of $x$. The index $n$ gives the order of the mode and equals the number of half cycles of sinusoidal variation of the fields in the $y$ direction.

The PE and PM modes are all hybrid modes, with the exception that the $n = 0$ order PE modes are transverse electric ($TE_{m0}$) modes. Two specific modes have been studied for device applications. The $PM_{11}$ mode (Fig. 4) is characterized by an electric field that is polarized predominantly in the $x$ direction but also has components in the $y$ and $z$ directions (Harris *et al.*, 1978; Yoneyama and Nishida, 1981). The cutoff condition for this mode is $\lambda_g = 2b$, where $\lambda_g$ is the guide wavelength. In a single-mode guide with $\lambda_g \lesssim 2b$, the plate spacing is less than one-half the free-space wavelength; thus the air-filled structure is beyond cutoff and the guided mode cannot radiate laterally into this structure. This characteristic has been used to suppress radiation from components such as bends and couplers in H guide (Yoneyama and Nishida, 1981).

Aside from the absence of radiation losses, the $PM_{11}$ hybrid mode has another important advantage: the $y$-directed electric field at the metal surfaces is small, and thus wall losses are low. However, at the present stage of development of electrooptic materials for phase shifting, material losses are dominant and the low wall losses of the $PM_{11}$ mode cannot be exploited. Furthermore, the $PM_{11}$ mode is not optimally suited for phase-shifting applications, because the $y$ and $z$ components of the electric field can lead to interference in the electrooptic phase shift or to unwanted amplitude modulation [violation of requirement (3)].

The other H-guide mode that has been analyzed and exploited for device

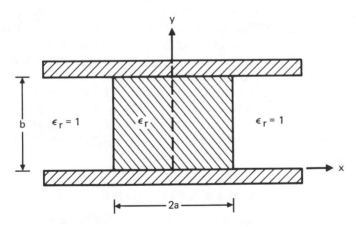

FIG. 3 Dimensions and coordinate system for H guide.

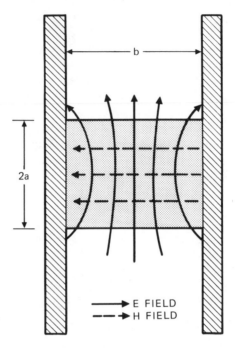

FIG. 4  Field lines for the $PM_{11}$ mode in H guide.

applications is the $TE_{10}$ mode (Cohn, 1959; Cohn and Eikenburg, 1962; Tischer, 1959). The electric field for this mode is directed normal to the metallic plates; no $x$ or $z$ component is present. Wall losses are larger than for the hybrid modes, but are still less than the dielectric losses in typical electrooptic materials. One important feature of the $TE_{10}$ mode is that the electric field is not a function of $y$, and thus there is no cutoff with decreasing plate spacing for decreasing frequency. Thus, the $TE_{10}$ mode can radiate laterally in or out of the dielectric medium. This feature may be a disadvantage in constructing components with radiation losses such as bends and couplers, but it is useful for intentionally radiating devices such as scanning antennas. The absence of a cutoff frequency for the $TE_{10}$ mode suggests that single-mode operation should be possible with a wide bandwidth of operation. This will be discussed further in Section V.

## IV.  Properties of the $TE_{10}$ Mode in H Guide

The $TE_{10}$ mode in H guide satisfies all the requirements specified, and is the mode used for our device investigation. The field components of all even-order TE modes (i.e., $TE_{10}$, $TE_{30}$, ... ) are as follows (Cohn, 1959). The factor $\exp[j(\omega t - \beta z)]$, which is common to all of these expressions, has been eliminated for the sake of brevity.

(1)  Region 1 ($-a \leq x \leq a$):

$$E_{y1} = -E_0(\omega\mu_0/k_d) \cos k_d x, \tag{23a}$$

$$H_{x1} = E_0(\beta/k_d) \cos k_d x, \tag{23b}$$

$$H_{z1} = jE_0 \sin k_d x. \tag{23c}$$

(2)  Region 2 ($x \leq -a$):

$$E_{y2} = -E_0(\omega\mu_0/k_d) \cos k_d a \exp[k_2(a + x)], \tag{24a}$$

$$H_{x2} = E_0(\beta/k_d) \cos k_d a \exp[k_2(a + x)], \tag{24b}$$

$$H_{z2} = -jE_0 \sin k_d a \exp[k_2(a + x)]. \tag{24c}$$

(3)  Region 3 ($x \geq a$): The field components are the same as those in region 2 except that $\exp[k_2(a + x)]$ is replaced by $\exp[k_2(a - x)]$ and the sign of $H_2$ is changed.

The normalized $x$-direction propagation constants for the dielectric region $k_d a$ and the outer regions $k_2 a$ are related by the equation

$$k_2 a = k_d a \tan k_d a. \tag{25}$$

Note that $(k_2 a)^{-1}$ can be considered as the lateral extent of the evanescent fields outside the dielectric, normalized to the width of the sample. The $x$-direction propagation constant for the dielectric region is related to the free-space wavelength $\lambda_0$ by

$$\pi^2(2a/\lambda_0)^2(\epsilon_r - 1) = (k_d a/\cos k_d a)^2. \tag{26}$$

For the $TE_{10}$ mode, $k_d a$ is constrained to be within the interval $0 \leq k_d a \leq \pi/2$. The next higher mode ($TE_{20}$) is an odd-order mode and is not likely to be excited using the coupling schemes described in Section VI. However, in our experiments the $TE_{30}$ even-order mode could be excited. Equations (25) and (26) also apply for this mode, with $\pi \leq k_d a \leq \frac{3}{2}\pi$. In general, the $TE_{m0}$ cutoff condition can be written as

$$2a/\lambda_0 = (m - 1)/2\sqrt{\epsilon_r - 1}. \tag{27}$$

Whereas the cutoff of the TE modes is independent of the plate separation $b$, the PM hybrid mode cutoff does depend on this parameter. Thus, for given experimental conditions, $b$ can be chosen small enough to suppress the lowest-order PM mode ($PM_{11}$) and yet not cause the TE modes to cut off.

The cutoff criterion for the hybrid modes has been derived by Moore and Beam (1957):

$$\frac{b_c}{\lambda_0} = \frac{1}{2}\left(\frac{1 + \tan^2 k_d a}{1 + \epsilon_r \tan^2 k_2 a}\right)^{1/2}, \tag{28}$$

where $b_c$ is the maximum plate separation that provides suppression of all hybrid modes.

We shall now review a number of other properties of the $TE_{10}$ mode in H guide, with emphasis on those relevant for phase-shifting applications. The expressions given also apply to the higher-order even TE modes ($TE_{30}$, $TE_{50}$, ... ). Our discussion is based on the analyses of Cohn (1959) and Cohn and Eikenberg (1962). In both of these works, the dielectric material is assumed to be isotropic. However, the results also apply to uniaxial or biaxial materials if they are oriented with a principal crystalling axis in the direction of the microwave field ($E_y$). Note that the analysis of the *hybrid* modes must be altered when considering an anisotropic dielectric in any orientation. The $PM_{11}$-mode properties in a $LiNbO_3$ H guide have been analyzed by Mortz and Stafsudd (1980).

A. GUIDE WAVELENGTH

The guide wavelength $\lambda_g$ is given by $\lambda_g = 2\pi/\beta$, where $\beta$ is determined from

$$\beta^2 = -k_d^2 + \omega^2 \mu_0 \epsilon_r \epsilon_0. \tag{29}$$

We thus obtain

$$\lambda_g/\lambda_0 = [\epsilon_r - (\epsilon_r - 1)\cos^2 k_d a]^{-1/2}. \tag{30}$$

It is convenient to define a normalized phase parameter $B$:

$$B = [(\beta/k)^2 - 1]/(\epsilon_r - 1), \tag{31}$$

where $k = 2\pi/\lambda_0$. The phase parameter is useful in the description of dispersion characteristics because it is normalized in the range [0, 1]. Combining Eqs. (30) and (31), we find the simple relation

$$B = \sin^2 k_d a. \tag{32}$$

In plotting dispersion characteristics, it is also convenient to define a normalized frequency $V$ given

$$V = (2\pi a/\lambda_0)\sqrt{\epsilon_r - 1}. \tag{33}$$

Using Eq. (26), we may also write

$$V = k_d a/\cos k_d a. \tag{34}$$

It is interesting to compare the dispersion of the $TE_{10}$ mode in an H guide with that of the $TE_{10}$ mode in a dielectric-filled rectangular metal waveguide.

In the latter the phase constant is

$$\beta_{rect} = (2\pi/\lambda_0)[\epsilon_r - (\lambda_0/\lambda_c)^2]^{1/2},$$

where $\lambda_c$ is the cutoff wavelength for the empty guide. By using this relationship and the definition of $V$ in Eq. (33), it can be shown that

$$B_{rect} = 1 - \pi^2/4V^2_{rect}.$$

This relationship is identical to the asymptotic variation of the phase parameter for the $TE_{10}$ mode in an H guide:

$$B = 1 - \pi/4V^2 \qquad (V \gtrsim 10).$$

For smaller values of $V$, rectangular waveguide is more dispersive than H guide, owing to the presence of cutoff at $V = \pi/2$.

## B. POWER DISTRIBUTION

An important consideration in the design of dielectric waveguide structures is the distribution of power between the confining dielectric and the outer medium (air in our case). The fraction of the power that flows in the dielectric is a good measure of the "confinement" of the power. The total axial power flow is

$$P_t = P_d + P_a, \tag{35}$$

where $P_d$ is the power flow in the dielectric, given by

$$P_d = \int_0^b \int_{-a}^a E_{y1} H^*_{x1} \, dx \, dy, \tag{36}$$

and $P_a$ is the power flow outside the dielectric, given by

$$P_a = 2 \int_0^b \int_a^\infty E_{y3} H^*_{x3} \, dx \, dy. \tag{37}$$

The fractional power flow in the dielectric is defined as

$$\rho = P_d/P_t. \tag{38}$$

Using Eqs. (23) and (24), we find that

$$\rho = \frac{\sin k_d a \cos k_d a + k_d a}{\cot k_d a + k_d a}. \tag{39}$$

The quantity $\rho$ can also be obtained from a theorem derived for step-index optical fibers (Krumbholz et al., 1980):

$$\rho = B + (V/2)(dB/dV), \tag{40}$$

where $B$ and $V$ are given in Eqs. (32) and (34). By substitution into Eq. (40), we again obtain Eq. (39).

## C. NORMALIZED PHASE SHIFT

The expressions for induced phase shift presented earlier are applicable only if the carrier propagates as an unconfined plane wave in the electrooptic crystal. If the carrier propagates as the $TE_{10}$ mode in an H guide, then Eq. (5) must be modified to account for the guiding properties. We start by writing

$$\Delta\phi_{TE} = L \, \Delta\beta = L(d\beta/dn) \, \Delta n, \tag{41}$$

when $\Delta n$ is the induced change in microwave refractive index and $\beta$ is given by Eq. (29), or equivalently by

$$\beta = (1/a)[\pi^2(2a/\lambda_0)^2\epsilon_r - (k_d a)^2]^{1/2}. \tag{42}$$

In Eqs. (41) and (42) we have suppressed the tensor notation for $\beta$, $n$, and $\epsilon_r$. Depending on the crystal orientation, $\epsilon_r$ is given by either $\epsilon_o = \epsilon_{11} = \epsilon_{22} = n_o^2$ or $\epsilon_e = \epsilon_{33} = n_e^2$, where the subscripts o and e refer to the ordinary and extraordinary directions in the crystal. Using Eqs. (42), (41), and (30), we find that

$$\Delta\phi_{TE} = \frac{\lambda_g}{\lambda_0} \sqrt{\epsilon_r} \left(\frac{k_d a + \sin k_d a \cos k_d a}{k_d a + \cot k_d a}\right)\left(\frac{2\pi L}{\lambda_0} \Delta n\right). \tag{43}$$

A more useful quantity for device design is $\Delta\phi_R$, the phase shift $\Delta\phi_{TE}$ normalized to that of an unconfined plane wave in the same material (i.e., the high-frequency limiting value of $\Delta\phi_{TE}$):

$$\Delta\phi_R = \Delta\phi_{TE}/\Delta\phi_0, \tag{44}$$

where $\Delta\phi_0 = 2\pi L \, \Delta n/\lambda_0$. From Eqs. (43) and (44) we find that

$$\Delta\phi_R = \frac{\lambda_g}{\lambda_0} \sqrt{\epsilon_r} \left[\frac{k_d a + \sin k_d a \cos k_d a}{k_d a + \cot k_d a}\right]. \tag{45}$$

## D. NORMALIZED ABSORPTION COEFFICIENT

The relationship between the intensity (or power) absorption coefficient and loss tangent given in Eq. (22) is valid only for plane-wave propagation in the dielectric medium. If the wave propagates in the $TE_{10}$ mode in an H guide, then both wall and dielectric losses must be taken into account. For the purpose of this analysis we shall neglect wall loss, because the dielectric loss of available materials is expected to be greater.

The dielectric loss per unit length in the $z$ direction, $W_d$, is given by

$$W_d = \sigma_d/2 \int_0^b \int_{-a}^a |E_{y1}|^2 \, dx \, dy, \tag{46}$$

where $\sigma_d$ is the conductivity of the dielectric material:

$$\sigma_d = \omega\epsilon_0\epsilon_r \tan \delta. \tag{47}$$

The attenuation coefficient is defined as

$$\alpha_{TE} = W_d/P_t. \tag{48}$$

Using Eqs. (23a), (30), (47), and (35), we find that

$$\alpha_{TE} = \frac{\lambda_g}{\lambda_0} \sqrt{\epsilon_r} \left[ \frac{k_d a + \sin k_d a \cos k_d a}{k_d a + \cot k_d a} \right] \left( \frac{4\pi}{\lambda_0} \sqrt{\epsilon_r} \tan \delta \right). \tag{49}$$

A more useful quantity for device design is $\alpha_R$, the absorption coefficient $\alpha_{TE}$ normalized to that of a plane wave in the same material (i.e., the high-frequency limiting value of $\alpha_{TE}$):

$$\alpha_R = \alpha_{TE}/\alpha_I, \tag{50}$$

where $\alpha_I$ is given by Eq. (22). Measured values of $\alpha_I$ for LiNbO$_3$ and LiTaO$_3$ at 94 GHz are given in Table II. From Eqs. (49) and (50) we find that

$$\alpha_R = \frac{\lambda_g}{\lambda_0} \sqrt{\epsilon_r} \left[ \frac{k_d a + \sin k_d a \cos k_d a}{k_d a + \cot k_d a} \right], \tag{51}$$

which is identical to the expression for $\Delta\phi_R$ [Eq. (45)]. Note also that

$$\alpha_R = \Delta\phi_R = (\lambda_g/\lambda_0) \sqrt{\epsilon_r} \rho. \tag{52}$$

## E. IMPEDANCE

For a lossless transmission line the impedance is purely resistive, and is thus denoted by $R$. An expression for the impedance can be obtained from a voltage–power relationship:

$$R = [V(x = 0)_{max}]^2/2P_t, \tag{53}$$

where

$$V(x = 0)_{max} = E_{rf}b = E_0(\omega\mu_0/k_d)b \tag{54}$$

and $P_t$ is given by Eq. (35). By substitution into Eq. (53) we find that

$$R = 120\pi \frac{\lambda_g}{\lambda_0} \frac{b}{a} \frac{k_d a}{k_d a + \cot k_d a}. \tag{55}$$

## V. TE$_{10}$-Mode Parameters for LiNbO$_3$ H Guide

As mentioned earlier, we selected LiNbO$_3$ for our experiments because of its relatively large values of electrooptic figure of merit and its low values of absorption coefficient at 95 GHz. For all our experiments, the crystals were cut so that the crystalline $z$ axis was normal to the metal conductors. The carrier electric field is thus polarized in the extraordinary direction, and the relevant values of index of refraction and dielectric constant are $n_e$ and $\epsilon_e$, respectively. Because the modulating field is also normal to the metallic conductors, the appropriate electrooptic coefficient is $r_{333}$.

We have calculated the mode cutoff properties of a LiNbO$_3$ H guide with $\epsilon_r = \epsilon_e = 27$ using Eqs. (27) and (28); the results are plotted in Fig. 5. We do not show the TE$_{20}$-mode cutoff, because this mode is unlikely to be excited using the coupling schemes to be described. We see that the lowest-order (PM$_{11}$) hybrid mode is cut off for $b/\lambda_0 \leq 0.1$ and the higher-order TE modes are cut off for $2a/\lambda_0 \leq 0.2$. In selecting an optimum waveguide cross section for experiments, it is important for optimum mechanical strength that the cross-sectional area be as large as possible, which corresponds to the condition $b_c/\lambda_0 \simeq 0.1$ and $2a/\lambda_0 \simeq 0.2$. In our experiments we used waveguides with two different cross sections:

(1)  $2a = 0.55$ mm, $b = 0.30$ mm;
(2)  $2a = 0.55$ mm, $b = 0.40$ mm.

The normalized dimensions for each case are plotted as single points in Fig. 5, assuming $\lambda_0 = 3.18$ mm ($f = 95$ GHz). We see that case (1) is optimum

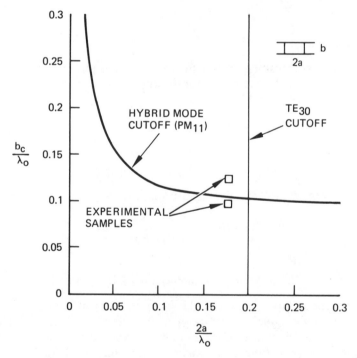

FIG. 5  Mode cutoff properties of H guide using Z-cut LiNbO$_3$. $\epsilon_r = \epsilon_e = 27$.

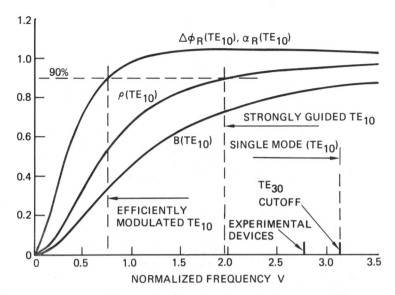

FIG. 6 Dispersion characteristics of the $TE_{10}$ mode in H guide using Z-cut $LiNbO_3$. $\epsilon_r = \epsilon_e = 27$.

for single-mode operation and case (2) allows the lowest-order hybrid mode to propagate. In our experiments, however, the input radiation is orthogonally polarized to the dominant polarization of the $PM_{11}$ mode, so this mode is unlikely to be excited.

In Fig. 6 we plot the normalized phase parameter $B$, the fractional power confinement $\rho$, and the normalized attenuation coefficient $\Delta\phi_R$ and phase shift $\alpha_R$ as a function of the normalized frequency $V$ for the $TE_{10}$ modes in Z-cut $LiNbO_3$ ($\epsilon_r = 27$). We also indicate the values of $V$ corresponding to the chosen sample width and the $TE_{30}$-mode cutoff.

The dispersion characteristics shown in Fig. 6 are useful for determining the bandwidth of the $TE_{10}$ mode. Note first that the $TE_{10}$ mode has no low-frequency cutoff, so by strict definition the operating bandwidth extends from $V = 0$ to the $TE_{30}$-mode cutoff at $V = 3.14$. However, for small values of $V$, most of the power flows outside the dielectric, resulting in weak guiding or "loose confinement." We shall consider the $TE_{10}$ mode to be strongly guided, or "closely confined," when at least 90% of the power is propagating in the dielectric. From Fig. 6 we see that the strongly guiding regime corresponds to $V \geq 2$, so that the range of operation for strongly guided, single-mode operation is $2 \leq V \leq 3.14$, corresponding to a 35% bandwidth.

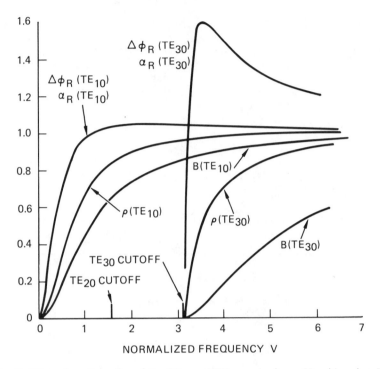

FIG. 7 Dispersion properties of the $TE_{10}$ and $TE_{30}$ modes in an H guide using Z-cut LiNbO$_3$. $\epsilon_r = \epsilon_e = 27$.

The rapid rise of $\Delta\phi_R$ (and $\alpha_R$) in Fig. 6 suggests a second definition for bandwidth. If efficient phase shifting is more important than power confinement, we can consider the $TE_{10}$ mode to be "efficiently modulated" when the induced phase shift is at least 90% of its limiting high-frequency value. In this case, the efficiently modulated regime corresponds to $V \geq 0.8$, so that the range of operation for efficiently modulated, single-mode operation is $0.8 \lesssim V \lesssim 3.14$, corresponding to a 75% bandwidth.

In Fig. 6 we note that the curve for $\Delta\phi_R$ (and $\alpha_R$) has a broad maximum at a value greater than unity, corresponding to a phase shift (or attenuation) larger than that of an infinite plane wave in the medium. This maximum is more pronounced for higher-order modes, as shown in Fig. 7. The reason for the maximum is as follows (Schweig, 1982). At very low frequencies, most of the fields are outside the dielectric and are thus subject to little modulation (or attenuation). As the frequency is increased, the fields become more concentrated in the dielectric. In a ray-optic picture, this situation corresponds to rays that bounce back and forth between the boundaries of the

## TABLE IV

### Properties of the $TE_{10}$ Mode in H Guide at 95 GHz Using Z-Cut $LiNbO_3$[a]

| Case | Cross section $2a \times b$ (mm) | Guide width $2a/\lambda_0$ | Distribution parameter $k_d a$ | Frequency $V$ | Guide wavelength $\lambda_g/\lambda_0$ | Phase parameter $B$ | Power confinement $\rho$ | Phase shift, attenuation coefficient $\Delta\phi_R, \alpha_R$ | Impedance $(\Omega)$ |
|------|------|------|------|------|------|------|------|------|------|
| | | | | | Normalized parameters | | | | |
| (1) | $0.55 \times 0.30$ | 0.172 | 1.14 | 2.73 | 0.210 | 0.826 | 0.948 | 1.04 | 18.7 |
| (2) | $0.55 \times 0.40$ | 0.172 | 1.14 | 2.73 | 0.210 | 0.826 | 0.948 | 1.04 | 24.8 |

[a] $\epsilon_r = \epsilon_e = 27$.

guide. The path followed by the ray is thus actually longer than if it were propagating along the guide axis, and therefore the phase shift (or attenuation) per physical length of guide is higher. At high frequencies, the guide cross-sectional dimensions become large compared to wavelength, and the mode can be thought of as a plane wave propagating parallel to the axis.

In Table IV we summarize the $TE_{10}$-mode properties for $LiNbO_3$ H guides with the cross-sectional dimensions given previously. All calculated values are based on equations given in Section IV. Several of these properties deserve comment. First, although the phase parameter is only $\sim 80\%$ of its high-frequency value, the fractional power confinement is very large and the normalized phase shift (and absorption coefficient) actually exceeds unity by a small amount. Thus, for our experimental conditions, the $TE_{10}$ mode has properties nearly identical to those of an unconfined plane wave in the $LiNbO_3$ sample. Second, we see that the evanescent field depth is a small fraction of the guide width, suggesting that the conducting plates need to extend only a small distance outside the dielectric region. This property is clearly related to the close mode confinement characterized by the large value of $\rho$. Finally, the indicated impedance values are reasonably well suited for impedance matching to other circuit elements.

## VI. $LiNbO_3$ H-Guide Phase Shifters at 95 GHz

On the basis of the analysis and calculations that we have presented, we see that $LiNbO_3$ H guide is a promising guiding structure for electrooptic phase shifting at 95 GHz. In this section we shall describe the construction and evaluation of several $LiNbO_3$ H-guide phase shifters (Klein, 1982), which differ primarily in their means of coupling between the $TE_{10}$ mode in a WR-10 hollow metal waveguide and the $TE_{10}$ mode in a $LiNbO_3$ H guide.

### A. END-COUPLED DEVICE

The first device utilizes end coupling, and is shown in Fig. 8. The $LiNbO_3$ sample is mounted on a flat, polished metal base, which is also the bottom portion of the hollow metal waveguide. The sample extends into the metal waveguide; the latter tapers down to the same height as the $LiNbO_3$ and to twice the width (over a length of 2 mm). The polished upper plate of the H guide is isolated from the surrounding metal structure by a 1.25-mm air gap and is held in place by spring loading. The overall length of the upper plate is 2.75 cm. A photograph of the completed device is shown in Fig. 9.

Careful preparation of the $LiNbO_3$ sample is an important requirement in this device. Samples are produced by first grinding and polishing the large faces of a Z-cut wafer to the required height of the waveguide (either 0.30 or 0.40 mm). The surfaces are made optically flat to optimize the contact with

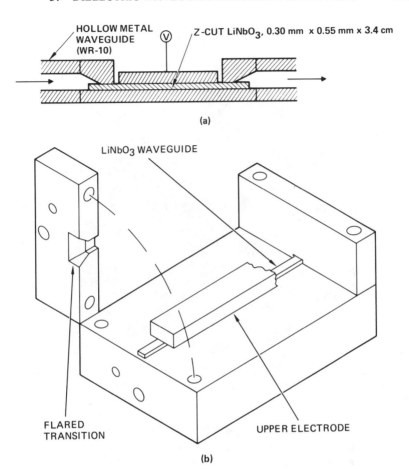

FIG. 8 LiNbO₃ phase shifter using end coupling: (a) cross section, (b) exploded view.

the metal conductors. The wafers are next cut slightly oversize in length and width. The resulting pieces are then stacked on a lap, and the side and end faces are ground to the required dimensions.

The modulation performance and insertion loss of this device were measured by insertion into the active arm of a waveguide phase bridge (see Section II.C for a discussion of this technique). Our device measurements were made at a modulation frequency of 1 kHz using the smaller cross section sample (0.3 mm in height). For a peak applied voltage of 1000 V, the measured amplitude modulation was 67% and the induced phase shift was 56° peak to peak, in good agreement with theory, based on the electrooptic coefficients measured by Klein (1981).

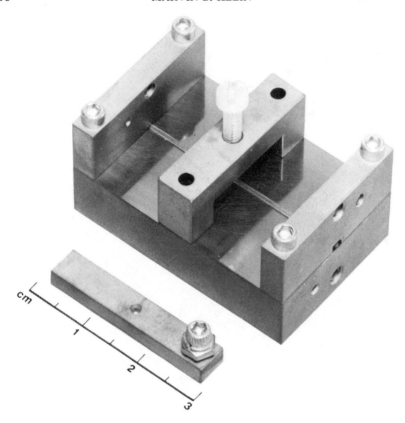

FIG. 9  Photograph of an end-coupled phase shifter with the upper plate removed.

The measured insertion loss of the device was typically $\sim 12$ dB. If we assume that only the $TE_{10}$ H-guide mode is excited (as expected from the spatial uniformity of the input field), then the calculated absorption loss is only $\sim 3$ dB. The excess loss is attributed to reflections and radiation from each transition and corresponds to $\sim 4.5$ dB/transition. It is important to reduce this coupling loss for practical device applications. To this end we varied the taper geometry and the penetration of the $LiNbO_3$ sample into the taper, with little or no reduction in the insertion loss. From these studies and measurements of radiated and reflected power, we conclude that most of the coupling loss is due to radiation from the region of the gap in the upper conductor. One means of reducing this radiation would be to install a choke joint that would act as a short circuit to microwave fields but still provide dc isolation.

## B. PROBE-COUPLED DEVICE

The second phase shifter that we designed and tested follows the design of an electrooptic phase shifter described and demonstrated at a carrier frequency of 200 MHz by Cohn and Eikenberg (1962). The key element of this device is that probe coupling was used to excite the $TE_{10}$ mode in an H-guide structure. This device is illustrated in Fig. 10. The $TE_{10}$ mode was launched using a current filament that extends across the parallel conducting plates. The current filament was energized from a coaxial line feeding through the upper conducting plate, with only the center conductor crossing the parallel-plate line and contacting the opposite wall. For optimum coupling, half holes were machined into the ends of the ferroelectric ceramic to allow passage of the coaxial line center conductor. The electrooptic material (or nonlinear dielectric) was a mixed-composition ferroelectric ceramic containing 35% $PbTiO_3$ and 65% $SrTiO_3$. This material has a Curie point near room temperature, leading to a large enhancement in both the dielectric constant and the electrooptic figure of merit. The measured dielectric parameters at 200 MHz are $\epsilon_r \simeq 5000$ and $\tan \delta \simeq 0.05$, and the measured dielectric nonlinearity corresponds to an electrooptic figure of merit $R \simeq 1.5 \times 10^{-3}$ cm/V. These favorable material properties led to a measured device phase shift of 348° at 22 kV/cm, with only 3 dB insertion loss.

On the basis of its large phase shift and low insertion loss, it is worthwhile

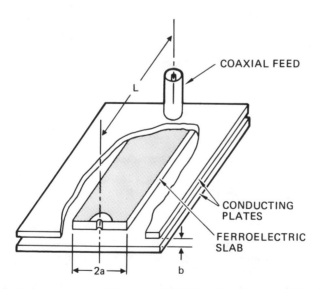

FIG. 10 Probe-coupled phase shifter used at 200 MHz. [From Cohn and Eikenberg (1962). Reprinted from *IRE Trans. Microwave Theory Tech.* **MTT-10.** © 1962 IEEE.]

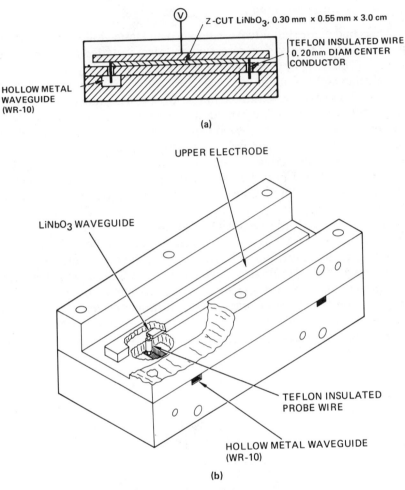

FIG. 11   LiNbO₃ phase shifter using probe coupling: (a) cross section, (b) cutaway view.

to consider scaling this device to operation at 95 GHz. However, it is important to note first that the same ferroelectric ceramic cannot be used at frequencies in the millimeter band, owing to its excessive absorption. More specifically, we note from Eq. (22) that the absorption coefficient $\alpha_I$ varies as $\lambda_0^{-1} \tan \delta$, so that even if $\tan \delta$ is invariant with frequency, $\alpha_I$ will increase from its measured value of $0.12 \text{ cm}^{-1}$ at 200 MHz to the undesirable value of $60 \text{ cm}^{-1}$ at 95 GHz.

The device constructed for use at 95 GHz is illustrated in Fig. 11. A short section of Teflon-insulated wire with insulation removed at each end pro-

FIG. 12   Photograph of probe-coupled phase shifter with the upper plate removed.

vides the connection between the hollow metal waveguide and the LiNbO₃ H guide. One end of the wire extends 0.75 mm into the hollow metal waveguide. The center (insulated) section is a coaxial waveguide connection, and the opposite end extends along the end of the LiNbO₃ sample and contacts the upper H-guide plate. The hollow metal waveguide sections run straight through the device and can be terminated by adjustable shorts for tuning. A photograph of the completed device is shown in Fig. 12.

Our device performance measurements were made using the phase bridge technique described in Section II.C and the smaller cross section sample. For a peak applied voltage of 1000 V, the measured amplitude modulation was 70% and the induced phase shift was 58° peak to peak, in good agreement with theory, based on the electrooptic coefficients given by Klein (1981). The measured insertion loss was typically ~ 8 dB, of which ~ 3 dB results from $TE_{10}$-mode absorption loss. There are a number of possible techniques for reducing the coupling loss below its present value of $\sim 2.5$ dB/transition. These include variation of the depth of the probe into the metal waveguide and the use of half holes in the ends of the $LiNbO_3$ samples.

## VII. Other Electrooptic Devices

Besides the phase shifters just discussed, a number of other dielectric waveguide components using the electrooptic or nonlinear properties of ferroelectrics appear feasible. In this section we specifically consider scanning antennas, tunable filters, and frequency doublers. The use of $LiNbO_3$ as the active material in these devices generally leads to performance that is inadequate for most applications. However, other materials in their earlier stages of development for microwave applications (such as $BaTiO_3$ and SBN) appear more promising. Whenever appropriate, we shall compare device performance using $LiNbO_3$ and $BaTiO_3$. The major challenge in considering highly nonlinear materials (such as $BaTiO_3$ and SBN) at frequencies in the millimeter band is the reduction of absorption loss. The prospects for improved materials are discussed in Section VIII.

### A.   ELECTRONICALLY SCANNABLE ANTENNAS AND TUNABLE FILTERS USING PERIODIC STRUCTURES

Periodically perturbed dielectric waveguides have unique propagation properties that allow their use as scanning antennas and tunable stopband filters. The antenna applications arise from the excitation of leaky waves in certain operating regimes. These waves radiate laterally at one or more given angles, and the radiation angles can be controlled by a variety of means. The stopbands of periodically perturbed waveguides are due to mode coupling in the surface wave (nonradiating) regime; these stopbands can also be tuned.

A leaky wave antenna in an inverted strip dielectric waveguide has been demonstrated at a fixed frequency of 15 GHz by Itoh (1977); the stopband properties of the experimental devices were also studied. Dielectric waveguides (in rectangular rod form) have also been used for beam scanning via frequency sweeping (Klohn et al., 1978). Two disadvantages of the latter experiment are that fixed-frequency operation is not possible and beam spread in the $E$ plane is large because the waveguide height is small. In more recent work, Trinh et al. (1981) eliminated the $H$-plane spreading by using a trough guide (half-symmetric H guide) with flaring of the metal plates to produce a one-dimensional horn. Beam scanning was still obtained by frequency sweeping.

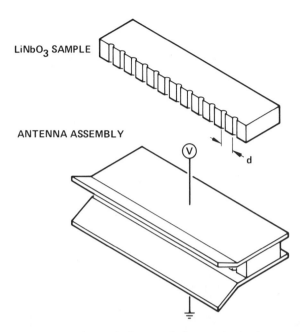

FIG. 13 LiNbO$_3$ sample with periodic perturbations and complete scanning antenna structure based on H-guide mounting (TE$_{10}$ mode).

In both of the scanning antennas described previously, the dielectric guiding medium was silicon and the periodic perturbations were produced using metallic strips or grooves on one wall. If we substitute LiNbO$_3$ for the silicon, the electrooptic properties of the LiNbO$_3$ can be used to provide beam scanning at a fixed frequency or tuning of the stopband center frequency. Figure 13 shows a sketch of the LiNbO$_3$ sample (with periodic indentations) and final H-guide antenna assembly.

The principles of operation of leaky wave antennas are reviewed by Collin and Zucker (1969) and Itoh (1981). Electromagnetic waves in a periodically perturbed waveguide can be described in terms of space harmonics whose phase constants are given by

$$\beta_m = \beta_0 + 2\pi m/d, \tag{56}$$

where $d$ is the perturbation period, $m$ is the order of the space harmonic, $\beta_0$ is the phase constant of the dominant ($m = 0$) space harmonic, and $\beta_m$ is the phase constant for the space harmonic of order $m$. If the waveguide perturbations are small, $\beta_0$ is generally very close to the phase constant $\beta_0(k)$ of the dominant mode in the unperturbed guide ($k$ is the free-space constant). The

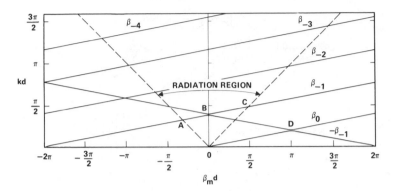

FIG. 14  Dispersion characteristics for the $TE_{10}$ mode in a $LiNbO_3$ H guide with periodic perturbations. The crystal is assumed to be Z-cut, so the refractive index $n_e = 5.2$. The stopbands due to mode coupling are shown only at points $B$ and $D$.

only significant deviations in $\beta_0$ occur in wave-coupling regions, as we shall explain.

In Fig. 14 we plot $\beta_m d$ as a function of $kd$ for the $m = 0$ through $m = -4$ forward-traveling space harmonics and the $m = -1$ backward-traveling space harmonic in a $LiNbO_3$ H guide with small periodic perturbations. The curve for $m = 0$ is obtained from the dispersion relation for the $TE_{10}$ mode obtained by combining Eqs. (26) and (29). Curves for higher spatial harmonics are obtained by shifting the curve for $\beta_0$ by multiples of $2\pi$. The coupling of forward- and backward-traveling waves at synchronous points such as $B$ and $D$ leads to forbidden values of $kd$ near these points. This behavior leads to the stopband properties of the perturbed structure to be discussed later. If $k$ is chosen so that $\beta_m$ satisfies the equation

$$|\beta_m/k| < 1, \tag{57}$$

then the radiation angle $\theta_m$ (measured from broadside) is given by

$$\sin \theta_m = \beta_m/k = \lambda_0/\lambda_g + m(\lambda_0/d). \tag{58}$$

The radiation region defined by Eq. (57) is indicated in Fig. 14. For example, as $kd$ increases from zero, the $m = -1$ space harmonic begins radiating in the backward direction ($\theta_{-1} = -\pi/2$) at point $A$. At point $B$, the radiation is broadside ($\theta_{-1} = 0$) and the corresponding value of $kd$ is $2\pi/n$, where $n = n_e \simeq 5.2$ for $LiNbO_3$ (Table II). The region in the immediate vicinity of point $B$ is not favorable for device operation, because mode coupling produces a large reflection. For larger values of $kd$ the structure radiates in the forward direction until point $C$ is reached, corresponding to $\theta_{-1} = \pi/2$.

As seen in Fig. 14, other negative space harmonics can also radiate, at correspondingly larger values of $kd$.

In the portions of Fig. 14 with $|\beta_m/k| > 1$, only surface waves can propagate. Near synchronous points such as $D$ (corresponding to $\beta_m d = \pi$ and $kd \simeq \pi/n$), mode coupling produces a stopband phenomenon. The bandwidth and reflectance depend on the strength and profile of the perturbations and on the interaction length (Itoh, 1981).

The electrooptic effect can be used to scan the beam angle in the radiation (leaky wave) region. To analyze the case of beam scanning, we consider the $m = -1$ space harmonic and assume sufficient mode confinement so that $\lambda_g \simeq \lambda_0/n$. Equation (58) then gives

$$\sin \theta_{-1} = n - \lambda_0/d. \tag{59}$$

The change in $\theta_{-1}$ induced by an electric field $E$ is

$$\Delta\theta_{-1} = [(dn/dE)/\cos \theta_{-1}]E. \tag{60}$$

Using Eq. (58), we find that

$$\Delta\theta_{-1} = (n^3 r/\cos \theta_{-1})E. \tag{61}$$

Thus, the scanning antenna described here is characterized by the same material figure of merit ($n^3 r$) as the bulk phase shifter described in Section II.A.

For a 20-kV/cm electric field in a LiNbO$_3$ device, $\Delta\theta_{-1} \simeq 10^{-2}$ rad. However, much larger scanning angles (on the order of 1 rad) are predicted for BaTiO$_3$ under the same conditions.

The electrooptic effect can also be used to tune the center frequency of the stopbands in the surface wave regime. As mentioned earlier, the stopband at $D$ corresponds to the condition $kd \simeq \pi/n$, or

$$\lambda_0 = 2nd. \tag{62}$$

The change in $\lambda_0$ induced by an electric field $E$ is

$$\Delta\lambda_0 = 2d \frac{dn}{dE} E. \tag{63}$$

Using Eq. (3), we find that

$$\Delta\lambda_0 = dn^3 rE. \tag{64}$$

The fractional change in $\lambda_0$ (of the frequency $\nu$) is thus

$$\Delta\lambda_0/\lambda_0 = \Delta\nu/\nu = \tfrac{1}{2}n^2 rE. \tag{65}$$

For a 20-kV/cm field in LiNbO$_3$ device, $\Delta\lambda_0/\lambda_0 \simeq 10^{-3}$. However, fractional tuning ranges approaching 100% are calculated for BaTiO$_3$.

## B. Electronically Tunable Resonant Filter

Fixed-frequency bandpass or bandstop filters have been demonstrated by coupling a wave traveling in a dielectric waveguide to a ring resonator (Knox, 1976; Rudokas and Itoh, 1976). The center frequency in this type of device is determined by the dimensions of the resonator structure and the refractive index of the material used. If the refractive index can be varied electrooptically, then the center frequency can be adjusted, resulting in a tunable filter. A diagram of such a device, along with the output frequency spectrum, is shown in Fig. 15. In this device only the resonator material needs to be electrooptic. The straight-waveguide sections can be fabricated from low-loss materials such as high-resistivity GaAs or Si.

To analyze the tunability of this device, we write the ring resonance condition as

$$m(\lambda_g/2) = \pi D, \qquad (66)$$

where $m$ is the order of interference and $D$ is the average ring diameter. If we assume tight mode confinement so that $\lambda_g \simeq \lambda_0/n$, then

$$m\lambda_0 = 2\pi nD. \qquad (67)$$

The change in $\lambda_0$ induced by a field $E$ is

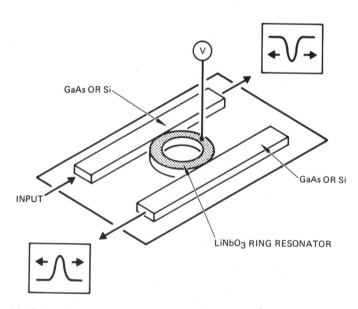

FIG. 15 Schematic diagram of a tunable ring-resonator filter. The output frequency spectrum is also shown.

$$\Delta\lambda_0 = \frac{2\pi D}{m} \frac{dn}{dE} E. \tag{68}$$

Using Eq. (3), we find that

$$\Delta\lambda_0 = (\pi D/m)n^3 rE. \tag{69}$$

The fractional change in $\lambda_0$ (or the frequency $\nu$) is thus

$$\Delta\lambda_0/\lambda_0 = \Delta\nu/\nu = \tfrac{1}{2}n^2 rE, \tag{70}$$

which is identical to the expression for the bandstop filter using a periodically perturbed waveguide [Eq. (65)].

Another useful performance parameter is the electric field required to tune by one linewidth. If the linewidth $\delta\nu$ (or resonator $Q$) is determined by dielectric loss, then

$$\delta\nu = \nu \tan \delta. \tag{71}$$

By combining Eq. (71) with Eq. (70) and using $\Delta\nu = \delta\nu$, we find that

$$E_{\delta\nu} = (2 \tan \delta)/n^2 r. \tag{72}$$

This field is equal to $\sim 20$ kV/cm for $LiNbO_3$ and $20-200$ V/cm for $BaTiO_3$. Although a wide tuning range is not possible with $LiNbO_3$, the ability to tune rapidly through more than one linewidth suggests that the same structure could be used as an amplitude modulator with a wide dynamic range.

## C. FREQUENCY DOUBLERS

There is an ongoing need for high-power pulsed sources in the millimeter and submilleter spectral regions. The highest frequency at which high-power conventional tube sources can be fabricated with reasonable efficiency is $\sim 50$ GHz. It is thus logical to consider frequency doubling (i.e., second-harmonic generation) in a bulk nonlinear material as a means of generating signals in the required band. The power conversion efficiency for frequency doubling increases with pump power, so that this mechanism is particularly favorable in high-power systems. If the doubling process can be made efficient, then the available efficient lower-frequency sources could be incorporated to make an overall system of high efficiency.

Materials useful for electrooptic phase shifting can also be used for nonlinear frequency doubling. Furthermore, the dielectric waveguide geometry is particularly favorable for devices, because the input power is concentrated in a small cross section over a large distance. The resultant increase in pump *intensity* and interaction length both lead to an increase in second-harmonic efficiency. Note also that H guide may not be the best

structure for frequency doubling because the upper conductor is not required. A more favorable structure is image guide, which is less complex and
has lower coupling losses.

The major technical issue in using any guiding structure for second-harmonic generation is the minimization of the mismatch between the phases
of the pump and second-harmonic waves. For plane-wave illumination
there is an inherent mismatch due to material dispersion, but if the sample is
mounted in a waveguide configuration, the dominant contribution to phase
mismatch is mode dispersion. The effect of mode dispersion is especially
strong if the structure is designed for single-mode operation.

The phase mismatch for second-harmonic generation is given by $\Delta\phi = \Delta\beta L$, where $\Delta\beta$ is the propagation constant mismatch:

$$\Delta\beta = \beta(2\omega) - 2\beta(\omega). \tag{73}$$

The requirement for efficient phase matching is $\Delta\phi < \pi$, or $\Delta\beta < \pi/L$. This
relation is equivalent to writing $L < L_c$, where $L_c = \pi/\Delta\beta$ is the coherence
length, i.e., the maximum interaction length useful in producing second-
harmonic power. The intrinsic contribution to $L_c$ from material dispersion
is given by

$$L_c = \lambda_0/4[n(2\omega) - n(\omega)]. \tag{74}$$

Using the far-infrared dispersion parameters of Bosomworth (1966) for
LiNbO$_3$, we obtain 8.1 cm for the coherence length in this material at
94 GHz for a wave polarized along the $c$ axis.

If the sample is in a waveguide configuration, the more general expression
for the coherence length is

$$L_c = \pi/[\beta(2\omega) - 2\beta(\omega)], \tag{75}$$

which is equivalent to

$$\frac{L_c}{\lambda_0(\omega)} = \frac{1}{4}\left[\frac{\lambda_0(2\omega)}{\lambda_g(2\omega)} - \frac{\lambda_0(\omega)}{\lambda_g(\omega)}\right]^{-1}. \tag{76}$$

Using the waveguide dispersion parameters presented in Section IV and the
material dispersion parameters of Bosomworth (1966), we plot $L_c/\lambda_0(\omega)$
and $L_c$ (at $\omega = 94$ GHz) in Fig. 16 for the TE$_{10}$ mode in both a rectangular
metal waveguide and an H guide. As before, the dielectric material is
assumed to be LiNbO$_3$ and the microwave electric field is assumed to be
polarized along the $c$ axis. The TE$_{30}$-mode cutoff for both structures is also
shown, as well as the value of normalized frequency corresponding to a
LiNbO$_3$-filled WR-10 waveguide at 94 GHz. The analogous curve for an
image guide is not included but is expected to be very similar to the plots
shown. It is clear that mode dispersion (and the resultant small coherence

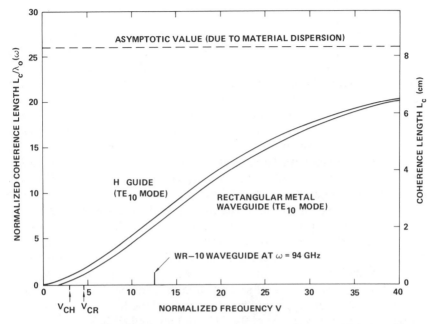

FIG. 16 Normalized coherence length and coherence length at 94 GHz in H guide and rectangular metal waveguide using Z-cut LiNbO$_3$. $V_{CH}$ and $V_{CR}$ are the cutoff frequencies for the TE$_{30}$ mode in H guide and rectangular metal waveguide, respectively.

lengths) precludes the use of a single-mode guide for frequency doubling. The effect of mode dispersion could be mitigated by coupling the signal at $2\omega$ into a different waveguide mode (Anderson *et al.*, 1971), but this does not seem practical in most cases.

The only viable alternative to obtaining coherence lengths approaching the intrinsic value due to material dispersion is to operate in a multimode structure. With careful transition design, it is possible to couple efficiently into the fundamental mode and take advantage of the low dispersion at high values of normalized frequency.

Recently, Ahn *et al.* (1983) demonstrated second-harmonic generation in LiNbO$_3$ and LiTaO$_3$ at 35 GHz. The sample filled the cross section of a $\sim$ 1-cm length of rectangular metal waveguide that was single mode in its unfilled portions. The coherence length in the dielectric-filled portion was $\sim$ 3 cm. For an input power of 24 kW, the signal power at 70 GHz was $\sim$ 10 W, corresponding to a conversion efficiency of $\sim$ 0.04%. For these same conditions we calculate that conversion efficiencies near 100% are obtainable in BaTiO$_3$ and other highly nonlinear materials.

## VIII.  Prospects for Improved Materials

The electrooptic devices discussed in Sections V and VI were primarily modeled using $LiNbO_3$ as the active material. The selection of this material is based on its small absorption coefficient, moderate nonlinearity, and ready availability. Although phase shifts on the order of 1 rad have been demonstrated (see Section V), it is desirable in many cases to have devices that have $\pi/2$ or $\pi$ phase shifts and require smaller modulating voltages. Similarly, the use of $LiNbO_3$ in the other proposed devices leads to inadequate performance for most applications. There is thus a critical need for improved materials with large electrooptic coefficents and small losses. Among the possible candidates are $BaTiO_3$ (Boyd et al., 1971) and SBN (Ho et al., 1981), materials discussed briefly in Sections I and II.C. These materials have large dielectric constants at microwave frequencies and thus have large electrooptic coefficients. However, the absorption losses are substantial in these materials and limit the interaction length in practical devices. In such materials a useful phase-shifting performance parameter is the phase shift *per absorption length*, i.e., Eq. (5) with $L = 1/\alpha_I$. The phase-shifting performance of several candidate materials in their present stage of development is plotted as a function of length in Fig. 17. We have assumed a carrier frequency of 94 GHz and an applied field of 20 kV/cm. On each curve we also show the phase shift corresponding to the limiting

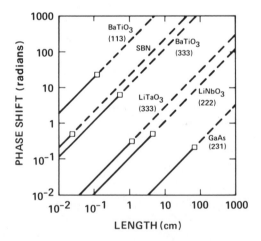

FIG. 17  Induced phase shift vs length for several electrooptic materials and orientations. The box on each curve corresponds to the condition $L = 1/\alpha_I$, where $\alpha_I$ is the absorption coefficient for the given material and orientation.

length $L = 1/\alpha_1$. We note that certain materials with large electrooptic coefficients have particularly large absorption losses (e.g., SBN), leading to a phase shift per absorption length only slightly larger than that in $LiNbO_3$ or $LiTaO_3$. We have also plotted the phase shift for GaAs in Fig. 17. Intrinsic or Cr-doped samples of this material are highly transmitting at millimeter and submillimeter wavelengths (Afsar and Button, 1983), and it is thus very attractive for a variety of active and passive components using dielectric waveguides. Although the electrooptic coefficient $r_{231}$ of GaAs is small, electrooptic applications requiring smaller phase shifts may favor use of the material.

The most promising material at present appears to be $BaTiO_3$; for both electrooptic orientations, phase shifts per absorption length greater than $\pi$ radians are calculated. At present, commercially available samples of this material are expensive and inconsistent in quality; any improvements in this material leading to lower absorption losses and improved crystal quality should lead to significantly improved device performance.

## IX. Conclusion

We have described the design and construction of a number of electrooptic control devices in the dielectric waveguide geometry, as well as the performance limitations imposed by presently available materials. The preferred waveguide structure in most cases is H guide, although image guide may be preferable for frequency doubling.

The devices we constructed required the assembly and alignment of a number of carefully machined discrete components. In future devices (especially at high frequencies) it is highly desirable to integrate all components onto a single chip. One integrated form of H guide is shown in Fig. 18. To fabricate this structure, a wafer of $LiNbO_3$ is contacted to a metal base. The wafer is then polished to the desired thickness and metalized. Ion beam etching is then used to form the waveguide. Although the upper electrode does not extend past the edge of the dielectric, the perturbation to the H-guide mode properties is expected to be small if the mode is well confined. An image guide can also be fabricated using the same technique, by omitting the metalization step.

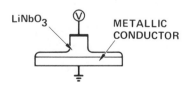

FIG. 18   $LiNbO_3$ H guide for integrated circuit applications.

174     MARVIN B. KLEIN

ACKNOWLEDGMENTS

I would like to acknowledge technical discussions with W. B. Bridges, E. Schweig, and J. F. Lotspeich and expert technical assistance by R. E. Johnson and R. H. Sipman. This work was supported in part by Subcontract No. 67-471915 with the California Institute of Technology under Prime Contract N00014-79-C-0839 with the Office of Naval Research.

REFERENCES

Afsar, M., and Button, K. J. (1983). *IEEE Trans. Microwave Theory Tech.* **MTT-31**, 217–223.
Ahn, B. H., Clark, W. W., Schurtz, R. R., and Bates, C. D. (1983). *J. Appl. Phys.* **54**, 1251–1255.
Anderson, D. B., Boyd, J. T., and McMullen, J. D. (1971). *Proc. Symp. Submillimeter Waves* (J. Fox ed.), pp. 191–210. Polytechnic Press, New York.
Bosomworth, D. R. (1966). *Appl. Phys. Lett.* **9**, 330–331.
Boyd, G. D., and Pollack, M. A. (1973). *Phys. Rev.* **B7**, 5345–5359.
Boyd, G. D., Bridges, T. J., Pollack, M. A., and Turner E. H. (1971). *Phys. Rev. Lett.* **26**, 387–390.
Cohn, M. (1959). *IRE Trans. Microwave Theory Tech.* **MTT-7**, 202–208.
Cohn, M., and Eikenberg, A. F. (1962). *IRE Trans. Microwave Theory Tech.* **MTT-10**, 536–548.
Collin, R. E., and Zucker, F. J. (1969). "Antenna Theory," part II. McGraw-Hill, New York.
DeFonzo, A. P., Lee, C. H., and Mak, P. S. (1979). *Appl. Phys. Lett.* **35**, 575–577.
Devonshire, A. F. (1954). *Adv. Phys.* **3**, 85–130.
Fetterman, H. R., Hu, C., Barch, W. E., and Parker, C. O. (1975). *Lincoln Lab. Q. Tech. Summary Rep.* (June), pp. 30–33.
Garrett, C. G. B. (1968). *IEEE J. Quantum Electron.* **QE-4**, 70–84.
Garrett, C. G. B., and Robinson, F. N. H. (1966). *IEEE J. Quantum Electron* **QE-2**, 328–329.
Harris, D. J., Lee, K.-W., and Reeves, J. M. (1978). *IEEE Trans. Microwave Theory Tech.* **MTT-26**, 998–1001.
Ho, W., Hall, W. F., Neurgaonkar, R. R., DeWames, R. E., and Lim, T. C. (1981). *Ferroelectrics* **38**, 833–836.
Itoh, T. (1977). *IEEE Trans. Microwave Theory Tech.* **MTT-25**, 1134–1138.
Itoh, T. (1981). In "Infrared and Millimeter Waves," vol. 4 (K. J. Button and J. C. Wiltse, eds.), pp. 199–273. Academic Press, New York.
Jacobs, H., and Chrepta, M. (1974). *IEEE Trans. Microwave Theory Tech.* **MTT-22**, 411–417.
Kaminow, I. P. (1974). "An Introduction to Electrooptic Devices." Academic Press, New York.
Klein, M. B. (1979). *Proc. Fourth Int. Conf. Infrared Millimeter Waves Their Appl, Miami Beach, Florida,* pp 280–281.
Klein, M. B. (1981). *Int. J. Infrared Millimeter Waves* **2**, 239–246.
Klein, M. B. (1982). *Int. J. Infrared Millimeter Waves* **3**, 587–595.
Klohn, K. L., Horn, R. E., Jacobs, H., and Freibergs, E. (1978). *IEEE Trans. Microwave Theory Tech.* **MTT-26**, 764–773.
Knox, R. M. (1976). *IEEE Trans. Microwave Theory Tech.* **MTT-24**, 806–814.
Krumbholz, D., Brinkmeyer, E., and Neumann, E.-G. (1980). *J. Opt. Soc. Am.* **70**, 179–183.
Lee, C. H., Antonetti, A., and Mourou, G. (1977). *Opt. Commun.* **21**, 158–161.
Lee, C. H., Mak, S., and DeFonzo, A. P. (1978). *Electron. Lett.* **14**, 733–734.
Li, M. G., Cao, W. L., Mathur, V. K., and Lee, C. H. (1982). *Electron. Lett.* **18**, 454–456.

Miller, R. C. (1964). *Appl. Phys. Lett.* **5**, 17–19.
Moore, R. A., and Beam, R. E. (1957). *Proc. Nat. Electronics Conf.,* vol. 12, pp. 689–705.
Mortz, M. S., and Stafsudd, O. M. (1980). *Proc. 1980 Eur. Conf. Opt. Syst. Appl. (Utrecht),* SPIE vol. 236, pp. 38–44.
Rudokas, R., and Itoh, T. (1976). *IEEE Trans. Microwave Theory Tech.* **MTT-24,** 978–981.
Schweig, E. (1982). Dielectric Waveguides For Millimeter Waves. Ph.D. Thesis, California Inst. Tech., Pasadena, California.
Tischer, F. J. (1959). *Proc. Inst. Elec. Eng.* (Suppl. 13), **106B,** 47–53.
Trinh, T. N., Mitra, R., and Paleta, R. J. (1981). *Proc. Int. Microwave Symp., Los Angeles, California,* pp. 20–22.
Vinogradov, E. A., Prisova, N. A., and Kozlov, G. V. (1970). *Sov. Phys. Solid State* **12,** 605–607.
Yariv, A. (1975). "Quantum Electronics," second edition. Wiley, New York.
Yoneyama, T., and Nishida, S. (1981). *IEEE Trans. Microwave Theory Tech.* **MTT-29,** 1188–1192.

CHAPTER 4

# Millimeter-Wave Propagation and Remote Sensing of the Atmosphere

*Edward E. Altshuler*

*Rome Air Development Center*
*Electromagnetic Sciences Division*
*Hanscom Air Force Base, Massachusetts*

## I. Introduction

The propagation characteristics of electromagnetic waves in the millimeter-wave region of the spectrum are of particular interest because the waves have a strong interaction with lower atmospheric gases and particulates. The effects of this interaction are twofold: on the one hand, atmospheric absorption, scattering, and refraction limit the performance of millimeter-wave systems; on the other hand, this interaction allows the propagated wave to be used as a diagnostic tool to probe lower atmospheric structure.

Millimeter waves are generally considered to cover wavelengths in the range from about 2 cm down to 1 mm (15–300 GHz). These limits are based both on wavelength and on the nature and magnitude of the interaction between the wave and the atmosphere. Although the interaction with the atmosphere does not change abruptly at these limits, it does become weaker at longer wavelengths and stronger at shorter wavelengths. Thus the concepts presented here can generally be extended to either slightly longer or slightly shorter wavelengths. We shall often refer to the "window regions" of the millimeter-wave spectrum. These are considered the low attenuation regions between the gaseous absorption resonances. In particular, wavelengths between the 1.35-cm water vapor resonance and 5-mm oxygen resonance, the 5- and 2.5-mm oxygen resonances, and the 2.5-mm oxygen resonance and 1.6-mm water vapor resonance compose the window regions. Low attenuation regions also exist at wavelengths longer than 1.35 cm and shorter than 1.6 mm.

The physics of the interaction between millimeter waves and the atmosphere is extremely complex, so many facets are considered beyond the scope of this book. However, an effort is made to provide the reader with a general understanding of the mechanisms of this interaction; if in-depth details are required, they can be obtained from the references. Likewise, complex mathematical expressions are used only to illustrate concepts that are considered important and cannot be satisfactorily explained otherwise. Many of the figures contain information that is "typical" or "average," that is, it is intended to provide the reader with an estimate of the magnitude of an interaction. More quantitative results are available from the cited references.

## A. PROPAGATION EFFECTS

Atmospheric gases and particulates often have a profound effect on millimeter waves and thus limit the performance of many millimeter-wave systems. Because the densities of these gases and particulates generally decrease with altitude, the effects of the atmosphere on the propagated wave are strongest very close to the earth and tend to diminish at higher altitudes. For this reason millimeter waves propagating above the tropopause, the altitude at which the temperature remains essentially constant with increasing height ($\sim 10-12$ km), are assumed to be unaffected. For the clear atmosphere there are essentially two types of interactions. The stronger interaction is the absorption–emission produced by oxygen and water vapor. The other interaction takes place with the refractivity structure of the atmosphere. This refractivity structure is often divided into two categories: gross structure and fine structure. The gross structure assumes the atmosphere is a horizontally stratified continuum characterized by a refractivity that normally decreases slowly with increasing altitude, and is assumed to be wavelength independent; that is, it affects all wavelengths from microwaves through millimeter waves in the same way. For the refractivity fine structure, the atmosphere is viewed as an inhomogeneous medium consisting of small pockets of refractive index that vary both temporally and spatially.

Because these pockets have different sizes, this interaction is wavelength dependent. Although the terrain is not actually part of the atmosphere, it does interface with the atmosphere and can affect millimeter-wave propagation. Thus, multipath propagation produced by the terrain and diffraction by prominent obstacles on the earth's surface are also reviewed.

Atmospheric particulates range in size from microns to close to 1 cm in diameter. For particles very small compared to wavelength, the only significant propagation effects are those of absorption and emission. As the particles become larger with respect to wavelength, scattering effects become pronounced. In Section II we shall review the effects of the clear atmosphere and then of atmospheric particulates on millimeter-wave propagation. We show that although the effects of the clear atmosphere are generally not as severe as those due to atmospheric particulates, they cannot be disregarded, even in the window regions and especially at short millimeter wavelengths.

B. REMOTE SENSING

Because millimeter waves are strongly affected by both atmospheric gases and particulates, which are in turn a function of the meteorological properties of the atmosphere, it is possible to obtain meteorological information by observing the behavior of the propagated wave. This technique was first employed at IR wavelengths and proved very successful for clear sky conditions. At IR, however, atmospheric particulates such as cloud or fog tend to mask out parameters of interest and thus limit the use of IR for remote sensing during overcast conditions. Millimeter waves are less affected by cloud and fog, so it is still possible to infer information on, for example, temperature and water vapor under adverse weather conditions. Three techniques for remotely sensing the atmosphere are reviewed: line-of-sight transmission, radiometry, and radar. Similar millimeter-wave sensors have been used to probe the earth and ocean; however, these applications are considered separate from the atmosphere and are not addressed in this chapter.

C. APPLICATIONS

Potential applications of millimeter waves have been considered for many years (Altshuler, 1968; Skolnik, 1970). However, for most applications other than remote sensing, atmospheric effects have always imposed limitations on system performance. The principal applications of millimeter waves have been in the areas of communications, radar, and remote sensing. Probably the first application for which millimeter waves were considered was communications. The discovery of the circular electric waveguide mode $TE_{01}$ about four decades ago prompted the use of these wavelengths for a waveguide communication system, because waveguide attenuation for that mode decreases with decreasing wavelength. It is believed that the most significant contribution resulting from this effort was not so much the system itself but the research that was directed toward the development of a whole new line of millimeter-wave equipment and techniques required for this application — i.e., sources, amplifiers, detectors, and waveguide components. Through the years consideration was often given to utilizing millimeter waves for point-to-point communications. However, proposed systems never materialized, principally because they were not economically competitive with those at longer wavelengths; they also lacked reliability because of atmospheric effects and inferior components.

In the past decade, the need for new types of communication systems and the need to alleviate increasing spectrum congestion have led to a reappraisal of millimeter waves. The availability of large bandwidths makes this region of the spectrum particularly attractive for high data rate earth-to-space communication channels. Furthermore, high-gain, high-resolution antennas of moderate size and light-weight, compact system components are indeed applicable for space vehicle instrumentation. Millimeter waves provide an excellent means for obtaining secure communication channels. For satellite-to-satellite links, where all propagation is above the absorptive constituents of the lower atmosphere, narrow-beamwidth antennas may be operated at a wavelength where atmospheric attenuation is very high (i.e., $\lambda \sim 5$ mm); thus the signal is confined by the antenna to a narrow cone and then absorbed by the lower atmosphere before it reaches the earth. Another application for secure communications is ship-to-ship and short terrestrial links; in these cases attenuation is sufficiently high at millimeter waves to allow a detectable signal only over short distances. Finally, point-to-point radio-relay systems that were previously not considered feasible are now being reexamined. Recent studies have shown that attenuation in the lower atmosphere can be combatted by using very short hops and diversity techniques; also, satisfactory system performance can now be obtained with solid-state components quite economically.

Because the angular resolution of an aperture is inversely proportional to wavelength, antennas can have higher resolution at millimeter wavelengths than at longer wavelengths. Furthermore, because range resolution is a function of bandwidth, improved range resolution is also possible. With this enhanced resolution, higher-precision radars for tracking, guidance, mapping, and target detection are now realizable. As in the case of communications, millimeter-wave radars also have far-reaching applications for aircraft and spacecraft because of the high resolution that can be obtained with relatively compact systems.

In the area of remote sensing there are numerous applications for millimeter-wave sensors. Although in this chapter we address only remote sensing of the atmosphere, millimeter-wave sensors also have application in geology, hydrology, agriculture, oceanography, and astronomy.

An attempt has been made to identify most of the applications for which millimeter waves are being considered, and the future looks bright. Furthermore, it must be remembered that in any emerging field the benefits derived from unforeseen applications often outweigh the benefits initially foreseen.

## II. Atmospheric Effects on Propagated Waves

Atmospheric gases and particulates may severely alter the properties of millimeter waves. For the clear atmosphere the most pronounced effect is absorption due to the gases, oxygen and water vapor. However, refraction, scattering, diffraction, and depolarization effects are also reviewed, and it is shown that under special conditions their impact on the propagated wave can be significant. For an atmosphere containing particulates, the absorption and scattering by rain seriously limit propagation at millimeter wavelengths. However, the effects of smaller, water-based particulates are also significant, particularly at shorter millimeter wavelengths.

### A. CLEAR ATMOSPHERE

We shall consider an atmosphere "clear" if it is free of atmospheric particulates. Thus the only interaction that takes place is that between the propagated wave and the atmospheric gases (and terrain). We shall see that for very low elevation angles the gross structure of the refractive index causes the propagated wave to be bent and delayed, whereas the fine structure of the refractive index scatters the propagated wave and may produce scintillations. The gases, oxygen and water vapor, absorb energy from the wave and reradiate this energy in the form of noise. Finally, if the wave is propagating close to the earth's surface, part of the energy may be reflected from the surface, or it may be diffracted by prominent obstacles on the surface and then interfere with the direct signal. If some form of reflection or scattering takes place, the propagated wave may also become depolarized. All of these effects are discussed in detail in Sections II.A.1–II.A.5.

### 1. *Refraction*

The lower atmosphere is composed of about 78% nitrogen, 21% oxygen, and 1% water vapor, argon, carbon dioxide, and other rare gases. The densities of all these gases, except for water vapor, decrease gradually with height; the density of water vapor, on the other hand, is highly variable. The index of refraction of the atmosphere is a function of the temperature, pressure, and partial pressure of water vapor. Because the refractive index $n$

is only about 1.0003 at the earth's surface, it is often expressed in terms of a refractivity $N$, where

$$N = (n - 1) \times 10^6. \tag{1}$$

The radio refractivity can be approximated by the following theoretical expression, which has empirically derived coefficients (Smith and Weintraub, 1953):

$$N = 77.6P/T + 3.73 \times 10^5 e/T^2, \tag{2}$$

where $P$ is the atmospheric pressure (in millibars), $T$ the absolute temperature (in degrees Kelvin), and $e$ the water vapor pressure (in millibars). This expression consists of two terms; the first is often referred to as the dry term, and the second is often referred to as the wet term, because it is a function of water vapor. Equation (2) neglects dispersion effects and in principle does not hold near absorption lines; however, in practice it is assumed valid down to a wavelength of about 3 mm and is often considered acceptable down to wavelengths of even 1 mm, particularly in the window regions. The refractivity decreases approximately exponentially with height and is often expressed as

$$N(h) = N_s e^{-bh}, \tag{3}$$

where $h$ is the height above sea level (in kilometers), $N_s$ the surface refractivity, $b = 0.136/\text{km}$, and $N(h)$ is the refractivity at height $h$ (in kilometers).

As mentioned earlier, the gross behavior of atmospheric refractivity affects millimeter waves in much the same way as microwaves; because effects such as bending, time delay, and ducting are amply treated elsewhere (Bean and Dutton, 1968), they will only be summarized here. The refractive index fine scale structure is too complex to be treated by simple refraction theory. Because of the variability in the scale sizes of the refractive inhomogeneities, the effects are wavelength dependent, and thus the interaction is different for millimeter waves than for microwaves.

a. *Bending.* The angular bending is due primarily to the change of the index of refraction of the atmosphere with height. Because the refractivity generally decreases with height, the wave passing obliquely through the atmosphere is bent downward. Thus the apparent elevation angle of a target tends to appear slightly higher than its true elevation angle, and this difference is called the angle error. For a horizontally stratified atmosphere the angle error is zero at zenith and increases very slowly with decreasing elevation angle. This error becomes appreciable at low elevation angles, and for a standard atmosphere it approaches a value of about 0.07° near the horizon. For illustration we plot in Fig. 1 the angle errors of targets at

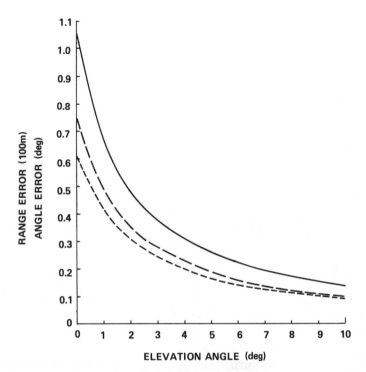

RANGE ERROR (100m)
ANGLE ERROR (deg)

ELEVATION ANGLE (deg)

FIG. 1 Typical tropospheric refraction angle and range errors: ———, range error; — —, angle error ($h = \infty$); – – –, angle error ($h = 90$ km). CRPL reference atmosphere, 1958. (From Altshuler and Mano, 1982. Reproduced by permission of the U.S. Dept. of Commerce, National Technical Information Service, Springfield, Virginia 22161.)

altitudes of 90 km and infinity (radio source) for an atmosphere with a surface refractivity of 313 $N$ units that decreases exponentially with height.

For a typical atmosphere the refractivity decreases at a rate of about 40 $N$ units/km near the surface and then more slowly at higher altitudes; this produces superrefraction. As the gradient of this decrease in refractivity becomes larger the wave is bent more toward the earth's surface, and when the gradient decreases at a rate of 157 $N$ units/km the wave travels parallel to the surface; this condition is called ducting. For still steeper gradients the wave will actually be bent into the earth. There are times when the decrease in refractivity is less than 40 $N$ units/km and the wave is bent less than normal toward the earth; this is called subrefraction.

In actuality we have a curved ray passing over a curved earth. It is sometimes easier to visualize a straight ray over a curved earth (or a curved ray over a flat earth). For a vertical negative gradient of 40 $N$ units/km, it can

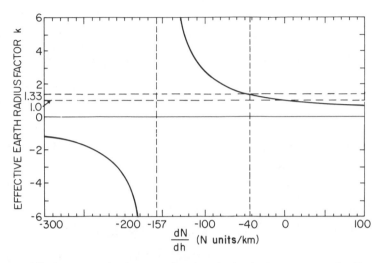

Fig. 2   Effective earth radius factor as a function of refractive index gradient $dN/dh$: ducting region, $dN/dh < -157$ N units/km, $K < 0$; supperrefraction region, $-157$ N units/km $< dN/dh < -40$ N units/km, $0 < K < \frac{4}{3}$; subrefraction region, $-40$ N units/km $< dN/dh$, $\frac{4}{3} < K < 0$. (From Hall, 1979. © Peter Peregrinus Ltd.)

be shown that an earth with an effective radius about $\frac{4}{3}$ that of the true earth with the wave propagating in a straight line is equivalent to the true case — the curved ray path over the curved earth. The ratio of the effective earth radius to the true earth radius is often designated $k$. The behavior of $k$ as a function of the refractive index gradient is illustrated in Fig. 2 (Hall, 1979). If the refractivity gradient is less than 40 $N$ units/km, then $k$ decreases and reaches unity for the case of no gradient. As the gradient becomes more positive, $k$ approaches zero. When the negative gradient is more negative than 40 $N$ units/km, $k$ is larger than $\frac{4}{3}$ and approaches infinity for the special case of ducting, for which the gradient reaches 157 $N$ units/km. For still larger negative gradients, $k$ becomes negative, because the flat earth for $k = \infty$ becomes curved in the opposite sense.

   b.   *Time Delay.*   Time delay occurs primarily because the index of refraction of the atmosphere is greater than unity, thus slowing down the wave, and to a lesser extent because of the lengthening of the path by angular bending. For radar and navigation systems the range is determined from time delay measurements; thus the additional time delay produced by the troposphere results in a corresponding range error. This error causes the target to appear further away than its true distance. For a typical atmosphere the range error is slightly larger than 2 m in the zenith direction and increases very slowly with decreasing elevation angle. The range error becomes much larger at lower elevation angles, and for a standard atmosphere it approaches a value of about 100 m near the horizon. For illustration, we plot in Fig. 1 the range error for an atmosphere with a surface refractivity of 313 $N$ units that decreases exponentially with height.

c. *Refraction Corrections.* As seen in Fig. 1, the angle and range errors become appreciable for very low elevation angles. It has been shown that both errors are strongly correlated with the surface refractivity, and for many applications adequate corrections based on a linear regression on $N_s$ are possible (Altshuler, 1970; Bean and Dutton, 1968). More accurate corrections can be obtained by measuring the vertical refractivity profile and calculating the corrections. The principal limitation of this approach is that a horizontally stratified atmosphere is usually assumed, and this is not always valid for the long distances traversed at low elevation angles. It has been shown that the range error is correlated with the brightness temperature of the atmosphere, and techniques to take advantage of this dependence have been proposed (Elgered, 1982; Gallop and Telford, 1975; Schaper *et al.*, 1970; Westwater, 1976; Wu, 1979).

One of the most effective methods for obtaining angle error corrections involves the use of "targets of opportunity." These may be either calibration satellites or radio sources, the angular positions of which are normally known to an accuracy of the order of microradians. In principle, the angular error of the calibration source is measured prior to the target being tracked. If the target is in the same general direction as the calibration source and the atmospheric refractivity does not change appreciably with time, then the correction can be determined directly. It is more difficult to use this technique to obtain range error corrections, because the true range of a satellite is generally not known to sufficient accuracy for calibration and, because radio sources are detected passively, no range information is available. However, Mano and Altshuler (1981) and Altshuler and Mano (1982) have derived an expression for the range error as a function of angle error and show that range error corrections can be obtained from a set of angle error corrections.

d. *Tropospheric Multipath.* We have seen in Section II.A.1.a that for a superrefractive troposphere the rays are bent toward the earth and that in extreme cases the waves can follow the earth's curvature (ducting). The same type of refractivity structure can also produce tropospheric multipath, in which rays are essentially reflected from layers set up by these large negative gradients in refractive index. The mechanism that produces tropospheric multipath has been modeled by Rummler (1979), who used both two- and three-ray models to show how tropospheric multipath can affect line-of-sight propagation. The model is heavily dependent on the refractive index structure, the separation of transmitter and receiver, and the location of each with respect to the layer. Generally, the likelihood of multipath occurring increases as the separation between transmitter and receiver increases or as the transmitter and receiver come closer to the layer. Multipath effects may be lessened by employing antenna beamwidths that are narrow enough that the layer is well outside the main beam.

Experiments to study multipath propagation have been conducted by many investigators, using two different techniques. In one method a very short pulse is transmitted and the time delays between the direct signal and the multipath signals are measured. Time delays are usually of the order of nanoseconds, and the amplitudes of the delayed signals generally decrease with increasing delay. An equivalent technique makes use of a swept frequency; the delays are not measured directly but are obtained through a Fourier transform of the transfer function. Stephanson (1981) has summarized the results obtained from these experiments in a very complete review paper.

Of particular interest are the experimental results obtained by Sandberg (1980). These are in good agreement with Rummler's three-ray model, which seems to give a more complete description of the multipath phenomenon, in terms of both the time delay of the signal and its angle of arrival, than the two-ray model. The effects of tropospheric multipath on line-of-sight propagation will be discussed in more detail in Section III.A.

e. *Scintillations.* So far we have discussed the effects of the gross refractivity structure of the atmosphere on the propagated wave. The atmosphere also has a fine scale refractive index structure, which varies both temporally and spatially and thus causes the amplitude and phase of a wave to fluctuate; these fluctuations are often referred to as scintillations. The refractive index structure is envisioned to consist of pockets of refractive inhomogeneities that are sometimes referred to as turbulent eddies and may be classified by size into three regions: the input range, the inertial subrange, and the dissipation range. The two boundaries that separate these regions are the outer and inner scales of turbulence, $L_0$ and $l_0$, respectively. These are the largest and smallest distances for which the fluctuations in the index of refraction are correlated. A meteorological explanation of how these pockets are generated is beyond the scope of this book; however, in simple terms, large parcels of refractivity, possibly of the order of hundreds of meters in extent, continually break down into smaller-scale pockets. These pockets become smaller and smaller until they finally disappear. The very large pockets in the input range have a complex structure, and at the present time there is no acceptable formulation of the turbulence properties of this region. Pockets having scale sizes of less than about 1 mm have essentially no turbulent activity, and for all practical purposes the spectrum of the covariance function of the refractive index fluctuations, $\phi_n(k)$, equals zero. The inertial subrange bounded by $L_0$ and $l_0$ has a spectrum

$$\phi_n(k) = 0.033C_n^2 k^{-11/3} \tag{4}$$

for $2\pi/L_0 < k < 2\pi/l_0$, where $C_n$ is the structure constant and $k$ the wave number (not to be confused with the ratio $k$ of the effective earth radius to the true earth radius used earlier).

The phase fluctuations arise from changes in the velocity of the wave as it passes through pockets of different refractive index. As the wavelength becomes shorter, the changes in phase increase proportionally. The amplitude fluctuations arise from defocusing and focusing by the curvature of the pockets. A detailed explanation of millimeter-wave scintillations is provided by Strobehn (1968) and Fante (1980).

## 2. Absorption and Emission

Atmospheric gases can absorb energy from millimeter waves if the molecular structure of the gas is such that the individual molecules possess electric or magnetic dipole moments. It is known from quantum theory that, at specific wavelengths, energy from the wave is transferred to the molecule, causing it to rise to a higher energy level; if the gas is in thermal equilibrium it will then reradiate this energy isotropically as a random process, thus falling back to its prior energy state. Because the incident wave has a preferred direction and the emitted energy is isotropic, the net result is a loss of energy from the wave. The emission characteristics of the atmosphere may be represented by those of a blackbody at a temperature that produces the same emission; therefore the atmospheric emission is often expressed as an apparent sky temperature. Because absorption and emission are dependent on the same general laws of thermodynamics, both are expressed in terms of the absorption coefficient. Using Kirchhoff's law and the principle of conservation of energy, one can derive the radiative transfer equation, which describes the radiation field in the atmosphere that absorbs and emits energy. This emission is expressed as

$$T_a = \int_0^\infty T(s)\gamma(s) \exp\left(-\int_0^\infty \gamma(s')\, ds'\right) ds, \tag{5}$$

where $T_a$ is the effective antenna temperature, $T(s)$ the atmospheric temperature, $\gamma(s)$ the absorption coefficient, and $s$ the distance from the antenna (ray path). In simpler terms

$$T_a = T_m(1 - e^{-\gamma s}), \tag{6}$$

where $T_m$ is the atmospheric mean absorption temperature within the antenna beam. Solving for the attenuation, we obtain

$$A = \gamma s = 10 \log[T_m/(T_m - T_a)], \qquad (7)$$

where $A$ is measured in decibels. The only atmospheric gases with strong absorption lines at millimeter wavelengths are water vapor and oxygen. The absorption lines of $O_3$, CO, $N_2O$, $NO_2$, and $CH_2O$ are much too weak to affect propagation in this region.

a. *Water Vapor.* The water vapor molecule has an electric dipole moment with resonances at wavelengths of 13.49, 1.64, and 0.92 mm (22.24, 183.31, and 325.5 GHz) in the millimeter-wave region. In general, the positions of these resonances, their intensities, and their linewidths agree well with experimental data. There are, however, serious discrepancies between theoretical and experimental absorption coefficients in the window regions between these strong lines; experimental attenuations are often a factor of two to three times larger than theoretical values (URSI Commission F Working Party, 1981). Although the cause of the discrepancy is not known, indications are that either the line shapes do not predict enough absorption in the wings of the resonances or there is an additional source of absorption that has not yet been identified. It should be mentioned that there are over 1800 water vapor lines in the millimeter-wave–infrared spectrum, 28 of which are above 0.3 mm. Because the wings of these lines contribute to the absorption in the window regions, very small errors in the line shapes could significantly affect the overall absorption. In an effort to overcome this problem, several workers have introduced an empirical correction term to account for the excess attenuation (Crane, 1980a; Waters, 1976).

In addition to the uncertainty of the absorption coefficient of water vapor, there is also the problem of water vapor concentration. The amount of water vapor in the lower atmosphere is highly variable in time and altitude and has densities ranging from a fraction of a gram per cubic meter for very arid climates to 30 g/m³ for hot and humid regions; for this reason it is very difficult to model. A plot of the water vapor absorption as a function of wavelength is shown in Fig. 3 for a density of 7.5 g/m³. Because the attenuation is linearly proportional to the water vapor density, except for very high concentrations, attenuations for other water vapor densities are easily obtained.

b. *Oxygen.* The oxygen molecule has a magnetic dipole moment with a cluster of resonances near a wavelength of 5 mm (60 GHz) and a single resonance at 2.53 mm (118.75 GHz). Although the more than 30 lines near

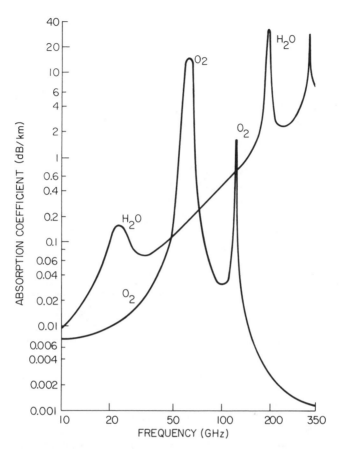

FIG. 3 Absorption coefficients for water vapor and oxygen. $P = 1$ atm, $T = 20°C$, $P_w = 7.5$ g/m$^3$. (From Smith, 1982. © American Geophysical Union.)

a wavelength of 5 mm are resolvable at low pressures (high altitudes), they appear as a single pressure-broadened line near sea level owing to a large number of molecular collisions. Even though the magnetic dipole moment of oxygen is approximately two orders of magnitude weaker than the electric dipole moment of water vapor, the net absorption due to oxygen is still very high, simply because it is so abundant. The fact that the distribution of oxygen throughout the atmosphere is very stable makes it easy to model. A plot of oxygen attenuation as a function of wavelength is shown in Fig. 3 along with that of water vapor. Note the importance of water vapor attenuation at very short wavelengths.

### 3. Scattering

The principal effect of the pockets of refractive index on the propagated wave is to produce scintillations, as described in Section II.A.1. When the pockets are of the order of a wavelength in size, they can also scatter the signal. At millimeter wavelengths and for line-of-sight paths, this scattered field is generally very weak compared to the direct signal and is not considered significant. There are, however, special cases for which the scattered field is important, and these are considered next.

a.   *Troposcatter.*   Signals propagated at angles very close to the horizon can be received at distances far beyond the horizon; this mode of propagation is called tropospheric scatter propagation, or simply troposcatter. There is still not complete agreement on the mechanism that produces the scattered field; however, two models in particular have been used through the years. Probably the most widely held view is that of scattering from turbulent pockets originally proposed by Booker and Gordon (1950). Another view is that the scattering takes place from very thin layers of refractivity (Friis *et al.*, 1957). More recently, Crane (1981) has suggested that both of these models may be valid depending on the wavelength, the scatter angle, and the thickness of the layer. By using very-high-power radars in a backscatter mode, it has been possible to observe thin turbulent layers (Crane, 1973; Hooke and Hardy, 1975). Crane believes that if the wavelength $\lambda$, scatter angle $\theta$, and vertical outer scale of turbulence $L_{0v}$ satisfy the condition

$$\lambda/2 \sin \tfrac{1}{2}\theta > L_{0v}, \tag{8}$$

then the layer model is valid and the mechanism is predominantly one of partial specular reflection from the layer. This model applies best for longer wavelengths. When Eq. (8) is not satisfied, the mechanism is believed to be predominantly volume scattering from the pockets that compose the layer. Because the outer scale of turbulence and scattering angle are typically of the order of kilometers and 1°, respectively, it is seen from Eq. (8) that troposcatter at millimeter wavelengths is indeed volume scattering from turbulent pockets. Most turbulence theories assume a Kolmogorov energy spectrum with a wavelength dependence of $\lambda^{5/3}$. This in turn leads to a scattering cross section per unit volume proportional to $\lambda^{1/3}$. The elevation angle dependence is proportional to $\theta^{11/3}$. Although the scattered signal and antenna gain increase with decreasing wavelength, this increase is limited by the aperture-to-medium coupling loss and atmospheric attenuation. These limitations will be reviewed in Section III.C.

b.   *Terrain Multipath.*   Electromagnetic waves scattered from the earth's surface may interfere with the direct signal; this is called multipath propagation. The extent of multipath is dependent on the geometry of the transmitter and receiver with respect to the surface, their respective beamwidths and polarizations, and the dielectric constant and surface roughness of the terrain.

Let us first consider a surface that is relatively smooth with respect to wavelength. The reflection coefficients of vertically and horizontally polarized waves for a nonmagnetic surface are

$$\Gamma_v = \frac{\varepsilon \sin \alpha - (\varepsilon - \cos^2 \alpha)^{1/2}}{\varepsilon \sin \alpha + (\varepsilon - \cos^2 \alpha)^{1/2}}, \tag{9}$$

$$\Gamma_h = \frac{\sin \alpha - (\varepsilon - \cos^2 \alpha)^{1/2}}{\sin \alpha + (\varepsilon - \cos^2 \alpha)^{1/2}}, \tag{10}$$

where $\alpha$ is the grazing angle and $\varepsilon = \varepsilon' - j\varepsilon''$ the complex dielectric constant. Typical values of the magnitudes and phases of these reflection coefficients as a function of grazing angle are shown in Figs. 4 and 5 (Povejsil et al., 1961). For very small grazing angles the magnitudes approach unity and the phases approach 180°. As the grazing angle increases, the magnitude and phase of the vertically polarized wave fell off sharply and those of the horizontally polarized wave decrease very slightly. The sharp falloff of the

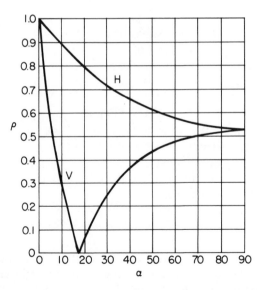

FIG. 4 Magnitude of reflection coefficient $\rho$ for average land as a function of incidence angle $\alpha$ ($\varepsilon' = 10$, $\sigma = 1.6 \times 10^{-3}$ S/m). (From "Airborne Radar" by Donald J. Povejsil, Robert S. Raven, and Peter Waterman. © 1961 by D. Van Norstrand Company, Inc. Reprinted by permission of Wadsworth Publishing Company, Belmont, California 94002.)

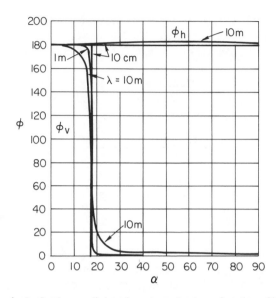

FIG. 5 Phase $\phi$ of reflection coefficient for average land as a function of incidence angle $\alpha$ ($\varepsilon' = 10$, $\sigma = 1.6 \times 10^{-3}$ S/m). (From "Airborne Radar" by Donald J. Povejsil, Robert S. Raven, and Peter Waterman. © 1961 by D. Van Norstrand Company, Inc. Reprinted by permission of Wadsworth Publishing Company, Belmont, California 94002.)

vertically polarized reflection coefficient can be explained as follows. For most surfaces, with the exception of very dry ground, $|\varepsilon| \gg 1$. With this approximation, Eq. (9) can be rewritten as

$$\Gamma_v = (\sqrt{\varepsilon} \sin \alpha - 1)/(\sqrt{\varepsilon} \sin \alpha + 1). \qquad (11)$$

Note that the numerator is $\sqrt{\varepsilon} \sin \alpha - 1$. Thus at some angle the numerator approaches zero; this is the Brewster angle, and if the terrain were a perfect dielectric ($\sqrt{\varepsilon}$ is real) then the reflection coefficient would actually go to zero. As the grazing angle approaches normal incidence the vertical reflection coefficient increases and finally equals the horizontally polarized reflection coefficient at normal incidence.

When the surface is relatively smooth the scattering is predominantly specular; that is, it can be considered coherent. As the surface becomes rougher a diffuse, incoherent component appears, and for a very rough surface the scattered signal is predominantly diffuse. The criterion usually applied for characterizing surface roughness is that introduced by Rayleigh. It is based on the phase difference of adjacent rays reflected from a rough surface; when the path difference between these rays increases to 90°, the surface is assumed to transform from smooth to rough. Obviously this transition is very gradual and should be interpreted as such. Mathematically the surface can be considered smooth when $h \sin \alpha < \frac{1}{8}\lambda$, where $h$ is the height of a surface irregularity. It must be emphasized that even at millimeter wavelengths, for which $\lambda$ is very small, typical surfaces tend to look smooth at very low grazing angles.

We have reviewed the general characteristics of multipath propagation. Now let us summarize multipath propagation in the context of millimeter waves. At longer wavelengths the reflection coefficient of a vertically polarized wave is significantly lower than that of a horizontally polarized wave, particularly in the vicinity of the Brewster angle, so microwave systems are often designed to operate with vertical polarization to minimize multipath interference. At millimeter wavelengths this polarization dependence is of less importance, because the reflection coefficients of vertically and horizontally polarized waves are comparable for most millimeter-wave applications. First of all, multipath effects at these short wavelengths will generally occur only at very small grazing angles, because most surfaces based on the Rayleigh criterion appear rough for larger grazing angles. Furthermore, because the dielectric constants of most surfaces tend to remain constant or decrease with decreasing wavelength, the Brewster angle increases and the reflection coefficient of the vertically polarized wave does not drop off as rapidly with increasing grazing angle. Therefore at millimeter wavelengths multipath is confined to much lower grazing angles than at microwave wavelengths and is thus less sensitive to polarization. In Fig. 6, Rayleigh's roughness criterion is plotted as a function of grazing angle. As the surface becomes rough with respect to wavelength, the grazing angle must become extremely small to have specular reflection.

## 4. *Diffraction*

Electromagnetic waves incident on an obstacle may be bent around that obstacle; this is known as diffraction. The extent of diffraction is dependent on the shape and composition of the obstacle, its position with respect to the direct path of the incident wave, and the wavelength. Classical diffraction theory has been used to treat simple shapes such as individual knife edges, rounded edges, and in some instances sets of these edges. An underlying assumption is that the knife edge is very sharp or the rounded edge very smooth with respect to wavelength. It is also often assumed that the edge is a perfect conductor, although solutions have been obtained for edges having finite conductivity.

Diffraction loss is often expressed as a function of the dimensionless Fresnel parameter $v$, which is in turn a function of the geometric parameters of the obstacle and path. For knife edge diffraction the Fresnel parameter can be defined as

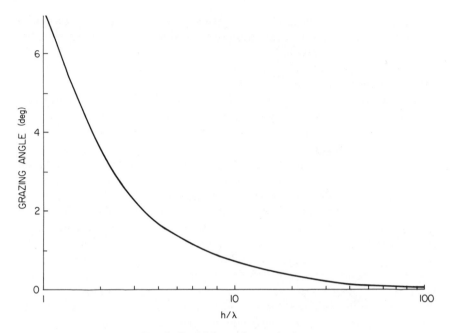

FIG. 6   Rayleigh roughness criterion.

$$v = h\left[\frac{2}{\lambda}\left(\frac{1}{d_1} + \frac{1}{d_2}\right)\right]^{1/2}, \tag{12}$$

where $h$ is either the height of the obstacle above the direct path or the distance of the obstacle below the path, and $d_1$ and $d_2$ are the respective distances of transmitter and receiver from the knife edge. For illustration, let us assume that the knife edge is midway between transmitter and receiver; then $d_1 = d_2 = \frac{1}{2}d$ and

$$|v| = 8h^2/\lambda d, \tag{13}$$

where $v$ is positive when the ray path is below the edge and negative when the ray path is above the edge. It is known that the diffraction loss is approximately zero for $v < -3$ and very high for $v > 3$, so $-3 \le v \le 3$ can be considered the region of interest.

Equation (13) can be expressed as

$$h = \sqrt{\tfrac{1}{8}\lambda d v^2} = \tfrac{1}{2}v\sqrt{\tfrac{1}{2}\lambda d}. \tag{14}$$

Because $d$ is generally on the order of kilometers and $\lambda$ on the order of millimeters, $h$ can only be on the order of meters. Thus from a practical standpoint only isolated obstacles such as small hills or buildings would produce diffraction effects at millimeter wavelengths.

## 5. Depolarization

Depolarization of an electromagnetic wave can occur when the incident wave is scattered and a cross-polarized component is produced along with the copolarized component. It is defined as

$$|\text{depolarization}| = 20 \log(|E_x|/|E_y|), \tag{15}$$

where $E_x$ and $E_y$ are the cross-polarized and copolarized components, respectively, and the depolarization is measured in decibels. Olsen (1981) has summarized in detail both the mechanisms that can produce depolarization during clear air conditions and some experimental observations of this phenomenon. He divides these mechanisms into two groups: those that are independent of the cross-polarized pattern of the antenna (a perfect plane-polarized wave) and those that are dependent on the cross-polarized pattern. In principle, depolarization can arise from scattering by refractive inhomogeneities or from terrain. For the plane-polarized wave it appears that depolarization due to refractive multipath is insignificant (Olsen, 1981) but that depolarization due to terrain multipath can be much stronger. For an antenna having a measurable cross-polarized pattern, it is believed that both atmospheric and terrain multipath mechanisms contribute to the depolarization of the wave. Although most experimental results of depolarization by the clear atmosphere have been obtained at centimeter wavelengths, there is no reason to believe that the same effects will not occur at millimeter wavelengths. However, because both atmospheric and terrain multipath may normally be weaker at millimeter wavelengths than at microwave wavelengths, the depolarization may not be as severe.

## B. ATMOSPHERIC PARTICULATE EFFECTS

In this section we discuss the degrees of absorption, scattering, and depolarization that may occur from atmospheric particulates. We shall see that rain is by far the most important of the particulates, for two reasons: the interaction of rain with millimeter waves is very strong and rain occurs more often than other particulates. Thus a detailed discussion of the rain–millimeter-wave interaction is presented.

## 1. Absorption and Scattering

Millimeter waves incident on atmospheric particulates undergo absorption and scattering, the degree of each being dependent on the size, shape, and complex dielectric constant of the particle and the wavelength and polarization of the wave. An expression for calculating the absorption and scattering from a dielectric sphere was first derived by Mie (1908):

$$Q_t = -\lambda^2/2\pi \ \text{Re} \sum_{n=1}^{\infty} (2n + 1)(a_n^s + b_n^s), \tag{16}$$

where $Q_t$ represents losses due to both absorption and scattering, and $a_n^s$ and $b_n^s$ are very complicated spherical Bessel functions that correspond to the magnetic and electric modes of the particle, respectively. $Q_t$ has the dimension of area and is usually expressed in square centimeters. Physically, if a wave with a flux density of $S$ W/cm$^2$ is incident on the particle, then $S \times Q_t$ is the power absorbed or scattered.

When the circumference of the particle is very small compared to wavelength, i.d., $\pi D \ll \lambda$, then the scattering and absorption losses can be represented by

$$Q_s = (\lambda^2/2\pi)(\tfrac{4}{3}\rho^6)|(n^2 - 1)/(n^2 + 2)|^2 \tag{17}$$

and

$$Q_a = (\lambda^2/2\pi)(2\rho^3) \ \text{Im}[-(n^2 - 1)/(n^2 + 2)], \tag{18}$$

where $\rho = kD/2 = \pi D/\lambda \ll 1$ and $n$ is the complex index of refraction.

Because $\rho$ is very small, the loss due to scattering, which is proportional to $\rho^6$, will be much smaller than that due to absorption, which is proportional to $\rho^3$. This condition is often referred to as the Rayleigh approximation, for which

$$Q_s \propto 1/\lambda^4 \quad \text{and} \quad Q_a \propto 1/\lambda.$$

Because the scattering loss is often assumed negligible, the total loss is proportional to the volume of the drop. Often the backscatter cross section (or radar cross section) is of interest:

$$\sigma = (\lambda^2\rho^6/\pi)|(n^2 - 1)/(n^2 + 2)|^2 = \tfrac{3}{2}Q_s. \tag{19}$$

The relationship with $Q_s$ arises because Rayleigh scatterers are assumed to have the directional properties of a short dipole and the directivity of a dipole in the backscatter direction is 1.5 times that of an isotropic source. As the drop becomes large with respect to wavelength, the Rayleigh approximation becomes less valid and the Mie formulation must be used.

## 2. Depolarization

Atmospheric particulates having a nonspherical shape will depolarize a wave (produce a cross-polarized component) if the major or minor axis of the particulate is not aligned with the $E$-field of the incident wave. The extent of the depolarization is a strong function of the size, shape, orientation, and dielectric constant of the scatterer. The depolarization defined in Eq. (15) arises because the orthogonal components of the scattered field undergo different attenuations and phase shifts. These differences are referred to as the differential attenuation and differential phase shift. An alternative definition related to depolarization is the cross-polarization discrimination, which is simply the reciprocal of the depolarization. In general, the depolarization increases as the particulate size and eccentricity increase. The depolarization also increases as the angle between the $E$-field of the incident wave and the major axis of the particulate increases up to approximately 45°, for which the depolarization passes through a maximum.

### 3. *Types of Particulates*

Rain is the most common particulate; drops range in size from a fraction of a millimeter to 6 or 7 mm. Sleet and snow, which are considered quasi-solid forms of water, are then treated. Because of their complexity of shape and composition only limited theoretical work has been done on them, and because they are rare events in most locations only limited experimental data have been obtained. Hail is frozen water, and does not occur very often. However, the losses due to hail can be calculated quite accurately, and are very small at millimeter waves because the complex dielectric constant of ice is small. Cloud, fog, and haze particulates are very similar in that they are all composed of very small water droplets suspended in air (clouds may also contain ice crystals) with diameters ranging from several microns up to about 100 $\mu$. Therefore, through most of the millimeter-wave region the Rayleigh approximation is valid for these particulates. Dust and sand particulates have size distributions comparable to that of clouds, but because their complex dielectric constants are low their interaction with millimeter waves is very weak.

a. *Rain.* Rain is an extremely complex phenomenon, both meteorologically and electromagnetically. From a meteorological standpoint it is generally nonuniform in shape, size, orientation, temperature, and distribution, thus making it very difficult to model. Electromagnetically, the absorption, scattering, and depolarization characteristics can be calculated only for very simple shapes and distributions. However, theoretical results do provide a qualitative understanding of the effects of rain on millimeter waves, and when they are combined with experimental data, empirical parameters can be derived and more quantitative results are possible.

Let us first examine the absorption and scattering characteristics of a single spherical raindrop. From Eq. (16) we can calculate the total cross section $Q_t$, which is the sum of the absorption cross section $Q_a$ and the

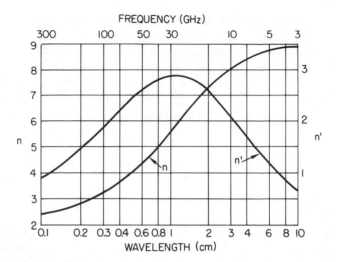

FIG. 7 Typical values of refraction indices of water at 20°C. (From Hogg and Chu, 1975. © 1975 IEEE.)

scattering cross section $Q_s$. This cross section is a strong function of the drop diameter and its complex index of refraction; the real and imaginary parts of the index of refraction of water are shown in Fig. 7. At millimeter wavelengths both parts decrease with decreasing wavelength, and we shall see that this is one of the reasons why the cross section of a drop eventually starts to decrease at shorter wavelengths.

In Fig. 8, the total cross section of a drop is plotted as a function of drop diameter for several wavelengths. When the drop is very small with respect to wavelength, the Rayleigh approximation is valid; it is seen from Eqs. (17) and (18) that the scattering and absorption cross sections $Q_s$ and $Q_a$ are proportional to $(D/\lambda)^6$ and $(D/\lambda)^3$, respectively. Because the loss due to scattering is negligible compared to that due to absorption, the total cross section is proportional to the volume of the drop. As the drop becomes larger both the scattering and absorption cross sections continue to increase, with the scattering cross section increasing more rapidly. Finally, after reaching a peak, the total cross section begins to level off, and would eventually approach a value of twice the geometric cross section of the drop when it is very large with respect to wavelength (Van de Hulst, 1957). Thus, as the drop becomes larger, the cross section, which is initially proportional to the drop volume, becomes proportional to the drop area.

The dependence of the cross section on wavelength is more complicated than that of size, because both the relative drop size and the complex index of refraction are changing. In Fig. 9, the cross sections are plotted as a

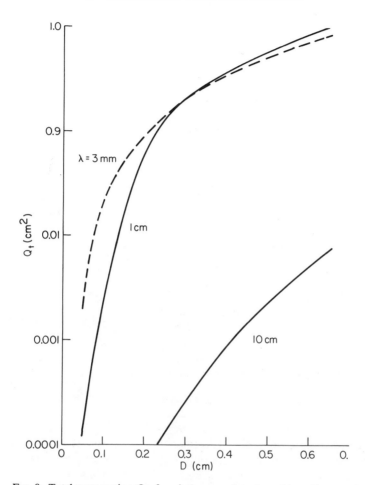

FIG. 8   Total cross section $Q_t$ of a raindrop as a function of drop diameter $D$.

function of wavelength for a number of drop sizes. The cross sections increase with decreasing wavelength, reach a peak, and then start to decrease very slightly for still smaller wavelengths. This behavior can be explained by referring to Figs. 7 and 8. Although the cross section increases as the drop becomes larger with respect to wavelength, the real and imaginary components of the index of refraction decrease as the wavelength becomes smaller, and this decrease eventually causes the total cross section to decrease.

We shall now consider the effect of the *shape* of rain drops on electromagnetic parameters. Whereas small drops tend to be spherical in shape, larger drops become oblate because of distortion due to air drag and are often

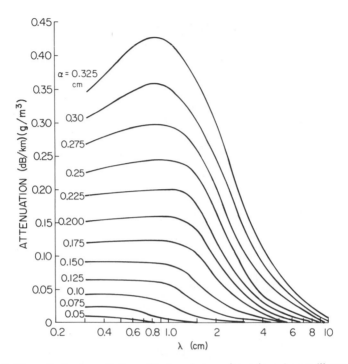

FIG. 9  Theoretical values of attenuation by raindrops for various drop radii, expressed in decibels per kilometer per drop per cubic meter. (From "Propagation of Short Radio Waves" by D. E. Kerr. © 1951, McGraw-Hill Book Company.)

modeled as oblate spheroids (Pruppacher and Pitter, 1971). The cross section of an oblate drop is generally larger than that of a corresponding spherical drop having an equal volume of water (Atlas *et al.,* 1953). The cross section is strongly dependent on the polarization of the wave, being larger when the polarization vector is aligned with the major axis of the spheroid and smaller when the polarization vector is aligned with the minor axis. We shall see that the most significant effect of a nonspherical drop is the depolarization it can produce.

  We have reviewed the absorption and scattering characteristics of individual raindrops and found that they are a very complex function of both the drop geometry and the index of refraction. Rain can be considered a collection of drops having diameters ranging from a fraction of a millimeter (mist) up to possibly 7 mm. To compute the attenuation of rain the cross sections of the drops must be calculated and then summed. Because the characteristics of a precipitation system are controlled largely by the air flow, the net result is a collection of drops that is continually varying both spatially and temporally; it is thus very difficult to model. It has been found that the meteorological parameter that is most easily measured and also most effectively characterizes rain is the rain rate. A number of investigators have shown that rain rate is correlated with drop size distribution; their results are summarized by Olsen *et al.* (1978). The attenuation can be expressed in the form

$$A = 0.4343 \int_0^\infty N(D)Q_t(D, \lambda) \, dD, \tag{20}$$

where $N(D) \, dD$ is the number of drops per cubic meter having diameters in the range $dD$, $Q_t$ is the total cross section of each drop, and the attenuation is measured in decibels per kilometer. Rain is often assumed to have an exponential distribution of drop diameters, so that

$$N(D) = N_0 e^{-\Lambda D}, \tag{21}$$

where $N_0$ are $\Lambda$ are empirical constants that are a function of the type of rain and more particularly the rain rate. The original rain attenuation calculations were done by Ryde and Ryde (1945); these were later refined by Medhurst (1965). Attenuations based on Medhurst's calculations are plotted as a function of rain rate in Fig. 10. These curves can be approximated by the expression

$$A = aR^b, \tag{22}$$

where $a$ and $b$ are numerical constants that are a function of wavelength and type of rain, $R$ is the rain rate in millimeters per hour, and the attenuation is measured in decibels per kilometer. Olsen *et al.* (1978) calculated rain attenuation as a function of rain rate using the Mie formulation and then performed a logarithmic regression to obtain values for $a$ and $b$. These values have been tabulated for frequencies from 1 to 1000 GHz, and although they are believed to provide a good approximation to the attenuation, it should be remembered that they are statistical and must be interpreted accordingly.

The depolarization characteristics of rain are very heavily dependent on the shape and orientation of the drops. Because light rain consists mostly of small drops and small drops tend to be spherical, depolarization effects are minimal. As the rain becomes heavier, the average drop size increases and the larger drops tend to become more oblate. The more oblate the drop, the larger the differential attenuation and phase between the orthogonal fields. However, it should be emphasized that these differentials do not in themselves produce depolarization; the incident field must also be tilted with respect to the axes of the drop. Because large oblates are easily canted by winds, their axes are not usually aligned with either horizontally or vertically

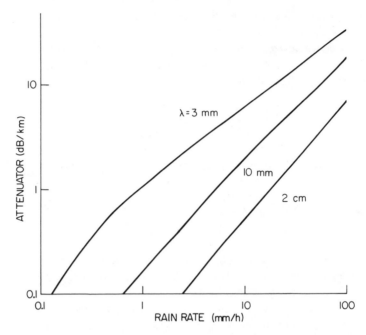

FIG. 10   Rain attenuation as a function of rain rate. (From Medhurst, 1965. © 1965 IEEE.)

polarized waves. Brussard (1976) has shown that the canting angle of oblate rain drops is a function of the average drop diameter and vertical wind gradients. Typically the canting increases with drop size and levels off for drops on the order of a few millimeters in diameter. It naturally increases with wind speed and usually becomes smaller with increasing height.

Representative curves of differential attenuations and phases of rain as a function of frequency for a number of rain rates have been prepared by Hogg and Chu (1975); these are presented in Figs. 11 and 12. Because the differential attenuation is proportional to the total attenuation, it increases with rain rate. It also initially increases with shorter wavelengths, as does the total attenuation, but then reaches a peak and eventually starts to decrease at very short millimeter wavelengths, mostly because the attenuation at very short wavelengths is produced primarily by the smaller drops and these tend to be spherical. The differential phase is affected mostly by the real part of the refractive index of water and is seen to decrease with shorter millimeter wavelengths. Thus, at millimeter wavelengths, differential attenuation is the dominant cause of depolarization.

b.  *Sleet, Snow, and Hail.*   These very complex forms of precipitation have attenuation characteristics that vary markedly at millimeter wave-

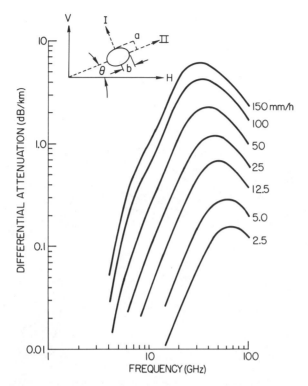

FIG. 11   Rain-induced differential attenuation. (From Hogg and Chu, 1975. © 1975 IEEE.)

lengths. We have seen that liquid water is a strong attenuator of millimeter waves; therefore sleet, which is a mixture of rain and snow, can also produce very high attenuations. In fact, these attenuations may exceed those of rain, because nonspherical shapes have been shown to produce higher attenuations than equivalent spheres and sleet particulates are often very elongated (Atlas *et al.*, 1953). The depolarization effects of sleet can be very strong if the flakes show a nonparallel preferential alignment with the *E*-field. Wet snow has characteristics very similar to those of sleet. As the snow becomes drier its composition approaches that of ice crystals, and because ice has a low imaginary index of refraction the absorption is very small. Losses due to scattering are small at longer millimeter wavelengths but may become appreciable at shorter wavelengths if the flakes are large.

   The effect of hail on millimeter waves is better understood than that of sleet or snow because there is less variability in its shape and composition. Because the imaginary part of the index of refraction of ice is about three orders of magnitude less than that of water, absorptive losses are negligible.

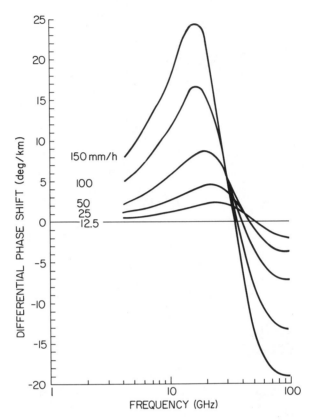

FIG. 12   Rain-induced differential phase shift. (From Hogg and Chu, 1975. © 1975 IEEE.)

The real part of the index of refraction is about one-fourth that of rain, so although scattering losses produced by hail are smaller than those of rain, they can be important, particularly at shorter millimeter wavelengths. If the hailstone is covered with even a very thin coat of water, its attenuation rises significantly and approaches that of a rain drop having an equivalent volume (Battan, 1973).

In summation, because it is extremely difficult to produce accurate models of sleet, snow, and hail, and because there are very few experimental attenuation data at millimeter wavelengths (or any other wavelengths), it is not possible to provide quantitative results. Although it is known that the absorption and scattering losses of sleet and wet snow are large and the scattering losses of dry snow and hail are significant, there are not presently any attenuation data for these particulates comparable to those for rain. However, from a practical standpoint, these particulates do not occur very often, and so their impact on millimeter-wave systems is not considered critical.

c. *Cloud, Fog, and Haze.* Meteorologically, cloud, fog, and haze are very similar because they consist of small water droplets suspended in air. The droplets have diameters ranging from a fraction of a micron for fog and haze to over a 100 $\mu$ for heavy fog and high-altitude clouds (Stewart, 1980). Haze may be considered a light fog, and of course the principal difference between cloud and fog is that clouds exist at higher altitudes and often contain ice as well as water particulates. Electromagnetically, cloud, fog, and haze can be treated identically. Because the droplets are very small with respect to wavelength, the Rayleigh approximation is valid, even at short millimeter wavelengths. Therefore, scattering losses are negligible and the attenuation, which is proportional to the volume of the drops, can be calculated from Eq. (18) and is seen to increase with decreasing wavelength. Because it is very difficult to measure fog or cloud water content, the attenuations produced by these particulates are not easily determined. Typical attenuations are plotted as a function of wavelength in Fig. 13 for several temperatures.

Fog is often characterized by visibility, which is much easier to measure than density; however, it should be emphasized that visibility is an optical parameter, a function of the scattering characteristics of the droplets, whereas the attenuation at millimeter wavelengths, as mentioned previously, is strictly a function of the fog density. Although there is a correlation between fog or cloud visibility and total liquid water content (Eldridge, 1966; Platt, 1970), this relationship must be used with caution because the correlation coefficient is strongly dependent on the type of fog. Fog is often

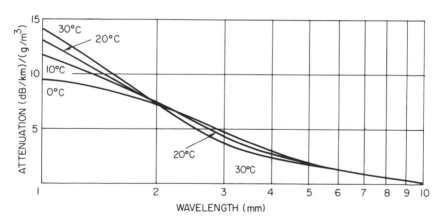

FIG. 13 Attenuation of cloud and fog as a function of wavelength.

divided into two types: radiation fog, which forms when the ground becomes cold at night and cools the adjacent air mass until it becomes supersaturated, and advection fog, which forms when warm, moist air moves across a cooler surface. The average drop size of an advection fog is usually larger than that of a radiation fog. Thus, if two fogs had the same liquid water densities but one consisted of relatively large drops, then that fog would have a much higher visibility. Another fog or cloud parameter not to be overlooked is temperature, because it has a strong influence on the complex index of refraction, particularly at longer millimeter wavelengths (Rozenberg, 1972). Once again, as with snow and hail, the overall effects of cloud and fog are not as severe as those of rain.

d. *Sand and Dust.* Sand and dust are both fine-grained, quartz-type particles that have diameters of a fraction of a millimeter, densities on the order of 2600 kg/m$^3$, and a relative dielectric constant of about $2.5 - 0.025i$ (Chu, 1979; Rafuse, 1981). The principal difference between sand and dust is that the average grain size of sand is larger. As a result, whereas larger wind-blown sand grains rise to a maximum height of only 2 m, heavy winds can carry fine dust particles to altitudes as high as 1 km. In general, the height to which the particles rise is proportional to the wind speed and inversely proportional to the particle size (Rafuse, 1981). Because sand and dust particulates are very small with respect to wavelength, even at millimeter wavelengths, scattering losses are negligible and the only losses are due to absorption. However, the imaginary part of the complex dielectric constant is very small; thus absorption losses can be considered minimal under naturally disturbed conditions. If a large amount of dirt or dust were to become suspended in the atmosphere as a result of an explosion, however, then significant attenuations might arise, particularly if the dust were moistened by the presence of water; these attenuations would last only for a duration of seconds (Gallagher *et al.,* 1980).

## III. Transmission Paths

There are essentially three types of transmission paths, each with specific characteristics with regard to millimeter-wave propagation. Two of these paths are line-of-sight: terrestrial and earth–space. The third is the transhorizon path. We shall now take the results of Section II and apply them to systems that would use these propagation paths.

### A. TERRESTRIAL LINE-OF-SIGHT PATHS

For terrestrial paths the atmospheric effects on propagation become more pronounced as the length of the path increases and the wavelength decreases. For short paths, attenuation is of principal concern; refraction and atmospheric multipath effects are unlikely, terrain multipath and diffraction problems arise only when transmitter and receiver are close to the surface, and depolarization and scintillations occur only under very extreme conditions of precipitation. For a clear atmosphere, gaseous absorption in the window regions is only a fraction of a decibel per kilometer at longer millimeter wavelengths but can become appreciable at shorter wavelengths. Sand and dust attenuations throughout the millimeter-wave spectrum are well below 1 dB/km, except for conditions of a large cloud that may be produced by an explosion. Haze and fog attenuations increase with decreasing wavelength as shown in Fig. 13, and although they become appreciable they are generally lower than water vapor absorption at very short wavelengths. Attenuation due to rain, sleet, and wet snow can be significant even for very short paths; this attenuation is lowest at longer wavelengths and gradually increases with decreasing wavelength, reaching a maximum at a wavelength of a few millimeters and then leveling off at still shorter wavelengths.

As the path becomes longer, attenuation effects become more severe; in addition, all of the other propagation effects mentioned in the previous paragraph are more likely to occur. The use of millimeter waves for applications requiring long terrestrial paths appears unlikely at this time because the attenuation would be prohibitive, except perhaps for a region having a relatively dry climate.

## 1. *Attenuation*

For many applications, particularly communications, it is important to be able to estimate the percentage of time that the path attenuation exceeds a certain value. To accomplish this one must first examine the climate of the region of interest. If the absolute humidity is known, the gaseous absorption can be estimated from the expression (CCIR, 1982a)

$$A = a + b\rho_0 - cT_0. \tag{23}$$

The coefficients are plotted in Fig. 14; $\rho_0$ is in grams per cubic meter and $T_0$ in degrees Celsius. This calculation is not very accurate at very short wavelengths, where there is a lack of agreement between the theoretical and experimental values of water vapor absorption. As mentioned in the previous paragraph, sand, dust, haze, and fog attenuations are low at longer wavelengths and generally small compared to water vapor absorption at very short wavelengths; only at wavelengths between about 2 and 3 mm is fog attenuation generally important.

Rain attenuation is by far the most serious propagation problem. Only for a few locations are rain attenuation statistics available. A procedure for

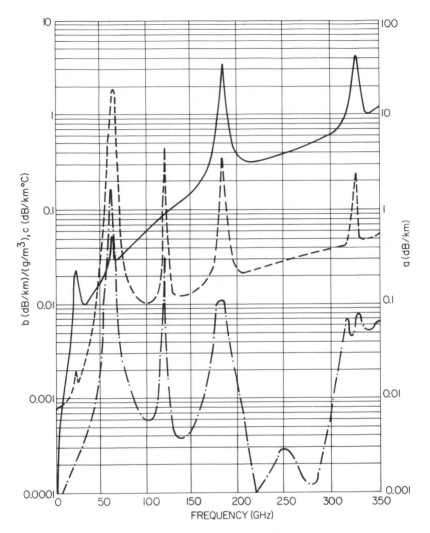

FIG. 14  Coefficients for computing specific attenuation. Specific attenuation $A = a + bp_0 - cT_0$ (dB/km), with $p_0$ in grams per cubic meter and $T_0$ in degrees Celsius: $---$, $a$ coefficient; ——, $b$ coefficient; $-\cdot-$, $c$ coefficient. (From CCIR, 1982a).

estimating rain attenuation for different climates has been outlined by Crane (1980b). A global model representing typical rain climates throughout the world was developed, based on rain data provided by the World Meteorological Organization, and from this model the percentage of time that the point rain rate exceeds a particular value can be estimated. However, rain ordinarily consists of cells of different sizes and generally is not homogeneous in the horizontal plane. Thus, it is necessary to derive a path average rain rate from the point values. By pooling worldwide rain statistics it was possible to obtain an empirical relationship between point and path average rain rates; this has been done so far for distances up to 22.5 km. For low rain rates the rain is usually widespread; widespread rain, however, may contain convective cells having a high rain rate, so on an average the rain rate along the total path will be higher than that at a point. For high rain rates the rain tends to be localized, so the average rain rate for the total path would ordinarily be lower than that at a point. As expected, the correction factor is heavily dependent on the path length.

In Section II.B.3.a the $aR^b$ relationship for attenuation as a function of rain rate was introduced. By using the path average rain rate concept of the previous paragraph it is possible to derive a correction term for the $aR^b$ expression. From Crane (1980b), we have

$$A(R_p, D) = aR_p^\beta \left( \frac{e^{u\beta d} - 1}{u\beta} - \frac{b^\beta e^{c\beta d}}{c\beta} + \frac{b^\beta e^{c\beta D}}{c\beta} \right), \quad d \le D \le 22.5 \text{ km}, \quad (24)$$

$$= aR_p^\beta \left( \frac{e^{u\beta D} - 1}{u\beta} \right), \qquad\qquad 0 < D \le d, \qquad (25)$$

where $R_p$ is in millimeters per hour, $D$ in kilometers, the specific attenuation $A(R_p, D)$ in decibels per kilometer, and

$$u = \ln(be^{cd})/d, \qquad b = 2.3R_p^{-0.17},$$

$$c = 0.026 - 0.03 \ln R_p, \qquad d = 3.8 - 0.6 \ln R_p.$$

For $D > 22.5$ km, the probability of occurrence $P$ is replaced by a modified probability of occurrence

$$P' = (22.5/D)P. \qquad (26)$$

For example, suppose that for a particular climatic region the rain rate exceeds 28 mm/h 0.01% of the time and 41 mm/h 0.005% of the time ($R_{0.01} = 28$ mm/h, $R_{0.005} = 41$ mm/h. The attenuation along a 22.5-km path that would be exceeded 0.01% of the time can be calculated directly from Eq. (24) or (25). To determine the attenuation that would be exceeded 0.01% of the time over a 45-km path, a new percentage of time $P' = (22.5/45)P = 0.005\%$ would be used, with a corresponding rain rate of $R_{0.005} = 41$ mm/h. Now the attenuation that would be exceeded 0.01% of the time for a 45-km path would be based on a rain rate of $R_{0.005} = 41$ mm/h for a 22.5-km path.

To improve the reliability of a terrestrial link, path diversity can be utilized. As mentioned previously, the heavy rain that has a severe impact on terrestrial link performance tends to be localized. By using redundant terminals, the probability of having a path free from heavy rain is increased. Ideally, the optimum separation of transmitter or receiver terminals (or both) is that for which the rain rates (and corresponding attenuations) for the pair of terminals are uncorrelated. This separation is a function of the climate and rain rate. Blomquist and Norbury (1979) have studied diversity improvement for a number of paths with lengths of about 3 – to 13 km and terminal separations of 4 – 12 km. Based on very limited data, they found that the diversity improvement increased as the terminal separation was increased from 4 to about 8 km; there was, however, no additional improvement as the separation was increased further. Because only limited statistical data on diversity advantage are presently available, it is not possible to determine optimal terminal spacings for most locations. However, there is sufficient evidence to indicate that a diversity mode of operation can significantly improve the performance of line-of-sight links.

## 2. Terrain Scatter and Diffraction

If the terminals of a line-of-sight path are close to the surface, then propagation losses due to both multipath and diffraction are possible. These mechanisms were described in Sections II.A.3.b and II.A.4, respectively, and it was seen that the multipath or diffracted signal can interfere with the direct signal; the net effect is a resultant signal that may vary in amplitude from zero intensity to twice the intensity of the direct signal. It should be emphasized that the interference is strongest when the multipath or diffracted signal is coherent. For multipath this occurs when the terrain is smooth with respect to wavelength; for diffraction it occurs from a prominent obstacle or a set of obstacles that happen to add constructively (or destructively).

Regarding terrain muiltipath, even though most terrains have surface irregularities much larger than 1 mm, we see from Eqs. (9) and (10) that as the grazing angle becomes small the surface becomes electromagnetically smooth (the reflection coefficient approaches unity) and large specular signals are possible. These signals interfere with the direct signal and can significantly degrade the performance of a line-of-sight system. Hayes et al. (1979) have made multipath measurements over grass and snow at wavelengths of 8.6, 3.1, and 2.1 mm and obtained height – gain curves with nulls in excess of 20 dB, the nulls being deeper for snow cover than for grass and also deeper at longer wavelengths. Measurements were made for both vertical and horizontal polarizations; because grazing angles were well below the Brewster angle, the height – gain curves were similar for both polarizations.

Interference effects produced by a diffracted signal should not be as large as those produced by multipath, because the obstacles that would diffract the wave are not likely to produce a coherent signal. As mentioned previously, if the surface irregularities of a prominent obstacle are rough with respect to wavelength, there are many uncorrelated, diffracted rays and the resultant signal consists of a diffuse signal superimposed upon a weak specular signal; if the surface is very rough the specular component will disappear. Also, it is seen from Eq. (14) that the direct path must be within meters of the top of the obstacle for a diffracted signal to appear. However, diffracted signals have been measured at frequencies as high as 28.8 GHz (Haakinson *et al.*, 1980), so this phenomenon is certainly possible at millimeter wavelengths under the "right" conditions.

### 3. Depolarization

As mentioned in Sections II.A.5 and II.B.2, depolarized signals may arise from either multipath or precipitation. A multipath ray obliquely incident on a paraboloidal receiving antenna can produce a cross-polarized component. Oblate rain drops canted with respect to the plane of the polarization vector of the incident wave also produce a cross-polarized component. The net effect is that the resultant signal is depolarized and the system performance compromised. Vander Vorst (1979) has summarized the effects of depolarization on line-of-sight links. Measurements on a 53-km terrestrial path have shown that depolarization produced by multipath has a more severe effect on link performance than that produced by rain (Rooryck and Battesti, 1975). The influence of rain caused only a very small degradation of the performance of a dual-polarized system with respect to a single-polarization system, whereas the same was not true for multipath.

### 4. Refraction and Atmospheric Multipath

Under normal atmospheric conditions an electromagnetic wave is bent towards the earth's surface. The amount of bending is proportional to the length of the path, so for long paths refractive bending corrections may be required, as discussed in Section II.A.1.c. Under abnormal atmospheric conditions—for example, those producing a sharp negative gradient in the refractivity as a function of height—the wave may become trapped (ducting) or a multipath signal may be produced by the "layer" arising from the refractivity structure. Although refraction and multipath effects are possible in principle, they are not considered important as far as millimeter-wave line-of-sight links are concerned, because, as mentioned previously, attenuation effects will normally prohibit the use of very long paths.

### B. EARTH-SPACE PATHS

For earth–space paths, propagation effects become more severe with decreasing wavelength. For elevation angles above about 6°, attenuation and emission from atmospheric gases and precipitation are of principal concern. In addition, backscatter and depolarization resulting from precipitation may also cause problems. For low elevation angles all of the problems associated with long terrestrial paths are present. The determination of propagation effects for slant paths is generally more difficult than for terrestrial paths. The modeling of a slant path under conditions of cloud and precipitation is particularly complicated, because the structure of the particulates is varying in both time and space. Experimentally, it is very expensive to place a millimeter-wave beacon on a satellite or aircraft, so other means of obtaining attenuation information are often used.

## 1. *Attenuation and Emission*

Atmospheric attenuation and emission are the most serious propagation problems for earth–space systems. Because the attenuation decreases the signal level and the emission (or effective noise temperature) sets a minimum noise level for the receiver, the only way to maintain the system signal-to-noise ratio is through increased transmitter power or a diversity mode of operation. High powers are not readily available at millimeter wavelengths and diversity systems are costly, so these options have their difficulties.

For clear sky conditions, the attenuation is a function of oxygen and water vapor density along the path. The vertical distributions of these gases are assumed to decrease exponentially with height and have scale heights of approximately 4 and 2 km, respectively. The zenith attenuation as a function of wavelength can be estimated from (CCIR, 1982b)

$$A_{90°} = \alpha + \beta\rho_0 - \xi T_0. \tag{27}$$

The coefficients $\alpha$, $\beta$, and $\xi$ are plotted in Fig. 15. The attenuation at elevation angles above about 6° can be calculated by multiplying the zenith attenuation by the cosecant of the elevation angle. For angles below 6° the attenuation is assumed to be proportional to the path length through the attenuating medium. This distance is given by (Altshuler *et al.*, 1978)

$$d(\theta) = [(a_e + h)^2 - a_e^2 \cos^2 \theta]^{1/2} - a_e \sin \theta, \tag{28}$$

where $\theta$ is the elevation angle, $a_e$ ⅔ the earth's radius (8500 km), and $h$ the scale height of combined oxygen and water vapor gases ($\sim 3.2$ km). Therefore,

$$A(\theta) = A(90°)d(\theta). \tag{29}$$

The zenith attenuation is plotted as a function of wavelength in Fig. 16 for a completely dry atmosphere and more typical atmospheres having surface absolute humidities of 3 and 10 g/m³ (Smith, 1982). For a dry atmosphere

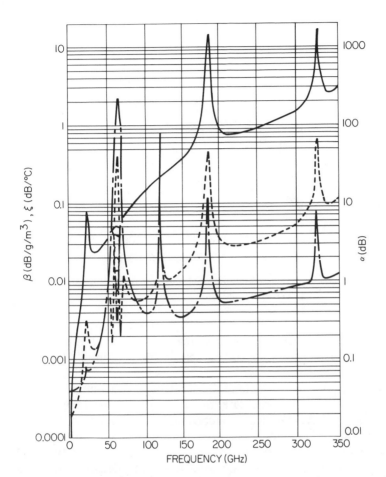

FIG. 15 Coefficients for computing zenith attenuation. Zenith attenuation $\tau_{90°} = \alpha + \beta\rho_0 - \xi T_0$ (dB), with $\rho_0$ in grams per cubic meter and $T_0$ in degrees Celsius: — · —, $\alpha$ coefficient; ——, $\beta$ coefficient; – – –, $\xi$ coefficient. (From CCIR, 1982b).

the total zenith attenuation in the window regions is only a fraction of a decibel per kilometer. However, this attenuation increases very sharply as the atmosphere becomes moist, particularly below a wavelength of a few millimeters, at which losses on the order of tens of decibels are possible. A plot of the apparent sky temperature (emission) is shown in Fig. 17 as a function of frequency for a set of elevation angles from the horizon to zenith and for an atmosphere having a water vapor density of 7.5 g/m³ (Smith, 1982). The sky temperature is relatively low for higher elevation angles and

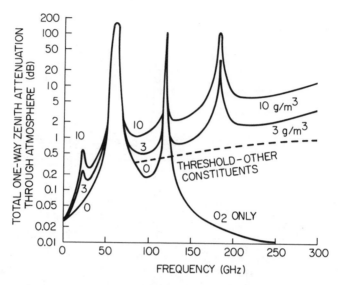

Fig. 16 Total zenith attenuation through atmosphere as a function of frequency. (From Smith, 1982. © American Geophysical Union.)

longer wavelengths and gradually increases for lower elevation angles and shorter wavelengths, approaching a terrain temperature of approximately 290 K.

For conditions of fog or cloud, the modeling of the atmosphere becomes increasingly difficult, particularly for slant paths close to the horizon. Fog and cloud models have been developed, however (Falcone et al., 1979), and in principle the attenuations can be calculated using the information provided in Fig. 13. It must be emphasized that these attenuations are only approximate, because neither the true liquid water density of the cloud nor the extent of the cloud is accurately known. Because the cloud particulates are in the Rayleigh region and scattering losses are negligible, the corresponding brightness temperature (emission) can be calculated from Eq. (6).

Several investigators have measured cloud attenuations at millimeter wavelengths. Altshuler et al. (1978) have presented cloud attenuation statistics at frequencies of 15 and 35 GHz based on 440 sets of measured data. They characterized sky conditions as clear, mixed clouds, or heavy clouds. Average attenuations as a function of elevation angle are shown in Fig. 18. They also demonstrated a reasonable correlation between the slant path attenuation and surface absolute humidity. Typical slant path attenuations extrapolated to zenith were 0.1 and 0.36 dB at 15 and 35 GHz, respectively. Lo et al. (1975) measured attenuations at wavelengths of 8.6 and 3.2 mm

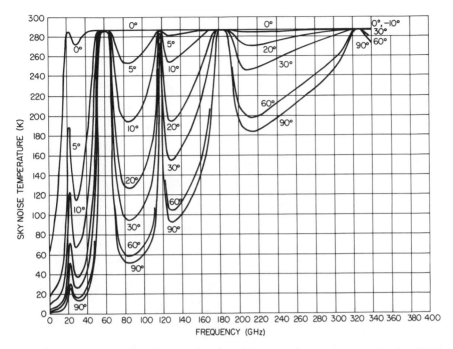

FIG. 17 Sky noise temperature as a function of frequency for a water vapor density of 7.5 g/m³. (From CCIR, 1982).

over a 6-month period, and obtained typical attenuations of 0.42 and 2.13 dB, respectively. Slobin (1982) has estimated average-year statistics of cloud attenuation and emission down to a wavelength of 6 mm for various climatically distinct regions throughout the United States.

As was the case for terrestrial paths, rain, sleet, and wet snow present the most serious propagation limitations on earth–space millimeter-wave systems. A number of investigators have derived models for predicting rain attenuation, and their results have been summarized by Ippolito (1981). All of the techniques assume that the slant path attenuation can be estimated by modifying the attenuation for a terrestrial path by an effective path-length parameter that is usually a function of the elevation angle and the type of rain. For example, the Crane (1980b) model for estimating rain attenuation for terrestrial paths, presented in Section III.A.1, can be modified for slant path attenuations. It is assumed that the rain rate has a constant value between the station height $h_0$ and the height $h$ of the 0°C isotherm. Precipitation above the 0°C isotherm consists of ice particles, so the attenuation is considered negligible. Although the height of the 0°C isotherm is variable, radar measurements have shown it to have a strong dependence on site

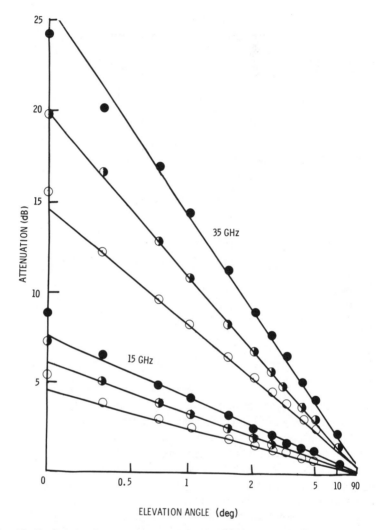

FIG. 18 Typical cloud attenuations as a function of elevation angle: ○, clear; ◑, mixed; ●, cloudy. (From Altshuler *et al.*, 1978. © Published by the American Geophysical Union.)

latitude and rain rate, which in turn are linearly related to the logarithm of the probability of occurrence $P$; Crane (1980b) has derived a set of curves for estimating the height $h$, and these are shown in Fig. 19. With $h$ and $h_0$ known, the effective path length $D$ through the rain can be calculated:

$$D = (h - h_0)/\tan \theta, \qquad \theta \geq 10°, \qquad (30)$$

$$= a_e \psi, \qquad \theta < 10°, \qquad (31)$$

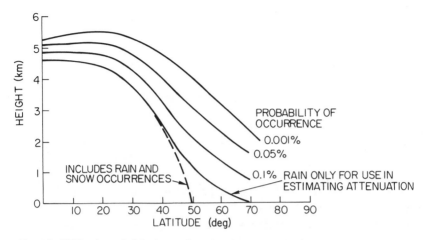

FIG. 19 0°C isotherm height for use in estimating the depth of the attenuating region on a slant path. (From Crane, 1980b. © 1980 IEEE.)

where

$$\psi = \sin^{-1}\left(\frac{\cos\theta}{h + a_e}\left\{[(h_0 + a_e)^2\sin^2\theta + 2a_e(h - h_0) + h^2 - h_0^2]^{1/2}\right.\right.$$
$$\left.\left. - (h_0 + a_e)\sin\theta\right\}\right), \tag{32}$$

$a_e$ is the effective earth radius (8500 km), and $\theta$ is the elevation angle.

The surface-projected attenuation $A(R_p, D)$ is calculated from Eq. (24) or Eq. (25). Then, the value for the slant path $A_s$ is estimated assuming a constant attenuation below $h$ by

$$A_s = LA(D)/D, \tag{33}$$

where

$$L = D/\cos\theta, \quad \theta \geq 10° \tag{34}$$
$$= [(a_e + h_0)^2 + (a_e + h)^2 - 2(a_e + h_0)(a_e + h)\cos\psi]^{1/2}, \quad \theta < 10°. \tag{35}$$

Several methods can be used to measure slant path attenuations. The most straightforward, but also most costly, is to place a millimeter-wave beacon in space. Measurements using satellite beacons have been made at wavelengths from approximately 30 to 10 mm and have been summarized by Ippolito (1981). A number of investigators have made attenuation measurements using the sun as a source (Altshuler and Telford, 1980; Davies, 1973; Wilson, 1969; Wulfsberg, 1967). When a radiometer is pointed at the sun, the noise power received consists of radiation from the sun and the atmosphere. The antenna temperature can be expressed as

$$T_a = T'_a e^{-\tau} + \int_0^\infty T(s)\gamma(s) \exp\left(-\int_0^\infty \gamma(s')\,ds'\right) ds, \qquad (36)$$

where $T'_a$ is the effective antenna temperature of the sun with no intervening atmosphere (in degrees Kelvin), $T(s)$ the atmospheric temperature, $\tau$ the total attenuation (in nepers), $\gamma(s)$ the absorption coefficient, and $s$ the distance from the antenna (ray path). In simpler terms

$$T_a = T'_a e^{-\gamma} + (1 - e^{-\gamma})T_m, \qquad (37)$$

where $T_m$ is the atmospheric mean absorption temperature within the antenna beam. The attenuation $\gamma$ appears in both terms on the right-hand side of Eq. (37). Because the second term is the emission, it can easily be canceled out by pointing the antenna beam toward and away from the sun. With the second term balanced off, Eq. (37) can be solved for $\gamma$, converted from nepers to decibels, and expressed as

$$A = 10 \log(T'_a/T_a). \qquad (38)$$

$T'_a$ is determined from a set of antenna temperature measurements made under clear sky conditions as a function of elevation angle; for these conditions the antenna temperature is proportional to the cosecant of the elevation angle, and it can be shown that $T'_a$ is equal to the slope of the line $\log T_a$ versus $\csc \theta$ (Wulfsberg, 1967).

Because the attenuation also appears in the emission term in Eq. (37), it is possible to determine the attenuation from an emission measurement. In this method the antenna must be pointed away from the sun or the moon (millimeter-wave radiation from all other natural sources is negligible), so the first term on the right-hand side of Eq. (37) can be considered zero and Eq. (37) reduces to Eq. (5). As before, the equation is then solved for the attenuation $\gamma$ and expressed in decibels as

$$A = 10 \log T_m/(T_m - T_a).$$

Another technique for estimating rain attenuation at millimeter wavelengths is from a measurement of the reflectivity factor of the rain. Attenuation is derived from established relationships between these parameters (Goldhirsh, 1979). When a single-wavelength radar is used, calibration errors and the uncertainty in the reflectivity–attenuation relationship, particularly for mixed phase precipitation, limit the accuracy of this technique. Uncertainties in the reflectivity–attenuation relationship can, however, be reduced by using a dual-wavelength radar and measuring the differential attenuation.

In summation, there are four methods for measuring slant path attenuations. The most direct method is to place a source in space. This allows measurements to be made over a very wide dynamic range. An additional advantage is that the polarization and bandwidth limitations imposed by the atmosphere can also be measured. One disadvantage is cost; and, depending on the satellite orbit, measurements may be possible at only one elevation angle, which may or may not be a drawback. Total attenuation can be measured very accurately and economically using the sun as a source over dynamic ranges approaching 25–30 dB. A disadvantage is that measurements can be made only in the direction of the sun and during the day.

Attenuation can easily be determined from an emission measurement on a continual basis and at any elevation angle. It must be emphasized that there are two major problems that limit the accuracy of this technique. The true value of $T_m$ is not always known; if $T_m - T_a$ is large this uncertainty is not serious, but if $T_m - T_a$ is small a large error may arise. Also, the emission is related only to the absorption, whereas the attenuation includes losses due to scattering in addition to those due to absorption. Therefore, in cases for which the Rayleigh approximation is not valid and scattering losses are appreciable, errors will arise. Techniques for correcting for the additional losses due to scattering have been investigated by Zavody (1974) and Ishimaru and Cheung (1980). For these reasons this method is not generally recommended for attenuations much above 10 dB.

Attenuations determined from radar reflectivity measurements have the limitations in accuracy discussed previously, and in general this technique is not suitable for losses arising from very small particulates such as fog, cloud, or drizzle. This method does, however, have the advantage of measuring attenuation as a function of distance from the transmitter. McCormick (1972) has compared attenuations derived from radar reflectivity with those obtained directly utilizing a beacon placed on an aircraft. Strickland (1974) has measured slant path attenuations using radar, radiometers, and a satellite beacon simultaneously.

## C. TRANSHORIZON PATHS

Very little is known about transhorizon propagation at millimeter wavelengths. Although a large number of troposcatter, tropospheric ducting, and diffraction experiments have been conducted over the years, they have been limited to wavelengths on the order of centimeters and longer. Based on the results that have been obtained, coupled with theoretical findings, empirical relationships for estimating path losses have been derived (CCIR, 1978). Under normal atmospheric conditions and for paths just slightly beyond the horizon, the diffracted field is the largest, particularly if the diffracting terrain approximates a knife edge. As the distance beyond the horizon is increased, the received signal consists of both diffracted and troposcatter components. For still longer distances, the diffracted field disappears and the received signal is entirely troposcatter. Finally, for very long paths, tropospheric ducting is possible; however, it must be emphasized that this is an anomalous mode of propagation that occurs only for the special refractivity structure described in Section II.A.1.a.

In principle, troposcatter, tropospheric ducting, and diffraction modes of propagation are all possible at millimeter wavelengths. Traditionally, transhorizon propagation has been used for paths hundreds of kilometers in extent. However, this is not practical at millimeter wavelengths, because these modes of propagation occur only for very low elevation angles and atmospheric attenuation is large, particularly for shorter millimeter wavelengths. Although the advent of satellite communications has reduced the need for transhorizon systems, there are still potential applications for paths only slightly beyond the horizon.

### 1. *Troposcatter*

The principal advantages of using millimeter wavelengths for troposcatter are smaller antennas and larger available bandwidth. Parl and Malaga (1982) have calculated the path loss for a troposcatter system using turbulence scatter theory and obtained the results plotted in Fig. 20, which shows the received signal level as a function of frequency for several water vapor pressures. The calculations are based on transmitting and receiving antennas with 15-ft. apertures placed 100 km apart. It is interesting to note that the received signal level increases slowly with frequency up to about 12 GHz, at which point the aperture-to-medium coupling loss and atmospheric attenuation begin to outweigh the additional antenna gain due to a fixed-size aperture and shorter wavelength. In addition, the signal level increases significantly with higher water vapor pressures, because the refractive structure constant $C_n^2$ becomes higher. Although signal levels above 12 GHz drop off with increasing frequency, they are still high enough to be easily detected. The presence of precipitation along the path has two effects: it enhances the scattering but also attenuates the signal. The additional scattering, however, is not always an advantage, because it can produce intersystem interference (Crane, 1981).

Because the troposcatter mechanism is one of scattering from a turbulent medium, the field across the receiving aperture is not always a plane wave, and thus the gains of the transmitting and receiving antennas do not necessarily correspond to the plane-wave antenna gains. This loss in gain is the aperture-to-medium coupling loss, defined as the ratio of the power that would be received if the plane-wave gain of each antenna were realized to the power actually received (Waterman, 1958). This aperture-to-medium coupling loss increases as the beamwidths of the antennas decrease; thus the net gain of a troposcatter system does not increase proportionally with increasing antenna aperture. Likewise, the high gains that can be achieved with

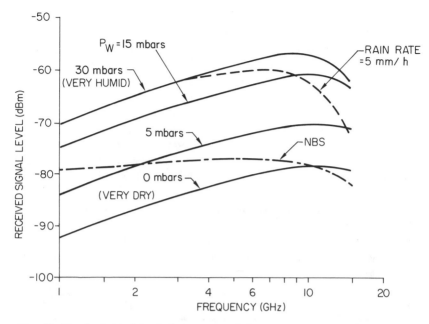

FIG. 20  Received signal level of troposcatter link compared to the National Bureau of Standards signal level. Based on 15-ft. antennas in a 100-km path. $P_T = 100$ W. (From Parl and Malaga, 1982.)

antennas of moderate size at millimeter wavelengths do not automatically translate into higher system gains. Unfortunately, the aperture-to-medium coupling loss is a complex function of the refractive index structure constant within the common volume, which is defined by the intersection of the antenna beams, and this is not easily determined; thus it is still not possible to predict this loss with any degree of confidence (Gough, 1968).

Another problem that arises from the nature of the scattering medium is fading of the received signal. Slow fades (of the order of hours) are usually attributed to major changes in the refractive index structure produced by changes in the weather. Fast fades (on the order of seconds) are attributed to the fact that the received field consists of many incoherent contributions from scatterers within the common volume.

Aperture-to-medium coupling loss and fading can be compensated for using diversity techniques. In this mode of operation the system is designed so that at least two essentially independent signals are received and then combined in an optimal way. The most common types of diversity are space (separate apertures), angle (separate feeds with common aperture), and frequency (separate frequencies). Empirical relationships for these separations have been derived and are summarized in CCIR (1978). At millimeter wavelengths, phased arrays are being considered for troposcatter systems. Because the amplitude and phase across this type of aperture can be controlled, more sophisticated diversity techniques are possible.

## 2. *Tropospheric Ducting*

Although it is possible to transmit over very long distances in a ducted mode, the fact that ducting occurs sporadically, depending upon location, time of day, season, etc., makes it of little use for transhorizon propagation applications. Ducting is usually of concern because it produces radio or radar holes (regions of weak signal level in the desired direction) and thus degrades system performance. Unfortunately, there is in general no way of preventing ducting, although changing the heights of the transmitter or receiver may sometimes help. Methods for detecting the presence of a duct or, better yet, for predicting the onset of a duct are being pursued. Because ducting is most likely to occur only for very low elevation angles, it should

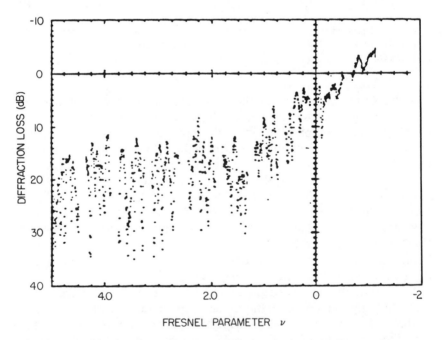

FIG. 21  Diffraction loss measured on a cluttered path with multiple diffraction. (From Haakinson *et al.*, 1980. © Reproduced by permission of the U.S. Dept. of Commerce, National Technical Information Service, Springfield, Virginia 22161.)

not normally be a problem at millimeter wavelengths, because atmospheric attenuation ordinarily prevents the use of these very low angles.

## 3. Diffraction

The diffraction mechanism described in Section II.A.4 may be a viable mode of propagation for paths slightly beyond the horizon, particularly if the diffracting terrain approximates a knife edge. However, diffraction effects are not limited to simple obstacles; for example, irregular terrain can produce a diffraction loss at frequencies as high as 28.8 GHz (Haakinson *et al.*, 1980). At millimeter wavelengths a statistical approach is used, and the diffraction loss is viewed as the superposition of many ray paths diffracted from these obstacles. The diffraction loss produced by trees on a smooth hill at 28.8 GHz is plotted in Fig. 21 (Haakinson *et al.*, 1980); this loss has a mean behavior of a single obstacle with random variations superimposed. Thus diffraction effects can occur at millimeter wavelengths, but are found to be of less importance than those at longer wavelengths. Also, theoretical solutions for natural terrains are not generally possible, because at these wavelengths the obstacles cannot be considered smooth.

## IV. Remote Sensing of the Atmosphere

Because of the strong interaction between millimeter waves and atmospheric gases and particulates, the propagated wave can serve as a diagnostic tool for sensing the meteorological properties of the atmosphere. In particular, it is possible to infer information on temperature, water vapor, turbulence structure, wind velocity, cloud composition and dynamics, and rain distribution and intensity. In addition, even a nonmeteorological application such as insect detection may prove to be a valuable use of millimeter-wave sensing.

Three types of systems are used for remote sensing: line-of-sight, radiometric, and radar. Line-of-sight systems consist of a transmitter and a receiver and probe the intervening atmosphere. Radiometric systems measure the atmospheric emission (or brightness temperature) of the atmosphere. Radar systems measure the reflective properties of atmospheric particulates or the refractive index structure. Radiometric and radar sensors may be either ground based (directed upward) or space-borne (directed downward). In addition, it is often advantageous to use a hybrid system that has both radiometric and radar channels.

## A. LINE-OF-SIGHT TRANSMISSION

In this section we shall review methods for inferring the composition and structure of the atmosphere between a transmitter and a receiver. For a clear atmosphere, two parameters can be inferred from millimeter-wave measurements: the refractive index structure constant $C_n^2$ and the wind velocity. A method for inferring average raindrop size is also described.

### 1. Refractive Index Structure Constant

As mentioned in Section II.A.1.e, the atmosphere has a fine scale refractive index structure that varies both temporally and spatially, and the strength of this structure is characterized by the parameter $C_n^2$. These variations produce amplitude and phase fluctuations across the wave front of the transmitted signal. In practice, the receiving antenna would be an array transverse to the propagated wave. It has been shown theoretically that the covariances of the fluctuations are related to the structure constant $C_n^2$; thus the wave front is sampled at each array element, the covariances are calculated, and $C_n^2$ is determined by inverting the integral that relates these quantities (Leuenberger et al., 1979; Shen, 1970). In principle, it should be possible to obtain the distribution of $C_n^2$ along the entire path; however, this has not yet been demonstrated, primarily because of the low degree of accuracy with which the integral can be inverted. Average values of $C_n^2$ for the whole path, however, have been obtained.

### 2. Wind Velocity

The procedure for determination of the wind velocity across the path is very similar to that used to obtain $C_n^2$. For this case one is interested in the shape of the time-lagged coverance function, determined by measuring the signal fluctuations at each array element and computing the covariances for a set of time delays. Once again the wind velocity is calculated by inverting the integral that relates the time-lagged covariance to $C_n^2$ (Leuenberger et al., 1979; Shen, 1970). As for the case of $C_n^2$, attempts to calculate the average wind velocity have been more successful than those to calculate the distribution of wind velocity along the path.

### 3. Raindrop Size Distribution

In Eq. (20) the rain attenuation is a function of the integral of the drop size distribution $N(D)$. In principle, if the path attenuation is measured at a number of wavelengths it is possible to invert this integral and determine the path-averaged raindrop size distribution $\overline{N(D)}$. Furuhama and Ihara (1981) measured copolar attenuations at 11.5, 34.5, and 81.8 GHz and phase variations between the frequencies 11.5 and 34.5 GHz and then inferred $\overline{N(D)}$ from these data. They compared the results obtained using several different inversion techniques with actual drop size distributions measured with a distrometer and found that the best results were obtained if the drop diameters were considered to be distributed exponentially. The inferred average drop size distribution was in good agreement with distrometer measurements over the range of drop diameters 1.5–3.5 mm, but the number of smaller and larger drop sizes was grossly overestimated. The authors conclude from these initial results that this technique appears feasible for remotely sensing the path-averaged raindrop size distribution.

## B. RADIOMETRIC SENSING

In Eq. (5) the atmospheric emission (or sky brightness temperature $T_a$) is a function of the kinetic temperature of the atmosphere $T(s)$ and the absorption coefficient of the absorbing gases $\gamma(s)$; $\gamma(s)$ is in turn a function of temperature, pressure, water vapor pressure, wavelength, and elevation (or nadir) angle. If the distribution of an absorbing gas is known, remains reasonably constant, and is sufficiently abundant, as is the case for oxygen, then Eq. (5) can be inverted and in principle a temperature profile $T(s)$ can be inferred from a set of brightness temperature measurements as a function of wavelength (or elevation angle).

Remote sensing of the atmosphere may be conducted with a ground-based system directed upward (Hogg, 1980) or a satellite or aircraft system directed downward (Staelin, 1981). For the downward-directed system an additional surface contribution must be added to Eq. (5). This is usually negligible for temperature profiles but it is very important in water vapor and liquid water measurements. The procedure for retrieving meteorological data from atmospheric emission measurements is similar for both ground-based and satellite systems. A concept that enables the best understanding of the inversion technique is that of the weighting function. Equation (5) can be written in the form

$$T_a = \int T(h)W(h, v)\, dh, \tag{39}$$

where $W(h, v) = \gamma(h) \exp[-\int \gamma(h')\, dh']$. The kernel of this integral, $W(h, v)$, is called the weighting function because it represents the contribution from each height interval to the total emission. The height distribution of the weighting function is a function of frequency; thus by varying frequency one can sample the emission from different heights. Typical weighting functions for temperature sensors directed downward and upward are shown in Figs. 22 and 23, respectively (Miner, 1972; Waters et al., 1975). The closer the frequency is to the peak of the absorption line, the closer the layer from which the emission is emanating, is to the antenna, because emissions from layers distant from the antenna are absorbed before reaching the antenna. Whereas the downward-directed weighting functions peak at different heights, those directed upward all peak at the surface, because the density of oxygen decreases with height and, therefore, in the upward direction emission is always greatest at the antenna.

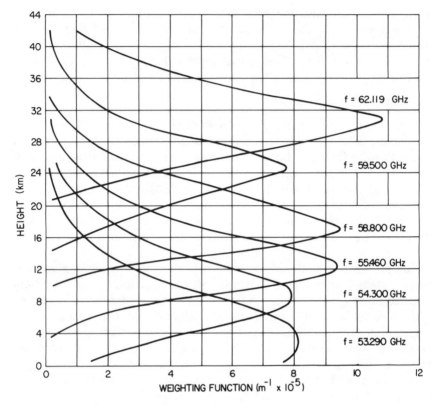

FIG. 22 Normalized temperature weighting functions for downward-directed temperature sensors for four frequencies (——) and the corresponding temperature profile (– – –). (From CCIR, 1982.)

Temperatures at discrete heights can be estimated from a set of brightness temperature measurements, because these temperatures are correlated with the temperature over the weighting function layer. Because the correlation is imperfect, the vertical resolution of the temperature profile is limited. There are, in addition, several uncertainties that further limit the accuracy of the inversion technique. First, this type of integral, often referred to as a Fredholm integral of the first kind, is unstable; that is, very small errors in the measurement of the brightness temperature, or in the accuracy with which the absorption coefficient of oxygen is known, can give rise to very large errors in the inferred temperature profile. Also, although wavelengths can be selected such that most of the absorption is due only to oxygen, contributions from water vapor and clouds further limit the accuracy of this technique. Methods for overcoming these limitations will be discussed later.

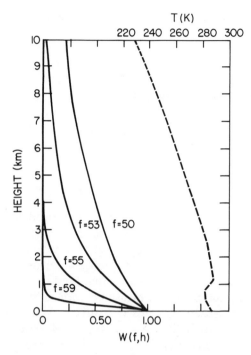

FIG. 23   Weighting functions for upward-directed temperature sensors (clear sky). (From Miner *et al.*, 1972. © American Geophysical Union.)

The possibility of inferring water vapor profiles from atmospheric emission measurements near the water vapor absorption line is complicated by the fact that the distribution of water vapor is not generally known, nor is it constant. Because of these uncertainties, water vapor profiles are only approximate; however, the integrated water vapor can be obtained more accurately. Because liquid water in clouds has different spectral characteristics than water vapor, it is possible to distinguish between liquid water and the integrated water vapor. Attempts have been made to map rainfall rates over the ocean using satellite radiometers, but so far they have not proved very successful (Viezee *et al.*, 1979; Wilheit *et al.*, 1977). The principal problem is a lack of spatial resolution owing to limitations on antenna size. Measurements over land are difficult because the antenna temperature measured by the radiometer is dominated by the brightness temperature of the land and is thus only slightly affected by the presence of rain. The resolution problem can be overcome by going to wavelengths on the order of 3 mm or even shorter. The measurement of rain over land presents a more serious problem; possibly a radar or IR channel may help to overcome this limitation.

To overcome the inversion limitations described previously, use is made of a priori data. The procedure usually followed is illustrated in Rosenkranz (1982), Staelin *et al.* (1976), Waters *et al.* (1975), Westwater (1965). It is often referred to as a statistical estimation technique that consists of a regression analysis of atmospheric temperature profiles and numerically calculated atmospheric emission. The atmospheric temperature, represented by the vector $T$, is related to the measured brightness temperatures $T_B$ by the matrix $D$:

$$T = D \cdot T_B. \tag{40}$$

The elements of the matrix are derived from a priori data. For example, a large set of previously measured radiosonde profiles that are representative of the region of interest are collected. Brightness temperatures for each of these profiles are then calculated for a set of frequencies. Using a least-squares regression approach, a $D$ matrix that produces the best agreement between the "true" temperatures measured with radiosondes and the inferred temperatures is derived. This $D$ matrix is then used to calculate a vertical temperature profile from a set of brightness temperature measurements. Different $D$ matrices may be used for different locations, seasons, etc. The complexity of the matrix may range from completely linear, with Gaussian statistics, to nonlinear. More sophisticated methods using spatial filtering concepts have also been employed (Rosenkranz, 1982).

### 1. Satellite Radiometry

Probably the most successful application to date of millimeter-wave remote sensing has been the determination of meteorological profiles from a satellite (Staelin, 1981). The groundwork for temperature sensing was provided by Meeks and Lilley (1963) when they calculated the weighting functions for oxygen. The idea of obtaining water vapor profiles was first proposed by Barrett and Chung (1962). These pioneering efforts were followed by measurements from a balloon (Barrett *et al.*, 1966), from an aircraft (Rosenkranz *et al.*, 1972), and finally from the Nimbus-5 and Nimbus-6 experimental satellites (Staelin *et al.*, 1977; Waters *et al.*, 1975). Millimeter-wave radiometers are now used on the Tiros-N and military DMSP operational satellites. It is expected that in the latter 1980s radiometers having as many as 15 or 20 channels will be employed.

*a. Temperature Profiles.* The caliber of temperature profiles that can be obtained from a satellite is illustrated by examining the results from the Nimbus-5 (Nimbus-E) Microwave Spectrometer (NEMS) (Waters *et al.*,

1975). A five-channel radiometer was employed; frequencies of 53.65, 54.9, and 58.8 GHz were used to sense the temperature, and frequencies of 22.235 and 31.4 GHz were used to sense water vapor and liquid water, respectively. Temperatures were retrieved for heights corresponding to pressure levels of 1000, 850, 700, 500, 400, 300, 250, 200, 150, and 100 mbars, so comparisons could be made between the inferred temperatures and those provided by the National Meteorological Center (NMC) obtained from radiosondes and commercial aircraft. Eighty-two pairs of temperature profiles obtained from NEMS and NMC were compared. The results show average temperature differences of 2.1 K for measurements conducted in June and 1.6 K for those conducted in December. These differences would probably have been slightly smaller if the effects of clouds had been taken into account (Staelin *et al.*, 1975). It should be emphasized that discrepancies are also likely to arise because the NEMS temperature is averaged over a surface area of approximately 200 × 300 km, whereas the radiosonde is a point measurement; also, the measurements were not taken at exactly the same time.

b. *Water Vapor and Liquid Water Content.* As mentioned previously, it is more difficult to infer water vapor profiles than temperature profiles from atmospheric emission measurements because of the spatial and temporal variations in the water vapor. However, approximate water vapor profiles and more accurate integrated water vapor abundances have been obtained. Although the 22.235-GHz water vapor resonance is the principal resonance below 50 GHz, it is relatively weak, and thus the total emission at that frequency can be influenced by liquid water emission. Cloud liquid water emission, however, is essentially nonresonant, so it can be distinguished from water vapor by a measurement off the water vapor line. The weakness of the water vapor line does, however, produce a serious problem, namely, the contribution from the earth's surface. Because land has an emissivity typically in the range 0.8–1.0, its brightness temperature is almost equal to its surface temperature. Thus, small contributions from the water vapor emission are dominated by the earth background and are not strong enough to be used for inferring water vapor data. For this reason water vapor and liquid water abundances can be inferred only from measurements over water that has an emissivity of about 0.45 at millimeter wavelengths and thus presents a cool background. The 183-GHz water vapor resonance may be better suited for measurements over land, because it is much stronger than the 22.235-GHz resonance. In addition, it has the potential of producing improved water vapor profiles, because it would result in better vertical resolution.

The state of the art of inferring water vapor and liquid water abundances using satellite radiometry, as for temperature, is best illustrated by reviewing the NEMS results (Staelin *et al.*, 1976). As mentioned previously, 22.235- and 31.4-GHz channels were used. The abundance of water vapor in the atmosphere was inferred from a linear combination of the brightness temperatures at the two frequencies. The coefficients of the brightness temperatures were obtained using a multidimensional regression analysis based on brightness temperatures computed for about 150 radiosonde profiles. The liquid water estimates were also based on a linear combination of the two brightness temperatures. The comparison of radiometric-derived and radiosonde water vapor and liquid water is difficult because they are not for identical regions, nor were they measured at exactly the same time.

With this limitation in mind, the accuracy of the water vapor estimates was tested by comparison with 31 sets of radiosonde data collected throughout the Northern Hemisphere during a 2-year period. It was concluded that integrated abundances of water vapor and liquid water had estimated NEMS accuracies of 0.2 and 0.01 g/cm², respectively. It is expected that improved results could be obtained by using additional frequencies.

## 2. *Ground-Based Radiometry*

From an applications standpoint, ground-based radiometric sensing has not progressed as rapidly as satellite radiometric sensing. Although, in principle, comparable meteorological profiles should be achievable with ground-based systems, particularly near the ground, they are not capable of readily obtaining global data. For temperature profiling there is a distinct difference between upward- and downward-directed systems, as was seen in Figs. 22 and 23. The upward-directed weighting functions are approximately exponential and all peak at the surface. For this reason the best accuracy is achieved at low altitudes and the accuracy gradually diminishes at higher altitudes. Temperature profiles are particularly important at very low altitudes because it is at these levels that temperature inversions often occur, and the ability to detect the presence of an inversion layer is very useful. Water vapor and liquid water ground-based systems are valuable because they are operable over land as well as water; this has not yet been achieved with satellite systems. Thus it is expected that ground-based radiometry will ultimately also be used for operational meteorological sensing.

a. *Temperature Profiles.* The procedure for obtaining temperature profiles with ground-based radiometers is very similar to that used for satellite radiometers. The state of the art of this technique can be illustrated by the controlled experiment conducted by Decker *et al.* (1978). Eighty-eight simultaneous rawinsonde and radiometer-derived profiles were measured under both clear and cloudy conditions at locations in Pt. Mugu,

California, and the Gulf of Alaska. A five-channel radiometer with temperature profiling frequencies of 52.85, 53.85, and 55.45 GHz and water vapor and liquid water frequencies of 22.235 and 31.65 GHz was utilized. Assuming the radiosonde temperatures are correct (not necessarily a valid assumption), the rms deviations between rawinsonde and radiometric-derived temperatures ranged from 1.1 K at heights within about 1 km of the surface to 2.8 K at a height of about 12 km. These accuracies compared well with the predicted accuracies and are considered acceptable for numerical weather prediction.

b. *Water Vapor and Liquid Water Content.* The same experiment was used to retrieve water vapor profiles and total precipitable water. The agreement between the rawinsonde and radiometric-derived water vapor profiles was not as good as predicted; it is believed that the uncertainty in the absorption coefficient of water vapor could be a significant source of error (Westwater, 1978). Although there was general agreement in the total precipitable waters of the two methods, the rms difference of 0.32 cm was about 0.1 cm larger than predicted. A problem that arises in inferring the liquid water content of a cloud is that a cloud temperature must be assumed and an error in this temperature can produce a large error in the estimated liquid water, which in turn produces an error in the vapor measurement. In an effort to obtain a more accurate integrated liquid water content, an absorption measurement was made in addition to the emission measurements (Snider *et al.,* 1980). The absorption was measured at a frequency of 28.56 GHz using the beacon on the COMSTAR satellite (Cox, 1978). By combining the absorption measurement with the brightness temperature, it is believed that a more accurate liquid water content can be obtained. Because it is very difficult to obtain independent liquid water contents of clouds, the expected improvement in utilizing an absorption measurement has not been confirmed. However, preliminary results seem to be consistent with the liquid contents that would be expected of certain types of clouds, so better accuracies are expected.

C. RADAR SENSING

In the earliest days of radar it was quickly recognized that some returns from the atmosphere were meteorologically induced, and since the early 1940s meteorological echoes have been classified and related to synoptic weather conditions. Most of the initial radar meteorological observations were conducted at longer wavelengths because of the availability of high-power radars; in addition, rain was of principal concern and centimeter wavelengths were optimal for rain studies. Later it was realized that cloud sensing could be more effectively accomplished at shorter wavelengths, because cloud droplets are typically less than 100 $\mu$ in diameter and are not easily detectable at centimeter wavelengths. This led to the development of the first millimeter-wave meteorological radar. The development of milli-meter-wave radars for cloud studies progressed rather slowly over the years, and only recently are newly designed millimeter-wave Doppler, dual-polar-ized radars being built to study cloud composition and dynamics.

Although ground-based millimeter-wave radars have not been used for rain studies because they are not as effective as longer-wavelength radars, they do have several features that make them attractive for space-borne precipitation measurements. The principal limitations imposed on space-borne radars are size and weight. To obtain a sufficiently small field of view and minimize ground clutter, very-high-resolution antennas are needed; these are available at millimeter wavelengths, and systems that would operate at these wavelengths are being planned.

Although millimeter-wave radars can be used to infer wind speed, they are not as effective as longer-wavelength systems because the returns are much weaker. Thus it is unlikely that they will be used for this application, unless perhaps the need arises for an airborne system for obtaining very-high-reso-lution maps of a localized region.

Finally, millimeter-wave radars have great potential in the area of ento-mology—the study of insects. The movements of insects have been the subject of much research, speculation, and controversy. More quantitative data on insect movement are desperately needed, and millimeter-wave radars are uniquely suited for this application.

In summation, it appears that the most promising applications for milli-meter-wave remote sensing radars are in the study of cloud dynamics and entomology. Space-borne millimeter-wave radars for sensing precipitation could prove useful when used in conjunction with longer-wavelength radars or millimeter-wave radiometers.

## 1. *Cloud Composition*

One of the earliest millimeter-wave radars was the AN TPQ-6, which was used for measuring the heights of cloud bases and tops (Donaldson, 1955). This ultimately evolved into the AN TPQ-11, an 8.6-mm-wavelength, pulse-modulated, two-dish radar capable of providing continuous records of cloud and precipitation height and density (Petrocchi and Paulsen, 1966). Information on wind shear, melting zone, and echo intensity in addition to cloud bases and tops and their trends could also be derived. Later a pseudo-noise coded Doppler 8.6-mm-wavelength radar was developed by Pasqualucci (1970) and used to measure drop size distributions in precipita-tion and vertical air velocity in both clouds and precipitation. A new

8.6-mm-wavelength Doppler radar with polarization diversity has been built (Pasqualucci, 1981) and is being used to study cloud microphysics, which includes liquid water content, ice crystal versus water drop concentrations, and air motion. More recently, it has been proposed that a Doppler phase-coherent radar be developed to operate at a frequency of 90 GHz (Lhermitte, 1981). An evaluation of the sensitivity characteristics of this radar indicates that it would have enormous potential for the study of clouds in their early stages of development.

## 2. Rain Distribution

Although ground-based rain sensing utilizing radar is generally best accomplished at centimeter wavelengths, a principal problem that arises when attempting this from space is that of antenna size (Atlas and Thiele, 1981). The cell sizes of rain are such that an instantaneous field of view of approximately a few kilometers is necessary. In addition, a narrow beam is needed so that effects of surface clutter can be minimized. For these reasons millimeter-wave radars are being considered for this application. If the radar were designed to operate at a wavelength of about 8 mm, then an antenna with a diameter of 3.4 m would provide 2-km resolution from an 800-km orbit. However, this system would have the limitation that accompanies attenuating radars, namely, the uncertainty that arises because the return signal is attenuated by the rain. This usually makes it necessary to assume that the rain rate is constant as a function of height, which is not necessarily the case.

It has been proposed that the SEASAT altimeter that operates at a wavelength of 2.2 cm (close to millimeter waves) be modified to measure rain rate (Goldhirsh and Walsh, 1982). This is considered a first-step approach, and although the assumption of a constant rain rate with height is considered a weak point, it is believed that meaningful rainfall data could be obtained with a very small investment. It would also provide an excellent opportunity to gain experience in using a millimeter-wave space-borne radar for measuring precipitation on a global scale.

As mentioned previously, there is some merit in considering a hybrid radar–radiometric system for this application. This combination would be particularly useful over the ocean, because the radiometer would provide data on the total attenuation and water content that would be very helpful in the interpretation of the radar returns. Measurements have been made with an airborne radar–radiometer system operating at 10 and 34.5 GHz (Yoshikado et al., 1981). This system was developed as a first step toward a satellite system. The aircraft measurements were compared with those from a ground-based C-band radar and the results were encouraging.

## 3. Wind Speed

Wind speed can be inferred from a measurement of the Doppler shift in the frequency of a scattered signal. This may be done with either a monostatic or a bistatic radar. Under clear sky conditions the transmitted signal is scattered by inhomogeneities in the refractive index structure; because these inhomogeneities are moving along with the wind, the Doppler shift produced by this motion is proportional to the wind speed. Under conditions of cloud or precipitation, the scattering is from particulates that are also moving along with the wind. With a monostatic radar, the horizontal components of the wind are generally measured by rotating the antenna in azimuth to obtain wind direction and varying the elevation angle to obtain wind height. The vertical wind component is measured with the antenna pointed toward zenith.

The bistatic system is not as versatile as the monostatic, because winds can be measured only in a fixed direction, unless either the transmitter or receiver is mobile or multiple transmitters and/or receivers are used. Another complication is that the vertical and horizontal components are not easily resolved. For the horizontal crosspath component the Doppler shift is zero when the common volume is at the center of the great circle path; it becomes positive when the common volume is moved to the side of the great circle path in the upwind direction and negative when the common volume is moved to the side in the downwind direction. Thus, the vertical component can be resolved only when the common volume is on the great circle path. The horizontal along-the-path component is very weak compared to the crosspath and therefore more difficult to measure.

Even though the bistatic system is much more limited than the monostatic system, because it can operate only in a fixed direction and does not have the resolution features of the monostatic radar, it has the distinct advantage of being a much more sensitive radar (Lammers, 1973); this is very important at millimeter wavelengths, where power is at a premium.

For the reasons stated, most of the radars that have been used as wind sensors have operated at longer wavelengths (Gage and Balsley, 1978), and the highest frequency at which wind sensing has been done is 15.7 GHz using a forward scatter system (Altshuler et al., 1968; Olsen and Lammers, 1973). At this time the only advantage in using millimeter waves for wind sensing seems to be the feature of being able to obtain higher resolution with smaller mobile antennas.

## 4. Insect Detection

In the early days of radar, unexplained returns from the clear air were classified as clutter or "angels" because these unwanted signals interfered with the detection of targets. Eventually, it was recognized that these sources of clutter were refractive index inhomogeneities and insects and birds (Atlas, 1959; Crawford, 1949; Hajovsky et al., 1966; Hardy and Katz, 1969). Before long, available radars were used to study insects (Frost, 1971; Glover et al., 1966; Richter et al., 1973; Riley, 1973). Radar has been considered an important tool in the field of entomology (Eastwood, 1967), and a workshop has been held to investigate the potential of radar for insect population ecology and pest management (Vaughn et al., 1979).

As in most emerging fields, existing radars designed for other applications have been used in entomology. The potential of radar for use in entomology now appears so important that the need for optimally designed ground-based and air-borne radars has been substantiated, and millimeter-wave radars are expected to play a key role in this application. Because the dielectric properties of insects depend mostly on body moisture, the maximum target cross-section should occur when the wavelength is comparable to the circumference of the insect. Many insects have sizes on the order of millimeters, so millimeter radars should be optimal, particularly for tracking single insects at short ranges. Longer-wavelength radars could still be used for tracking swarms of insects at long ranges. Because the operational aspects of pest management may encompass regions of millions of square miles in extent, air-borne radars will be needed and millimeter-wave systems have desirable characteristics for this application.

## V. Conclusion

The interaction of millimeter waves with atmospheric gases and particulates has been examined. Precipitation in general and rain in particular limit the performance of longer millimeter-wave systems. Systems operating at short millimeter wavelengths are significantly affected by high water vapor absorption in addition to precipitation, so applications in this region of the spectrum will of necessity be limited to very short paths. In the area of remote sensing, most applications will also use longer millimeter wavelengths.

The use of millimeter waves has probably not progressed as rapidly as had been originally anticipated. For many years, the more optimistically inclined envisioned millimeter waves revolutionizing the traditionally longer-wavelength communications and radar systems and traditionally shorter-wavelength remote sensing systems. When the discovery of the laser created a temporary lull in millimeter-wave research, the more pessimistically inclined feared that millimeter waves had passed from infancy to obsolescence without having experienced a period of fruitfulness. Finally, there were realists who recognized that cost is a major consideration and that millimeter-wave systems would reach the marketplace only when they could be shown to be competitive with systems that operate at longer or shorter wavelengths or to have unique properties such that needed applications could be realized only with millimeter waves. So far, history seems to be supporting the realists.

## REFERENCES

Altshuler, E. E. (1968). *Microwave J.* **11**, 38–42.
Altshuler, E. E. (1971). AFCRL Report 71-0419, AD73 1170, National Technical Information Service, Springfield, Virginia.
Altshuler, E. E., and Mano, K. (1982). Rpt. No. RADC-TR-82-7, National Technical Information Service, Springfield, Virginia.
Altshuler, E. E., and Telford, L. E. (1980). *Radio Sci.* **15**, 781–796.
Altshuler, E. E., Lammers, U. H. W., Day, J. W., McCormick, K. S. (1968). *Proc. IEEE* **56**, 1728–1731.
Altshuler, E. E., Gallop, M. A., and Telford, L. E. (1978). *Radio Sci.* **13**, 839–852.
Atlas, D. (1959). *J. Atmos. Terr. Phys.* **15**, 262–287.
Atlas, D., Kerker, M., and Hitschfeld, W. (1953). *J. Atmos. Terr. Phys.* **3**, 108–119.
Atlas, D., and Thiele, O. W. (1981). NASA Workshop Rpt. Precipitation Measurements from Space, Goddard Lab. for Atmos. Sci., Greenbelt, Maryland.
Barrett, A. H., and Chung, V. K. (1962). *J. Geophys. Res.* **67**, 4259–4266.
Barrett, A. H., Kuiper, J. W., and Lenoir, W. B. (1966). *J. Geophys. Res.* **71**, 4723–4734.
Battan, L. J. (1973). "Radar Observations of the Atmosphere." Univ. of Chicago Press, Chicago, Illinois.
Bean, B. R., and Dutton, E. J. (1968). "Radio Meteorology." Dover, New York.
Blomquist, A., and Norbury, J. R. (1979). *Alta Freq.* **66**, 185–190.
Booker, H. G., and Gordon, W. E. (1950). *Proc. IRE* **36**, 401–412.
Brussard, G. (1976). *IEEE Trans. Antennas Propag.* **24**, 5–11.
CCIR (1978). "Propagation Data Required for Transhorizon Radio Relay Systems." Rpt. 238-3, **V**, pp. 199–215.
CCIR (1982a). "Attenuation by Atmospheric Gases." Rpt. 719 (Mod I), **V**.
CCIR (1982b). "Attenuation and Scattering by Precipitation and Other Atmospheric Particles." Rpt. 721 (Mod I), **V**.
Chu, T. S. (1979). *Bell Syst. Tech. J.* **58**, 549–555.
Cox, D. C. (1978). *Bell Syst. Tech. J.* **57**, 1231–1255.
Crane, R. K. (1973). Tech. Rpt. 498, Lincoln Lab, MIT, Lexington, Massachusetts.
Crane, R. K. (1980a). ERT-Doc. P-A502, Environ. Res. and Technol., Inc., Concord, Massachusetts.
Crane, R. K. (1980b). *IEEE Trans. Commun.* **28**, 1717–1733.
Crane, R. K. (1981). *Radio Sci.* **16**, 649–669.
Crawford, A. B. (1949). *Proc. IRE* **37**, 404–405.
Davies, P. G. (1973). *IEE Conf. Publ.* **98**, 141–149.
Decker, M. T., Westwater, E. R., and Guirand, F. O. (1978). *J. Appl. Meteorol.* **17**, 1788–1795.
Donaldson, R. J. Jr. (1955). *J. Meteorol.* **12**, 238–244.
Eastwood, E. (1967). "Radar Ornithology." Methuen, London.
Eldridge, R. G. (1966). *J. Atmos. Sci.* **23**, 605–613.
Elgered, G. K. (1982). *IEEE Trans. Antennas Propag.* **30**, 502–505.

Falcone, V. J., Abreu, L. W., and Shettle, E. P. (1979). Rpt. No. AFGL-TR-79-0253, National Technical Information Service, Springfield, Virginia.

Fante, R. L. (1980). *Proc. IEEE* **68**, 1424–1442.

Friis, H. T., Crawford, A. B., and Hogg, D. C. (1957). *Bell Syst. Tech. J.* **36**, 627–644.

Frost, E. (1971). *Proc. Seventh Int. Symp. Remote Sensing Environ.* **7**, 1909–1915.

Furuhama, Y., and Ihara, T. (1981). *IEEE Trans. Antennas Propag.* **29**, 275–281.

Gage, K. S., and Balsley, B. B. (1978). *Bull. Am. Meteorol. Soc.* **59**, 1074–1093.

Gallagher, J. J., McMillan, R. W., Rogans, R. C., and Snider, D. E. (1980). U.S. Army Missile Command Rpt., RR-80-3, Redstone Arsenal, AL 35809, 100–108.

Gallop, M. A., Jr., and Telford, L. E. (1975). *Radio Sci.* **11**, 935–945.

Glover, K. M., Hardy, K. R., Konrad, T. G., Sullivan, W. N., and Michaels, A. S. (1966). *Science* **154**, 967–972.

Goldhirsh, J. (1979). *IEEE Trans. Geosci. Electron.* **17**, 218–239.

Goldhirsh, J., and Walsh, E. J. (1982). *IEEE Trans. Antennas Propag.* **30**, 726–733.

Gough, M. W. (1968). *IEE Conf. Publ.* **48**, 93–100.

Haakinson, E. J., Violette, E. J., and Hufford, G. A. (1980). NTIA Report 80-35, U.S. Dept. of Commerce, NTIA.

Hajovsky, R. G., Deam, A. P., and LaGrone, A. H. (1966). *IEEE Trans. Antennas Propag.* **14**, 224–227.

Hall, M. P. M. (1979). "Effects of the Troposphere on Radio Communication." Peregrinus Ltd., New York.

Hardy, K. R., and Katz, I. (1969). *Proc. IEEE* **57**, 468–480.

Hayes, D. H., Lammers, U. H. W., Marr, R. A., and McNally, J. J. (1979). *EASCON Rec.* **2**, 362–368.

Hogg, D. C. (1980). *Trans. Antennas Propag.* **28**, 281–283.

Hogg, D. C., and Chu, T. S. (1975). *Proc. IEEE* **63**, 1308–1331.

Hooke, W. H., and Hardy, K. R. (1975). *J. Appl. Meteorl.* **14**, 31–38.

Ippolito, L. J. (1981). *Proc. IEEE* **69**, 697–727.

Ishimaru, A., and Cheung, R. L. T. (1980). *Radio Sci.* **15**, 507–516.

Lammers, U. H. W. (1973). *Radio Sci.* **8**, 89–101.

Leuenberger, K., Lee, R. W., Waterman, A. T. (1979). *Radio Sci.* **14**, 781–791.

Lhermitte, R. (1981). *20th Conf. Radio Meteorol.* **20**, 744–748.

Lo, L., Fannin, B. M., Straiton, A. W. (1975). *IEEE Trans. Antennas Propag.* **23**, 782–786.

Mano, K., and Altshuler, E. E. (1981). *Radio Sci.* **16**, 191–195.

McCormick, K. S. (1972). *IEEE Trans. Antennas Propag.* **20**, 747–755.

Medhurst, R. G. (1965). *IEEE Trans. Antennas Propag.* **13**, 550–563.

Meeks, M. L., and Lilley, A. E. (1963). *J. Geophys. Res.* **68**, 1683–1703.

Mie, G. (1908). *Ann. Phys.* **25**, 377–445.

Miner, G. F., Thornton, D. D., and Welch, W. J. (1972). *J. Geophys. Res.* **77**, 975–991.

Olsen, R. L. (1981). *Radio Sci.* **16**, 631–647.

Olsen, R. L., and Lammers, U. H. (1973). *IEE Conf. Prop. Radio Waves 10 GHz* **98**, 197–201.

Olsen, R. L., Rogers, D. V., and Hodge, D. E. (1978). *IEEE Trans. Antennas Propag.* **26**, 318–329.

Parl, S. A., and Malaga, A. (1982). Personal communication, Signatron, Inc., Lexington, Massachusetts.

Pasqualucci, F. (1970). *14th Conf. Radar Meteorol.* **14**, 452–455.

Pasqualucci, F. (1981). *20th Conf. Radar Meteorol.* **20**, 646–648.

Petrocchi, P. J., and Paulsen, W. H. (1966). *12th Conf. Radar Meteorol.* **12**, 467–472.

Platt, C. (1970). *J. Atmos. Sci.* **27**, 421-425.
Povejsil, D. J., Raven, R. S., and Waterman, P. (1961). "Airborne Radar." Van Nostrand, Princeton, New Jersey.
Pruppacher, H. R., and Pitter, R. L. (1971). *J. Atmos. Sci.* **28**, 86-94.
Rafuse, R. P. (1981). Rpt. No. ESD-TR-81-290, Lincoln Lab., Lexington, MA.
Richter, J. H., *et al.* (1973). *Science* **180**, 1176-1178.
Riley, J. R. (1973). *Proc. IEE* **120**, 1229-1232.
Rooryck, M., and Battesti, J. (1975). Rpt. No. SP-113, European Space Agency, Paris, France, pp. 217-227.
Rosenkranz, P. W. (1982). *Radio Sci.* **17**, 245-267.
Rosenkranz, P. W., *et al.* (1972). *J. Geophys. Res.* **77**, 5833-5844.
Rozenberg, V. I. (1972). "Scattering and Attenuation of Electromagnetic Radiation Particles." Hydrometeorological Press, Leningrad, USSR.
Rummler, W. D. (1979). *Bell Syst. Tech. J.* **58**, 1037-1071.
Ryde, J. W., and Ryde D. (1945). General Electricity Company Rpt. No. 8670, Wembley, England.
Sandberg, J. (1980). *IEEE Trans. Antennas Propag.* **28**, 743-750.
Schaper, L. W., Jr., Staelin, D. H., and Waters, J. W. (1970). *Proc. IEEE* **58**, 272-273.
Shen, L. C. (1970). *Trans. Antennas Propag.* **18**, 493-497.
Skolnik, M. I. (1970). NRL Rpt. 2159, NRL, Washington, D.C.
Slobin, S. D. (1982). *Radio Sci.* **17**, 1443-1454.
Smith, E. K. (1982). *Radio Sci.* **17**, 1455-1464.
Smith, E. K., and Weintraub, S. (1953). *Proc. IRE.* **41**, 1035-1037.
Snider, J. B., Burdick, H. M., and Hogg, D. C. (1980). *Radio Sci.* **15**, 683-693.
Staelin, D. H. (1981). *Trans. Antennas Propag.* **29**, 683-687.
Staelin, D. H., *et al.* (1975). *J. Atmos. Sci.* **32**, 1970-1976.
Staelin, D. H., Kunzi, K. F., Pettyjohn, R. L., Poon, R. K. L. and Wilcox, R. W. (1976). *J. Appl. Meteorol.* **15**, 1204-1214.
Staelin, D. H., Rosenkranz, P. W., Borath, F. T., Johnston, E. J., and Waters, J. W. (1977). *Science* **197**, 991-993.
Stephanson, E. T. (1981). *Radio Sci.* **16**, 608-629.
Stewart, D. A. (1980). Rpt. RR-8D-3, Proc. of Workshop on Millimeter and Submillimeter Atmospheric Propagation Applicable to Radar and Missile Systems, U.S. Army Missile Command, Redstone Arsenal, AL, 77-82.
Strickland, J. I. (1974). *J. Rech. Atmos.* **8**, 347-358.
Strobehn, J. W. (1968). *Proc. IEEE* **56**, 1301-1318.
URSI Commission F Working Paper (1981). Radio Sci. **16**, 825-829.
Van de Hulst, H. C. (1957). "Light Scattering by Small Particles." Wiley, New York.
Vander Vorst, A. (1979). *Alta Freq.* **48**, 201-209.
Vaughn, C. R., Wolf, W., and Klassen, W. (1979). Radar, Insect Population Ecology, and Pest Management, NASA CP-2070, Wallops Flight Center, Wallops Island, Virginia.
Viezee, W., Shigeishu, H., and Chang, A. T. C. (1979). *J. Appl. Meteorol.* **18**, 1151-1157.
Waterman, A. T., Jr. (1958). *Proc. IRE* **46**, 1842-1848.
Waters, J. W. (1976). "Methods of Experimental Physics." **12B**, Chapter 23. Academic Press, New York.
Waters, J. W., Kunzi, K. F., Pettyjohn, R. L., Poor, R. K. L., and Staelin, D. H. (1975). *J. Atmos. Sci.* **32**, 1953-1969.
Westwater, E. R. (1965). *Radio Sci. J. Res.* NBS/USNC-URSI **69D**, 1201-1211.
Westwater, E. R. (1976). ESSA Tech. Rpt. IER 30-ITSA 30, National Technical Information Service, Springfield, Virginia.

Westwater, E. R. (1978). *Radio Sci.* **13**, 677–785.
Wilheit, T. T., Chang, M. S. V., Rao, E. B., Rogers, E. B., and Tyson, J. S. (1977). *J. Appl. Meteorol.* **16**, 551–560.
Wilson, R. W., (1969). *Bell Syst. Tech. J.* **48**, 1383–1404.
Wu, S. C. (1979). *IEEE Trans. Antenna Propag.* **27**, 233–239.
Wulfsberg, K. N. (1967). *Radio Sci.* **2**, 319–324.
Yoshikado, S., et al. (1981). *20th Conf. Radar Meteorol.* **20**, 287–294.
Zavody, A. M. (1974). *Proc. IEE* **121**, 257–263.

CHAPTER 5

# Technology of Large Radio Telescopes for Millimeter and Submillimeter Wavelengths

*J. W. M. Baars*

*Max-Planck-Institut für Radioastronomie*
*Bonn, Germany*

## I. Introduction

For the study of the physical processes in the universe the astronomer is totally dependent on the information contained in the electromagnetic radiation that accompanies physical processes. For centuries all information was gathered in the one-octave-wide optical window near the 0.5-$\mu$m wavelength. After the accidental discovery of radio radiation from our galaxy by Jansky in 1932 and the technical developments of the Second World War, radio astronomy quickly developed as an important branch of astronomy and astrophysics. The space era enabled detection of radiation for which the terrestrial atmosphere is totally opaque: infrared, ultraviolet, x-ray, and $\gamma$-ray observations added much to our knowledge of the universe.

The millimeter-wavelength region of the spectrum (30 – 300 GHz) has been opened up in the last dozen years, but the submillimeter-wavelength region (300 – 3000 GHz) is essentially unexplored. There are serious technical and operational difficulties for the development of millimeter-wavelength astronomy. First, the troposphere is only partially transparent in certain "windows" between the absorption bands of oxygen and water vapor. Good observation conditions are restricted principally to high and dry mountain sites. Second, the necessary technology of sensitive receivers and highly accurate reflector antennas is in an unsatisfactory state and requires state-of-the-art developments.

The detection of spectral-line radiation of the molecules of water vapor and ammonia in interstellar clouds in 1967 (Cheung *et al.*, 1968, 1969) at the rather long millimeter wavelength of 13 mm led astronomers to search for other molecules in the galaxy. Several new molecules were found with the millimeter telescope of the National Radio Astronomy Observatory (NRAO) at Kitt Peak, completed in 1968. The importance of the millimeter region for molecular astronomy, and consequently for the physics and chemistry of the interstellar medium, was established.

At the same time the need for larger and more accurate telescopes and for sensitive receivers was also clear. As had happened earlier for centimeter wavelengths, radio astronomers themselves became active in the development of the necessary technology. This has culminated at the present time in the advanced planning, construction, and early operation of several large telescopes for millimeter and submillimeter wavelengths.

For the purpose of this review we define a telescope to be *large* if at the wavelength of maximum gain $(\lambda_m)$ the reflector diameter $D \gtrsim 10^4 \lambda_m$. Following Ruze (1966), this implies a maximum gain of more than 83 dB, which occurs at $\lambda_m = 4\pi\varepsilon$, where $\varepsilon$ is the root-mean-square (rms) reflector surface deviation from the prescribed geometrical shape. The *quality factor* or *precision* of the reflector is now $D/\varepsilon \gtrsim 1.25 \times 10^5$ and the antenna beamwidth at $\lambda_m$ is $\theta_A \lesssim 25$ sec$^{-1}$.

This definition is rather stringent. For instance, the 100-m telescope at Effelsberg, which has the largest steerable dish in the world, barely meets our mark with its $\varepsilon = 0.6$ mm over the 75-m inner diameter. In Table I we have collected pertinent data on the most important existing and planned millimeter telescopes. The only existing telescope that is large according to our definition is the Caltech antenna (No. 11). All the new instruments (Nos. 12–19) easily meet our criterion. The special problems that must be solved in attaining such extreme specifications are the subject of this chapter. Examples will be taken from the new designs, mainly from the Max-Planck-Institut für Radioastronomie (MPIfR) 30-m millimeter telescope, the NRAO 25-m antenna, and the Caltech submillimeter telescope. This preference is based on two facts: first, these are the largest telescopes for the millimeter and submillimeter regions and, second, the MPIfR and Caltech telescopes are in an advanced stage of construction, so that actual achievements rather than design specifications can be presented.

The special problems of millimeter-wave astronomy are summarized in Fig. 1. Figure 1a indicates the atmospheric transmissivity, which is adequate

TABLE I

The Major Millimeter Telescopes[a]

| Institute | Location | Year | Height (m) | D (m) | ε (mm) | D/ε ($10^3$) | $\lambda_{min}$ (mm) | Remarks |
|---|---|---|---|---|---|---|---|---|
| 1. Aerospace Corp. | Los Angeles, CA | 1963 | 10 | 4.6 | 0.08 | 58 | 1.0 | open; source variability |
| 2. Crimean Observatory | Crimea, USSR | 1966 | 10 | 22.0 | 0.30 | 73 | 3.8 | open |
| 3. Univ. of Texas | McDonald Observatory | 1967 | 2030 | 4.9 | 0.08 | 61 | 1.0 | astrodome |
| 4. U.C. Berkeley | Hat Creek Observatory | 1968 | 1040 | 6.1 | 0.13 | 46 | 1.6 | open; $NH_3$, $H_2O$ |
| 5. NRAO | Kitt Peak Observatory | 1969 | 1940 | 11.0 | 0.14 | 78 | 1.8 | astrodome; many molecules (CO) |
| 6. CRAAM, Brazil | Itapetinga | 1972 | 800 | 13.7 | 0.34 | 40 | 4.3 | radome; $H_2O$ (southern hemisphere) |
| 7. Chalmers Univ. | Onsala, Sweden | 1976 | 10 | 20.0 | 0.18 | 110 | 2.3 | radome |
| 8. CSIRO, Australia | Sydney | 1977 | 10 | 4.0 | 0.09 | 44 | 1.2 | open |
| 9. Bell Laboratories | Holmdel, NJ | 1977 | 115 | 7.0 | 0.10 | 70 | 1.3 | open; offset paraboloid |
| 10. Univ. of Massachusetts | Amherst, MA | 1978 | 300 | 13.7 | 0.15 | 91 | 1.9 | radome |
| 11. Caltech | Owens Valley, CA | 1979 | 1220 | 10.4 | 0.035 | 300 | 0.5 | open |
| 12. Tokyo Observatory | Nobeyama, Japan | 1982 | 1350 | 45.0 | 0.2 | 225 | 2.5 | open |
| 13. MPIfR, Germany/ IRAM, Grenoble | Pico Veleta, Spain | 1983 | 2870 | 30.0 | 0.09 | 335 | 1.1 | open; operation by IRAM |
| 14. Caltech | Mauna Kea Observatory | 1985 (?) | 4000 | 10.4 | 0.015 | 690 | 0.2 | astrodome; advanced planning |
| 15. IRAM, Grenoble | Plateau de Bure, France | 1984/85 | 2500 | 15.0 | 0.05 | 300 | 0.6 | open; advanced design stage |
| 16. SERC, England/ ZWO, Netherlands | Mauna Kea Observatory | 1986 | 4000 | 15.0 | 0.05 | 300 | 0.6 | astrodome; design stage |
| 17. NRAO | Mauna Kea Observatory | — | 4080 | 25.0 | 0.07 | 355 | 0.9 | astrodome; project cancelled |
| 18. Astronomy and Space Research Center | Mt. Korek, Iraq | 1985 | 2100 | 30.0 | 0.09 | 335 | 1.1 | open; copy of No. 13 |
| 19. NRAO | Kitt Peak Observatory | 1983 | 1940 | 12.0 | 0.07 | 170 | 0.9 | astrodome; No. 5 with new reflector |

[a] $D$ is the reflector diameter, $\epsilon$ the rms reflector inaccuracy; $\lambda_{min} = 4\pi\epsilon$. References: Nos. 1–5, Cogdell et al. (1970), and the references therein; No. 6, Kaufmann et al. (1976); No. 7, Menzel (1976); No. 8, Eschenauer et al. (1977), Gardner et al. (1978); No. 9, Chu et al. (1978); Nos. 11, 14, Leighton (1978); No. 12, Nobeyama Radio Observatory Staff (1981); No. 13, Baars and Hooghoudt (1980), Baars (1981); No. 16, Shenton and Hills (1976); No. 17, NRAO Staff (1975, 1977).

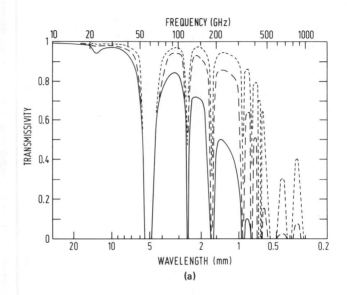

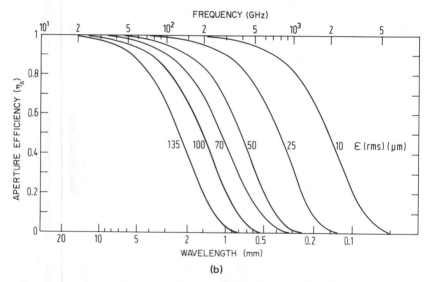

FIG. 1   (a) Atmospheric transmissivity vs frequency and wavelength for three values of the column height of precipitable water vapor: ——, 15 mm; — —, 3.3 mm; - - -, 1.3 mm. (b) Relative aperture efficiency $\eta_A$ vs frequency and wavelength for several values of the rms deviation $\varepsilon$ of the reflector surface from a perfect paraboloid.

only in certain "windows" of the spectrum. Because most of the absorption lines are of water vapor (those at 60 and 118 GHz are due to oxygen), much is to be gained by placing the telescope at a site with a small amount of precipitable water vapor. The three curves are valid for amounts of 15, 3.3, and 1.3 mm, which are representative for sites at sea level, at an altitude of 2–3 km, and at an altitude of 3–4 km, respectively. For a sufficiently low value of water vapor ($\lesssim 1$ mm), observations are feasible in the window near 350 $\mu$m (850 GHz). At lower wavelengths the atmosphere is totally opaque; observations are possible only from above the tropopause. We shall not discuss site problems in this chapter. It should be noted that only two telescopes are located at an altitude of about 2000 m (Nos. 3 and 5 in Table I). Most of the new large instruments are located well above 2000 m, which will add significantly to their usefulness.

The set of curves in Fig. 1b shows the relative aperture efficiency (the normalized relative absorption area of the reflector) as a function of frequency for several values of the rms deviation $\varepsilon$ of the reflector surface from a perfect paraboloid (Ruze, 1966). It is customary to consider $\lambda_{min} = 16\varepsilon$ as the minimum wavelength of the telescope. Normally, however, the beam shape and sidelobe level are acceptable to an efficiency of $\approx 0.3$; hence in Table I $\lambda_{min} = 4\pi\varepsilon$, at which maximum gain is reached. Thus, the telescope must have $\varepsilon < 100\ \mu$m for adequate performance in the 200–300-GHz window. For use up to 500 GHz we require $\varepsilon \lesssim 50\ \mu$m, and observations in the extreme window at 350 $\mu$m need a reflector with $\varepsilon = 10$–20 $\mu$m. Both Caltech and MPIfR intend to achieve this goal with a 10-m antenna.

## II. Natural Limits

The major limitation on the highest frequency at which a telescope may be used is the deformation of the structure under the influence of gravity, windforces, and differential thermal expansion. These result in a deviation of the reflector surface from the prescribed contour, which leads to a loss in aperture efficiency, and a variation in the pointing direction of the telescope, which leads to a loss in signal and/or an incorrectly derived source position.

von Hoerner (1967a, 1975) has identified several *natural limits* on the largest dimension of a telescope of a certain precision. These limits are indicated in Fig. 2, a plot of precision vs diameter. The stress limit at which the structure collapses under its own weight lies at about 600 m for the usual materials such as steel, aluminum, and concrete. Any material with a finite stiffness will deform under the influence of gravity. This gravitational deformation $\Delta_g$ is proportional to the mass times the length of the structural member divided by its stiffness. For the average structural member in the

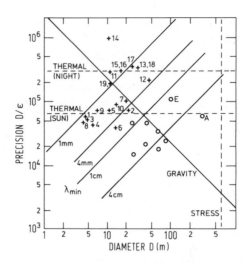

FIG. 2  Reflector precision vs diameter, with natural limits for stress, gravity, and thermals. The numbered crosses refer to the telescopes listed in Table I; the open circles represent centimeter-wavelength telescopes. E, Effelsberg telescope; A, Arecibo telescope.

reflector support structure, the cross section is proportional to its length, i.e., to the diameter of the reflector, $D$. Thus,

$$\Delta_g \propto (\rho/E)\, D^2, \tag{1}$$

where $\rho$ and $E$ are the density and Young's modulus of the material, respectively. For steel and aluminum, von Hoerner finds for the minimum wavelength in the gravitational limit:

$$\lambda_g \approx 70\,(D/100)^2, \tag{2}$$

where $\lambda_g$ is in millimeters and $D$ in meters. This limit applies to a classical structural design in which the stiffness is made large enough for the specified minimum wavelength. Only special tricks in design will make it possible to surpass this limit. The telescopes of Table I, along with several older instruments for longer wavelengths, are indicated in Fig. 2. The Arecibo telescope (A) can surpass the limit because its reflector surface is fixed with respect to the gravity vector. The fully steerable altazimuth-mounted 100-m Effelsberg telescope (E) achieves a precision five times greater than the gravitational limit, as the result of a basic design trick called *homology design,* which will be discussed in Section III. The new millimeter telescopes (Nos. 11–19 in Table I) lie well above the limit through the application of homology design.

The third natural limit is caused by deformations resulting from temperature differences in the structure. The thermal deformation $\Delta_t$ is proportional to the thermal coefficient of expansion $C_t$, the temperature difference $\Delta T$, and the length of the member:

$$\Delta_t \propto C_t D \, \Delta T. \qquad (3)$$

For steel the minimum wavelength in the thermal limit is

$$\lambda_t \simeq 6 \, \Delta T \, (D/100), \qquad (4)$$

where $\lambda_t$ is in millimeters, $T$ in degrees Kelvin, and $D$ in meters. For aluminum it is about twice as large.

The two cases indicated in Fig. 2 are for temperature differences of 4 K (sun) and 1 K (night). In the former case, thermal effects determine the limiting precision for $D \lesssim 40$ m; in the latter case, the thermal effect is more serious than the gravitational effect for $D \lesssim 12$ m. The values for the precision of the telescopes in Fig. 2 are either design values that disregard thermal effects or actual values measured at night. Several of the millimeter telescopes are clearly limited by the daytime temperature differences. If the new designs (Nos. 11–19 in Table I) are to achieve their design precision, tight control of the temperature uniformity of the structure will be needed. This problem will be treated in greater detail in Section IV.

It appears that the *large* telescopes for millimeter and submillimeter wavelengths considerably surpass the natural limits. We shall now review the methods by which this is achieved [see also Baars (1980)].

## III. Structural Design—Homology

The task of the structure of a telescope is to support both a reflector surface in a certain prescribed geometrical shape (normally paraboloidal) and a receiver system at the focal point of the reflector. The structure must have the strength and stiffness to do this for all azimuth and elevation angles and under the defined operating conditions regarding temperature and wind influences. Also, the specified performance must be accomplished in the most economical way, which, for nonexotic structures and/or materials, implies minimum weight. Finally, the structure must be laid out so that it survives the most extreme environmental conditions undamaged.

In a conventional design, the stiffness of the support structure is made high enough to keep the reflector paraboloidal to within a prescribed small fraction of the shortest operating wavelength for all attitude angles of the telescope. From this design method results the gravitational limit discussed in Section II [Eq. (2)]. Many telescopes designed along these lines have performed better than expected on the basis of structural stiffness. The reason is that the *best-fit paraboloid,* which is automatically found by optimizing the position of the receiver in the focal region, is better than the real deformations would suggest. The ultimate consequence of this effect was drawn by von Hoerner, who in 1965 proposed that the stiffness and geometry of the structural design be chosen so that, despite large absolute deformations, the best-fit paraboloid would remain of similar quality for all telescope angles. He coined the term *homologous deformations,* or *homology,* for a support structure that transforms the paraboloidal surface into other, equally well-fitting paraboloids when the elevation angle changes (von Hoerner, 1967a, 1969). Because the structure deflects, the parameters of the paraboloid will change; i.e., the focal length and the direction of the radio axis are changing. For each elevation angle, the optimum focus must be determined and the feed must be moved to that position.

Homology is achieved if all support points for the reflector panels possess equal stiffness, or, more appropriately, *equal softness.* The structural design problem is now that of finding a practical and economical structure between the many *soft* support points for the panels and the *hard* points where the whole structure is supported at the elevation bearings.

Von Hoerner (1967b) showed that an analytic solution to the problem exists and described a computer program that guides itself by iteration to a homologous structure. The procedure is described in general terms by von Hoerner (1977a). We shall not treat the method in detail here, but only comment on its practical realization. The requirement of equal softness points naturally to a structure of high symmetry and homogeneity. In a computer analysis only a section of a structure with a certain symmetry need be analyzed. This is indeed the case with the early homologous designs. Present-day computers and structural design programs such as ANTRAS and STRUDL can handle structures with hardly any symmetry requirements, allowing the analysis of one complete half of the reflector structure (which would almost always be sufficient).

Theoretically the homology calculation can result in a perfect structure, but practical constraints will impose limitations. A computational *homology imperfection* results from the following causes.

(1)   The ideal topology may be incompatible with the practical realization of the space frame.

(2)   Bar areas of the structural members will be chosen from the standard supply as offered by industry.

(3)   The stiffness of and tensions in the joints and other welded parts are difficult to determine and to account for in the computation.

In addition, deviation from homology is caused by the following *fabrication tolerances.*

(1)   The structural members will have a certain fabrication tolerance in area.

(2)   In the assembly a certain tolerance in the coordinates of the joints will occur.

Every homology design should include an estimation of the effects of these material and manufacturing tolerances.

The first telescope to employ the homology concept was the 100-m telescope of the Max-Planck-Institut für Radioastronomie, Bonn (Hachenberg *et al.*, 1973). For this telescope the design team of Krupp and M. A. N. did not use the iterative matrix procedure of von Hoerner, but applied a method of "trial–inspection of result–change of structure," using the standard structural programs. The success of the design is evident from Fig. 2. The telescope has been used with good performance at $\lambda = 7$ mm (Altenhoff *et al.*, 1980).

As examples of what can be achieved in the structural design of reflector antennas we choose three instruments of quite different appearance.

(1)   The proposed NRAO 25-m millimeter telescope (NRAO, 1975 and 1977). This instrument is based on an optimized, fully homologous structure of low weight. A perspective view is presented in Fig. 3. The rotationally symmetric reflector support structure is connected to an octahedron, which forms the basic structure between the reflector structure and the two elevation bearings. Computer optimization of this structure was extensive. The structural deformations in this telescope are presented in Table II. The main contribution to deviation from perfect homology is from the tolerance in the materials and misalignment, weight, and stiffness of the joints.

(2)   The 10-m submillimeter telescope of Caltech (Leighton, 1978). This structure is a tubular framework of vertical posts and connecting struts constructed in the form of a lattice of equilateral triangles (Figs. 4 and 11). It has a high degree of symmetry and redundancy, and therefore behaves quite homologously, although the structural calculation did not specifically use homology procedures. The joints are designed so that all strut forces pass through a point, thereby avoiding bending moments and resulting deformations. The fabrication tolerance is very small, and the main contribution to the deformation is due to the imperfection of the homology. The deformations in Table II do not include those of the mounting plate and the connection to the elevation bearings, but these are expected to be very small.

(3)   The 30-m millimeter telescope of MPIfR/IRAM (Baars and Hooghoudt, 1980; Baars, 1981). In the design of this telescope extreme symmetry was not used. A strict homology design procedure was not followed; rather, the structure was iteratively optimized to behave in a homologous fashion using the programs ANTRAS and STRUDL. A model of the elevation

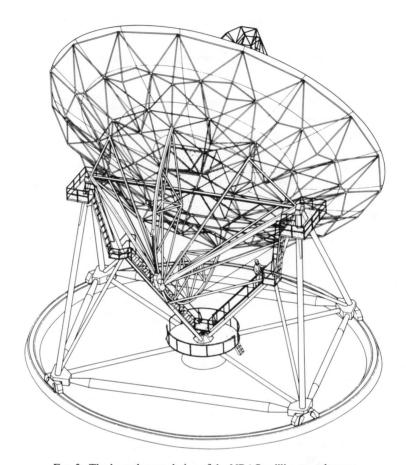

FIG. 3   The homologous design of the NRAO millimeter telescope.

structure is shown in Fig. 5. In this telescope the elevation axis is interrupted
to accommodate an equipment room between the elevation bearings. A
heavy square yoke, with counterweight arms to the rear, rotates on the two
elevation bearings. A cone-shaped extension from the four corners of the
yoke, realized by plates, is terminated in a disk (with a diameter of 14 m)
that supports the truss structure with 40 radial members. The members of
the cone are dimensioned so that the disk remains flat and round at all
elevation angles. Thus it forms an equal-softness support for the truss
section, which is designed homologously.

    The procedures for achieving homology are not as straightforward and
perfect here as in the NRAO design. The final design exhibits a quite

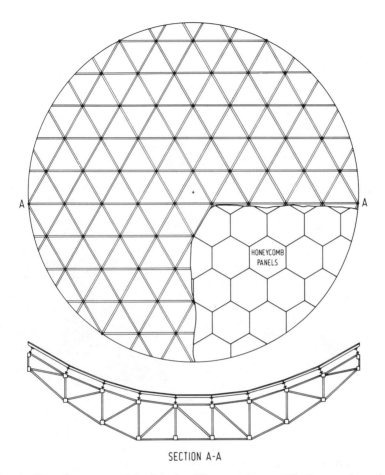

SECTION A-A

FIG. 4    The reflector part of the Caltech millimeter telescope. The hexagonal panels are connected to the support structure at three points.

substantial homology imperfection (Table II). It is interesting to note that a decisive improvement in the deformation was achieved by lateral displacement of the counterweight. This introduces a bending moment in the yoke that compensates for part of the gravity loading. The remaining gravitational deformation of the 30-m telescope in the horizon and zenith positions, assuming a setting angle of 55°, is shown in the contour plot of Fig. 6.

It has become customary to quote an effective tolerance of the reflector, $H$, assuming a perfect setting of the surface at an operational elevation angle of 45°–55° (von Hoerner and Wong, 1975). In setting the surface at 45°, based on a measurement in the zenith position, one introduces the calcu-

Fig. 5  Model of the elevation structure of the MPIfR/IRAM millimeter telescope. The yoke, supported on two elevation bearings, is extended by a cone-shaped extension to a disk 14 m in diameter, to which the support truss-structure is connected.

lated deformation between the zenith position and the setting angle. The accuracy of this procedure is limited by the fabrication tolerance (Table II). This can be circumvented only if measurement of the reflector profile at the setting angle is possible.

The weight of the MPIfR telescope is very much greater than that of the NRAO antenna, for two reasons. First, the optimized homology design of the NRAO structure minimizes the weight. Second, the MPIfR design has been laid out to withstand environmental loads (200-km/h wind, snow, and ice), whereas the enclosed NRAO antenna is not designed for such loading (the necessary material for the extreme loads must be placed in the astrodome, which weighs 370,000 kg).

TABLE II

STRUCTURAL DEFORMATIONS IN THREE MILLIMETER-TELESCOPES[a]

| Parameter | Caltech | NRAO | MPIfR/IRAM |
|---|---|---|---|
| Diameter (m) | 10.4 | 25.0 | 30.0 |
| Maximum deformation (mm) | $<1$ | $\sim 6$ | $\sim 6$ |
| Homology imperfection ($H_z$, $H_h$) ($\mu$m) | 10, 30 | 9, 2 | 60, 80 |
| Fabrication tolerance ($\mu$m) | ($<10$) | 12 | 20 |
| Total $H_z$, $H_h$ ($\mu$m) | 10, 30 | 17, 12 | 63, 82 |
| $H_\Phi$ for $20° < \Phi < 70°$ (after setting at 45°) | $<10$ | $<10$ | $<35$ |
| Weight on elevation axis (kg) (excluding counterweight) | $\sim 8000$ | 60.000 | 280.000 |

[a] $H_z$, $H_h$, and $H_\Phi$ are rms deviations from perfect contour in the zenith, horizon, and elevation angle $\Phi$ positions, respectively.

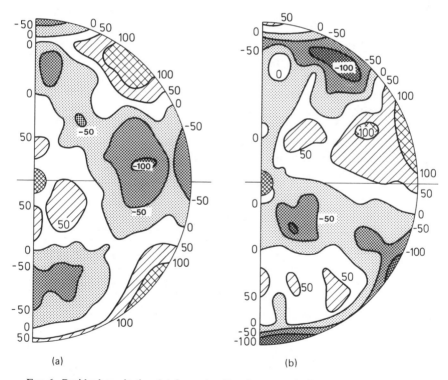

(a)                    (b)

FIG. 6 Residual gravitational deformation (in micrometers) of the 30-m MPIfR telescope in the (a) horizon (elevation 0°) and (b) zenith (elevation 90°) positions after perfect setting of the reflector surface at an elevation angle of 55°. $\varepsilon = 55$ $\mu$m.

## IV. Temperature Effects

We saw in Section II that temperature differences in the structure constitute a serious natural limit to the accuracy of the antenna. This becomes increasingly critical for structures that surpass the gravitational limit. From Eq. (4) it follows that, for telescopes in the 10–30-m-diameter range, the temperature must be uniform throughout the structure to better than 1 K to be able to operate at $\lambda \approx 1$ mm.

In Table III we have collected experimental data on temperature differences for several telescopes, which are placed either in the open air, in a closed radome, or in an astrodome with a slit (as is usual for optical telescopes). There is quite good agreement between the different measurements. It is clear that both the temperature gradients throughout the structure and the rate of change are much too large during a significant part of the day for an exposed telescope to function at millimeter wavelengths. The situation is much better at night, especially when wind helps to smooth the temperature difference. An astrodome will bring noticeable improvement only if the observation direction is restricted so that sunlight does not fall on the structure.

In a closed radome, stratification of the air would cause a large vertical gradient in temperature. To avoid this the air is moved around by ventilators at a typical "wind speed" of 10–20 km/h, providing a rather homogeneous and constant temperature regime. It should be noted, however, that the radome membrane is quite transparent to solar radiation, causing a temperature increase in the telescope structure on the side facing the sun. This has necessitated thermal insulation of the azimuth support pedestal to avoid large pointing errors (Rodman, personal communication).

TABLE III

MEASURED TEMPERATURE DIFFERENCES OF TELESCOPES

| Telescope | $\Delta T$ (K) | | | | $T$ (K/h) | |
|---|---|---|---|---|---|---|
| | Calm, sunny day | Covered sky, day | Calm, clear night | Windy or covered night | Clear night | Clear day |
| MPIfR, 100 m, open | 8 | 4 | 2–3 | 1 | 1 | 3 |
| NRAO, 43 m, open | 7 | — | 1.3 | — | 0.8 | 5 |
| NRAO, 11 m, astrodome | 5–7 | — | 1–3 | — | 1 | 3–5 |
| ESSCO, 13 m, radome | 4 | 1 | 2 | 2 | 1 | 2–5 |
| NRAO, 25 m | | | | | | |
|   Astrodome (study estimate) | 2–3 | — | 2 | — | 1 | 2–5 |
|   Radome (study estimate) | 2–3 | — | not used at night | | — | 2–5 |
| NRAO, test reflector panel | 5 | — | 1.2 | — | 0.8 | 5 |

The NRAO 25-m telescope will normally operate with an open slit; the radome door will be closed in strong winds and unfavorable thermal conditions. With the opening and closing of the slit, ambient temperature gradients will cause a thermal imbalance, with a calculated rms thermal deformation of the reflector of $\approx 40~\mu$m.

The thermal lag $\dot{T}$ is particularly harmful if the various sections of the telescope structure have different thermal time constants. For this reason, the NRAO design maintains as constant a wall-thickness in the structural members as possible and keeps the time constant of the entire structure at the rather low value of 20 min. The support structure of the Caltech antenna is very homogeneous, and consequently the $\dot{T}$ effect will be small.

A completely different situation arises with the MPIfR design. First, the telescope will be exposed and hence subject to all environmental influences, including temperature change and (partial) heating by direct sunlight. Second, the sections of the structure have very different thermal time constants. Extensive calculations of the temperature effects on reflector profile and pointing angle have been performed for this telescope. These calculations are summarized in Table IV. Similar calculations for the NRAO and Caltech antennas yield comparable results.

It is clear from Table IV that homogeneity of the temperature to $\approx 1$ K is

TABLE IV

CALCULATED TEMPERATURE EFFECTS FOR THE MPIfR 30-M MILLIMETER TELESCOPE

| Temperature load | Profile error (rms) ($\mu$m/K) | Pointing error (sec$^{-1}$/K) |
|---|---|---|
| Linear gradient across reflector support | 10 | 1.5 |
| Difference between front and back of reflector support | 17 | 0.5 |
| Difference between left and right or upper and lower half of support | 9 | 4.0 |
| Difference between truss support and yoke | 11 | — |
| Linear gradient along one leg of quadrupod or difference between legs | — | 1.1 |
| Randomly distributed differences | | |
| In truss structure | 10 | — |
| In truss structure and upper part of yoke and cone | 11 | — |
| In yoke and cone only | 3 | — |
| Difference between front and back of concrete pedestal | — | 0.1 |
| Difference between aluminum panels and steel support frame | 0.4 | — |

essential if the resulting errors are to play a minor role compared to gravity effects; and it is clear from Table III that an exposed telescope can achieve this only on a windy night. These considerations led to the decision to control the temperature of the 30-m telescope.

## V. Thermal Control of the 30-m Millimeter Telescope

The telescope and its cross section are shown in Fig. 7. The basic design —a compact structure on a concrete pedestal, with an equipment room between bearings—was the result of other considerations besides temperature uniformity. However, the structure lends itself well to the attachment of thermal insulation. The insulation is indicated in Fig. 7b by a zigzag line. It consists of panels 40 mm thick. No part of the structure apart from the reflector surface is directly exposed to the outside world. Insulation is also present between the steel truss support and the aluminum surface panels. In the closed room of the truss support, stratification would occur (as in a

Fig. 7a   The MPIf R/IRAM millimeter radio telescope, which has a diameter of 30 m and is located in the Sierra Nevada in Spain at an altitude of 2870 m. This photograph of November 1981 shows the nearly completed telescope during its first commissioning tests. (Photo courtesy MPIfR.)

radome). Therefore the air is constantly circulated by ventilators, assuring a uniformity of $\lesssim 0.5$ K. The heat produced by the ventilators is cooled away. The response of the total system to daily changes in outside temperature and to sudden heating by sunshine or loss of cooling has been analyzed by Baars, Greve, and Hooghoudt (1983).

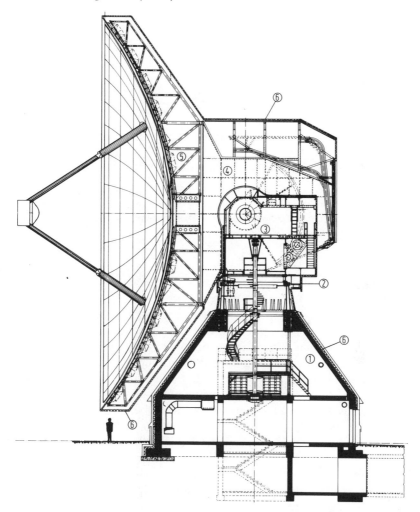

FIG. 7b  Cross-section of the 30-m telescope. The concrete pedestal (1) carries a 5-m-diam azimuth bearing (2). A two-story cabin (3) between the elevation bearings contains all drive systems and the radio-astronomy receivers. The yoke and cone section (4) is supported on elevation bearings and carries the reflector structure (5). The entire telescope is covered with thermal insulation (the zigzag line), which is also present between the reflector surface and its support structure (6).

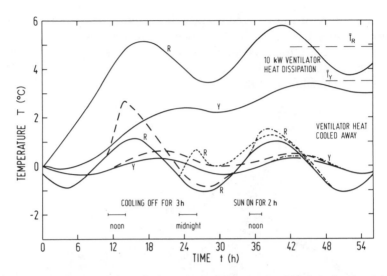

FIG. 8   Temperature of the reflector support (R) and yoke (Y) of the insulated 30-m telescope as a function of time $t$ for a sinusoidal ambient temperature (in degrees Celsius) $T_a = 2.5 \sin(\pi t/12)$. The dashed lines represent the reactions to disturbances indicated at the bottom of the figure. Regulated cooling keeps $T_R = T_Y$.

Figure 8 shows the temperature of the reflector support (R) and yoke (Y) as a function of time for several thermal loads. The solid lines represent the reaction of the telescope to a sinusoidal variation in outside temperature, with an average of 0°C, an amplitude of 2.5 K, and a maximum at noon and minimum at midnight. This is a realistic value for the average daily temperature variation at a high mountain site. The upper curves apply to the situation where the heat dissipation of the ventilators (10 kW) is not cooled away, resulting in an average temperature of 4.9 and 3.5°C for the reflector support and yoke, respectively, which is reached after a switch-on time of 60 h for the reflector support and 72 h for the yoke. The temperature difference between the reflector support and yoke fluctuates between 0.5 and 2.3 K. From Table IV it follows that the thermal deformation of the reflector support reaches about 25 $\mu$m, which is undesirably high.

Installation of a 10-kW cooling capacity in the ventilator system yields the lower curves, with an average of 0°C. The amplitude of the daily variation is 1.0 K for the reflector support and 0.3 K for the yoke, and the maximum difference between the two is 0.8 K. Figure 8 also shows the response to perturbations in the system. Switching off the cooling at 11:00 a.m. causes a fast rise in $T_R$ of almost 2 K, which takes more than 12 h to disappear. The rise in $Y_Y$ is limited to 0.3 K; it appears only after the cooling has been switched on again and takes 36 h to disappear. A similar switchoff at night causes the same initial temperature rise, but the temporal behavior at a later time is different, as shown in Fig. 8.

At $t = 35$ h (at 11:00 a.m.) the sun shines for 2 h on the rear side of the reflector support insulation, causing a $\Delta T = 5$ K between the front and back of the insulation. The rise in $T_R$ is $\approx 0.5$ K, with a delay of 1–2 h in reaching its maximum; $T_Y$ rises only $< 0.2$ K, with a large delay of 7 h.

Although the cooling away of the ventilator dissipation is already satisfactory and limits the thermal deformations to an insignificant value of $\approx 10$ $\mu$m rms, one further step is taken. By installing a cooling capacity of 20 kW (only slightly more expensive than a 10-kW system) the cooling power can be regulated to keep $T_R = T_Y$. Because the yoke has greater thermal inertia, the cooling will be regulated to keep the reflector support at the yoke temperature. The calculations show that even with a daily variation in outside temperature of 10 K (p.–p.), for which the maximum cooling capacity is not always sufficient, $T_R - T_Y < 0.5$ K.

To complete the thermal control, the subreflector support legs are also insulated and their temperature regulated to equal that of the yoke. By these measures this telescope will essentially be independent of temperature and sunshine effects. Its temperature uniformity will be better than 0.5 K throughout the day. From Table IV it follows that the resulting surface error is of the order of 10 $\mu$m and the pointing error is less than 1 ″. These are acceptable values in view of the overall specifications.

The cost of the insulation and temperature control system of the 30-m telescope amounts to about 10% of the price of the antenna. This percentage will probably be greater for smaller telescopes. In view of the increased availability during daytime, the small investment is warranted.

Insulation has been used for other telescopes. Caltech has found it necessary to insulate the mount of the 10-m telescope to avoid excessive pointing errors. The reflector support is still exposed. The Bell Laboratories telescope (No. 9 in Table I) has an insulated support structure. At Berkeley (No. 4 in Table I), the reflector has even been covered under a tent of grypholin. Such a cloth, which reflects 90% of the solar heat while transmitting $\approx 95$% at 86 GHz, will also cover the slit of the astrodome-enclosed British telescope (No. 16 in Table I). The large, exposed Japanese telescope (No. 12 in Table I), originally planned to operate without insulation, has now been fitted with insulating panels on the rear of the truss support. Finally, a significant improvement in the NRAO 43-m telescope after insulation of certain critical areas should be mentioned (von Hoerner, 1977b).

## VI.   Wind Effects

Wind exerts forces and moments on the telescope structure, leading to deformations of the reflector surface and to pointing errors. In a radome, wind effects are minimal, the only wind being that caused by the ventilator system in keeping the air mixed. Wind effects on a telescope in an astrodome will generally be less than for an exposed structure, but they are dependent on the wind direction with respect to the slit and rather hard to calculate.

An exposed telescope must be designed to survive the strongest wind likely to occur at the site. During operation wind velocities can be considerably lower. Thus the wind-induced deformations of the surface are generally small compared to the gravitational deformations. The 30-m telescope is designed to be fully operational for a wind speed of 50 km/h. The calculated deformations are quite dependent on the elevation angle. A maximum of 25 $\mu$m rms occurs at an elevation angle of about 60°, decreasing to < 10 $\mu$m at elevation angles of less than 40° and greater than 75°. The calculated deformation of the NRAO telescope is about 30 $\mu$m for the same wind speed. Because this telescope is in a radome – astrodome, the real deformations will be smaller.

Change of the pointing direction due to wind is often more serious than reflector deformation. For the MPIfR design, the calculations indicate a pointing error in 50-km/h wind of 2.5 " in the prime focus and 1 " in the secondary focus. The NRAO design would give a pointing error of 7 " if it were exposed. These errors are substantial compared to the beam width of about 10 ".

Wind effects scale with the wind velocity squared. Thus, typical median values of the errors will be about half the values quoted. The effects of gusts on pointing will be discussed in Section X.

## VII.   Radome, Astrodome, or Exposed?

This question is asked regularly by designers and users of radio telescopes. In fact, each of the three alternatives has certain advantages and disadvantages, and it can be difficult to decide on the best solution for a particular application. This is evident from Table I, which includes telescopes of all three kinds. In this section we attempt an objective comparison of radome- or astrodome-enclosed and exposed telescopes. This comparison is restricted to radio telescopes for millimeter and submillimeter wavelengths (typically in the frequency range 80 – 800 GHz) located at suitable sites, i.e., at high elevation in a relatively dry area. The criteria to be applied and the author's evaluations are presented in Table V.

An obvious (and at millimeter wavelengths, critical) disadvantage of the

## TABLE V

### Comparison of the Radome, Astrodome, and Exposed Options for Millimeter-Telescopes

| Parameter | Radome | Astrodome | Exposed |
|---|---|---|---|
| Gain loss | significant (10–40%) | generally avoidable (some with solar screen) | none |
| Wind influence | small (air circulation) | limited | full |
| Control of temperature | not trivial; air circulation | quite difficult; air circulation | more difficult; insulation |
| Means of avoiding snow/ice load | heating of radome | heating of astrodome | heating of antenna, insulation |
| Survival (storm, snow/ice) | similar for all three cases, because similar amounts of material are needed. | | |
| Cost | partially offset by cheaper antenna | high, especially with transparent door | limited, extra for insulation |
| Astronomical availability | depends only on clouds | added influence of wind | full influence of wind |
| Serviceability | always possible, easy | always possible with closed door | special measures necessary |
| Summary | higher availability at lower gain; easy operation | easy operation at full gain with lower availability; highest cost | full gain, but lowest availability and ease of operation |

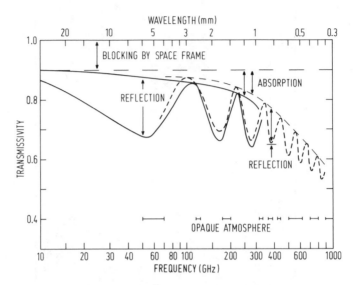

FIG. 9 Transmissivity of an ESSCO radome vs frequency and wavelength for two membrane materials: ——, Esscolam V; – – –, Escolam X-144-2. The three contributions are from the space frame, membrane absorption, and membrane reflection (the frequency of the minima can be changed by adjusting the membrane thickness). Atmospheric windows are indicated at the bottom of the figure (Courtesy A. Cohen, ESSCO).

completely closed radome is the loss in antenna gain due to absorption and reflection by the metal space frame that provides the mechanical rigidity of the radome and by the dielectric membrane. An extensive discussion of this subject can be found in the two-volume NRAO Report "A 25-Meter Telescope for Millimeter Wavelengths" (1975, 1977). In Fig. 9 we present the transmissivity of a radome as a function of frequency for two membrane materials. These curves were measured at the Electronic Space Systems Corporation (ESSCO) and kindly made available by Dr. A. Cohen.

The space frame has a *geometrical* blocking area of about 5% of the aperture area, causing a nominal gain loss of 10%. However, this loss can be partially compensated for by an optimized design of the subreflector support legs, because these are not exposed to environmental loads and hence can be made thinner. In the ESSCO antennas these legs are thin and ogival in cross section. Moreover, they penetrate the reflector surface almost at the edge, thereby minimizing the *spherical wave* blocking area. The resulting geometrical blocking of subreflector and support legs is only 2.5%. This and the space-frame blocking add up to 7.5%, which can be compared to 8% for the NRAO millimeter telescope alone. On the other hand, exposed supports can be made stiff enough without excessive blocking, as shown by the MPIfR design, which has 4.6% blocking despite the presence of 50-mm-thick thermal insulation on the support legs (see also Baars and Rusch, 1983).

The loss in the membrane has an *absorption* component, which increases monotonically with frequency, and a *reflection* component, which exhibits a periodicity with frequency. The period of this component can be changed by variation of the membrane thickness, allowing some optimization of the transmissivity with respect to the atmospheric windows. Because the atmospheric absorption lines of water vapor and oxygen are not harmonically related, this optimization works only over a restricted frequency range. The material Esscolam V (with which present-day ESSCO radomes — Nos. 7 and 10 in Table I — are equipped) exhibits a loss varying between 5 and 30% for frequencies up to 300 GHz. The newly developed Esscolam X-144-2 performs better, especially above 300 GHz (Fig. 9).

The membrane loss nevertheless becomes uncomfortably high above 300–400 GHz, and consequently the most accurate telescopes will be placed in an astrodome with an open slit (Nos. 14, 16, and 17 in Table I). The British design incorporates a "viewing window" over the slit made of sailcloth, loaded with $TiO_2$ and carbon black, which should reflect about 90% of the solar heat and transmit more than 90% of rf power up to 500 GHz (Shenton and Hills, 1976). The NRAO astrodome door will be transparent (it is essentially a piece of space-frame radome), and will be closed during daytime periods of unfavorable ambient thermal conditions and in strong winds.

The greatest advantage of a radome is the absence of wind effects on the antenna, resulting in lighter construction, a simple servo-drive system, and better pointing behavior. The astrodome with an open door will normally provide at least partial protection from wind effects.

We saw in Section IV that temperature differences most seriously limit the performance of millimeter and submillimeter telescopes. Temperature uniformity in a radome requires steady and substantial air movement. As mentioned in Section IV, the opening and closing of the slit of an astrodome causes a serious thermal imbalance. An exposed telescope has a good temperature uniformity at night in a light wind, but requires substantial insulation during sunshine.

To withstand extreme environmental conditions, the protection of a dome is not necessary (von Hoerner, 1975). Servicing and operating an instrument in an enclosed environment is both convenient and cost saving. Unfortunately, many weather influences, that a radome or astrodome easily copes with render observations at millimeter and submillimeter wavelengths impossible, thereby restricting the operational advantages of the enclosure. In the author's opinion there is no need to use an enclosure, in spite of its convenience. Telescopes of intermediate size ($D \lesssim 20$ m) for frequencies below 250 GHz might advantageously be placed in a radome. At higher frequencies, the high losses of present-day membrane materials require the use of an astrodome if an enclosure is necessary. An astrodome is expensive, especially the variety with "transparent" door. The NRAO estimates a cost for the astrodome as high as that of the telescope in it. The required performance can be achieved in a properly designed and locally insulated exposed telescope, as the MPIfR 30-m design has shown. Of course, the insulation increases the cost.

It is likely that very accurate submillimeter telescopes, using advanced materials (e.g., carbon fiber) and techniques will be designed to operate exposed in astronomically feasible conditions—a clear sky (sunshine during the daytime), no precipitation, and reasonable wind speed—and will be protected against extreme weather conditions by a simple astrodome or a removable shack.

## VIII.  Reflector Surface

The reflector surface is the most important part of a telescope. Its quality determines the sensitivity of the telescope. As illustrated in Fig. 1b, the rms deviation of the reflector from the prescribed shape must be 100 $\mu$m or less for satisfactory operation in the millimeter wavelength range, and the surface should have a ≈ 20-$\mu$m accuracy for use at the shortest wavelengths in the atmospheric windows near $\lambda = 350$ and 450 $\mu$m. Thus the demands placed on the manufacturing accuracy and gravitational and thermal stability of the reflector surface are very high.

There are two basic types of reflector surfaces.

(1)  A *contiguous surface* attached to the supporting backup structure. This, of course, is the type usually used for optical telescopes. The total surface has to be shaped in the required form, and the maximum size will be determined by the machine available for the shaping. The 4.9-m telescope of the University of Texas has an epoxy-resin reflector supported by an Invar backup structure. The parabolic shape was achieved by spinning the reflector while the epoxy was still in a fluid state. The reflector of the NRAO 11-m telescope was machined on a specially built machine from one piece of aluminum welded to an aluminum support structure (Fig. 10). As an indication of the difficulty of machining such a large piece to the specified 0.1 mm, we mention that the effects of the tides on the machine in the workshop near the Pacific Coast considerably limited the usable time for machining. Later it was determined that the biggest problem of the telescope was the "bimetallic" behavior of the steel mount and aluminum reflector with temperature variations. An improved reflector support and surface panels were installed in 1982 (No. 19 in Table I).

FIG. 10 The NRAO 11-m antenna at Kitt Peak. It has a surface machined out of one piece of aluminum. Built in 1967–1968, it was the first large millimeter telescope, with which many important astronomical discoveries were made. A new reflector was installed in 1982. (Photo courtesy NRAO, Green Bank, West Virginia.)

FIG. 11  The Caltech 10.4-m antenna, which has the most accurate reflector surface of any radio telescope. Note the hexagonal shape of the reflector panels. For the fourth reflector in the series an accuracy of 10 $\mu$m is envisaged. (Photo courtesy Caltech, Pasadena, California.)

(2)  A surface consisting of a set of *reflector panels* individually supported by adjustable screws on the backup structure. Almost all radiotelescopes have a reflector of this type, including most millimeter telescopes. Here the problem of achieving an accurate reflector profile has two facets. First, the panels must be fabricated with the correct shape and, second, their position on the support structure must be measured and adjusted to lie on the required contour.

An interesting combination of the two methods has been used in the Caltech telescopes (Leighton, 1978). Here the surface consists of 84 hexagonal panels (Fig. 11), all mounted on a support structure to form the 10-m-diam reflector. The complete surface is now machined to a parabolic form, using an ingenious method to be described in Section IX.

There exists a great variety of manufacturing methods for panels. Those used, or proposed, for millimeter and submillimeter telescopes are summarized in Table VI. Included in the table are the best presently obtained panel accuracies and examples of telescopes taken from Table I. From a purely manufacturing viewpoint the different types achieve similar accuracies. The differences are more a result of the effort put into achieving a given specifica-

TABLE VI

PARAMETERS OF REFLECTOR PANELS

| Type | Method of contour forming | Typical size $(l \times w \times h)$ (m) | Best fabrication accuracy $(\mu m, rms)$ | Example (No. in Table I) |
|---|---|---|---|---|
| 1. Cast aluminum plate with reinforcing ribs | machined on numerically controlled machine | $1.5 \times 0.8 \times 0.10$ (including ribs) | 30–40 | Bell (9), NRAO (17) |
| 2. Aluminum sheet on reinforcing ribs | stretched–formed on mold | $3.5 \times 0.7 \times 0.07$ (including ribs) | 30–40 | ESSCO (6, 7, 10, 19) |
| 3. Aluminum sheet with many pins on frame | preformed on mold, pins adjusted | $0.9 \times 0.8 \times 0.1$ (pin height) | 60 | CSIRO (8) |
| 4. Stiff aluminum honeycomb epoxied to sheet | stretched–formed on mold | $1.5 \times 0.8 \times 0.23$ (honeycomb height) | 20–30 | SERC (16) |
| 5. Flexible aluminum honeycomb on 15 pins | stretched–formed on mold, pins adjusted | $2.0 \times 1.0 \times 0.4$ (honeycomb height) | 20–30 | MPIfR (13) |
| 6. Aluminum honeycomb epoxied to carbon fiber skin | formed on mold (metal) formed on mold (pyrex) | $2.2 \times 1.1 \times 0.06$ $1.2 \times 0.9 \times 0.06$ | 65 5–15 | Tokyo (12) in 1979 MPIfR, IRAM (15) in 1982 |
| 7. "Minipanel" of aluminum honeycomb | formed on mold | $0.3 \times 0.3 \times 0.01$ | 10 | IRAM (15) (proposed) |
| 8. Aluminum honeycomb epoxied to aluminum sheet | in situ machining of honeycomb | 1.15 (hexagonal) | 10–20 | Caltech (14) |
| 9. Pyrex honeycomb | slumped on mold and ground | up to 7 (monolithic) | 2–5 | Optical Sciences Center, Tucson (proposed) |

tion than of the basic limitation of a certain method. Accuracies of 25- to 50-$\mu$m rms can be achieved in various ways. With increased experience the manufacturer learns to circumvent unexpected problems. This is illustrated by the improvement in a series of ESSCO reflectors (Nos. 6, 7, 10, and 19 in Table I), for which in a decade the accuracy has been improved almost by a factor of 5.

The carbon fiber skin used in the Japanese telescope (No. 6 in Table VI) is the first to be applied to a large ground-based antenna. The achieved accuracy is a factor of 2 better than originally specified. Carbon fiber is being increasingly considered for use, mainly because of its extremely low thermal expansion coefficient. The proposed "minipanels" (350 mm) for the IRAM telescopes (No. 15 in Table I) are supported on one point only, and could be made of aluminum honeycomb or a sheet of carbon fiber material.

The Optical Sciences Center of the University of Arizona, Tucson, is proposing to construct a lightweight mirror 7 m in diameter made of Pyrex honeycomb. Pyrex sheets are fused to the required size, the honeycomb (or "egg crate") core and back plate are added, and the whole is fused and "slumped" on a mold in an oven to form a single mirror of approximately the correct curvature (Carleton and Hoffmann, 1978). After one machine-grinding procedure the surface error will be less than 5 $\mu$m. The estimated weight of a 7-m mirror is $\geq 10^4$ kg. This is an interesting possibility for a telescope in the submillimeter range ($350 < \lambda \leq 1000 \ \mu$m), and it would also be useful for near-infrared work at wavelengths of 10 to 30 $\mu$m.

In several cases in Table VI, a mold is used to achieve the panel shape. The final accuracy of the panel is strongly dependent not only on the accuracy of the mold but also on the "working" of the panel once it is released from the mold. This working of the panel—essentially a release of stresses intro-duced in the manufacturing process—has been nicely corrected for in example 5 of Table VI. The multiple support points (about 500 mm apart) enable deviations in a panel with a relatively large scale length to be corrected, whereby an improvement of a factor of 5 to 10 in the rms value can be achieved.

Wind effects are similar to, but generally smaller than, gravity effects. Because gravity deformations of the panel will always be restricted by the designer to a fraction of the manufacturing tolerance, the even smaller wind-induced deformations will be insignificant.

The effects of temperature variation are much more serious. The panels distort under a temperature gradient across the height of the panel structure, which can arise from

(1) a rapidly changing ambient temperature environment,
(2) direct sunlight on the panel surface,
(3) radiation from the panel surface toward the clear, cold sky at night.

TABLE VII

THERMAL EFFECTS IN PANELS IN SUNSHINE

| Panel type | Measured $\Delta T$ (K) | Length $l$ (mm) | Height $h$ (mm) | rms deformation $\Delta z$ ($\mu$m) | |
|---|---|---|---|---|---|
| | | | | Calculated | Measured |
| Cast aluminum plate with ribs | 0.4 | 1900 | 71 | 40 | 33 |
| Aluminum sheet on ribs | 0.3 | 1500 | 75 | 18 | 20 |
| Thick aluminum honeycomb | 1.0 | 1500 | 225 | 30 | — |
| Thin aluminum honeycomb | 0.6 | 500 | 40 | 8 | 12 |

The temperature differences to be expected in telescopes are given in Table III. The groups at NRAO and MPIfR have measured the resulting temperature differences across different types of panels and calculated and/or measured the resulting deformation. The results are summarized in Table VII.

The deformation $\Delta z_m$ at the center of a plate fixed at the four corners is approximately given by

$$\Delta z_m \simeq 0.25 C_t \, \Delta T \, l^2/h, \tag{5}$$

where $C_t$ is the thermal conductivity ($C_t = 2.34 \times 10^{-5}$/K for aluminum) and $\Delta T$ the gradient across the height $h$ of the panel, with a largest linear dimension $l$ between support points. Because the deformation of the panels is consistently in one direction ($\Delta T$ will have the same sign across all panels), the rms $\Delta z \simeq \frac{1}{3} \Delta z_m$.

The measured $\Delta T$ in Table VII appears to be typical for a situation in which the rise in average temperature of the panel due to solar heat is limited to $\lesssim 5$ K by an appropriate white paint on the surface. The calculated and measured deformations are within values that are acceptable for a millimeter telescope. However, with the surface accuracies required in the submillimeter region ($\varepsilon \lesssim 20 \, \mu$m), the thermal deformation of these types of panels would constitute a very significant contribution to the overall surface error.

## IX. Measuring and Setting the Reflector Profile

In Section VIII we saw that the reflector of most telescopes consists of panels individually attached to the backup structure. To achieve an accurate reflector, the panels must be placed along the prescribed contour with an accuracy similar to the manufacturing accuracy of the panels. Thus we need measurement methods for the determination of the position of a point on the panel with an accuracy of better than 100 $\mu$m or even, for the newer instruments, 25–50 $\mu$m — i.e., a precision of 1–3 $\times$ $10^{-6} D$ (where $D$ is the reflector diameter). This is beyond the capability of the geodetic method (theodolite and tape or pentaprism and tape) that have been used for the large centimeter wavelength telescopes and achieve a precision of about $10^{-5}D$. These methods have been reviewed by Findlay (1971). Here we shall limit ourselves to a brief description of some new methods that are being developed in conjunction with some of the projects listed in Table I.

First, the shape of the panels must be checked and optimized, as in examples 2, 3, and 5 of Table VI. This can be achieved by comparison with an accurate template, or by use of an optical level instrument or, most conveniently, a sufficiently large three-dimensional measuring machine. Such machines allow repeatable measurements with an accuracy of $\lesssim 5$ $\mu$m. An original application of this principle has been used at NRAO (Fig. 12). The absolute calibration of such machines is particularly difficult.

A method developed by Leighton (1978) deserves special attention. In it the problems of panel manufacture, panel measurement, and total reflector measurement are simultaneously attacked and solved in a physically straightforward and technically refined way. The system is sketched in Fig. 13. The basic idea is to employ the geometric characteristic of a

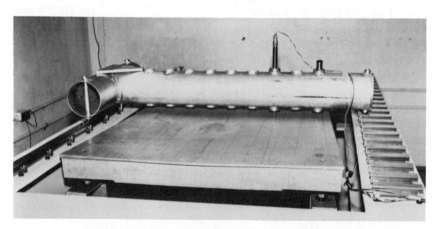

FIG. 12 The NRAO solution for the fast measurement of individual surface panels. A bar with nine accurate depth gauges (only one is installed here) is stepped along the long side of the panel at 20 reproducible points. The measurement accuracy is about 5 $\mu$m after calibration. The panel is supported separately. (Photo Courtesy NRAO, Green Bank, West Virginia.)

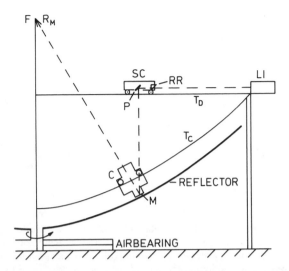

FIG. 13 Principle of Leighton's measurement method used for the Caltech antennas. The ray of the laser interferometer (LI) is deflected 90° by the pentaprism (P) on the slave cart (SC) running on the horizontal directrix template ($T_D$), hits the mirror (M) on the cart (C) moving on the parabolic template ($T_C$), and returns on its path after reflection by the retroreflector ($R_M$) located in the focus (F) of the parabola. The retroreflector (RR) on the SC delivers the reference signal for the LI.

parabola (path length FMP is constant for any position of M along a parabola). Along a parabolic template $T_C$ runs a cart with a mirror for the measuring system, to be replaced later by a cutting head for shaping the surface. The vertical position of the tract $T_C$ can be adjusted to a parabola. The control signal for this is derived from the measuring signal of the laser interferometer, which will be zero if M is on the parabola. After $T_C$ has been brought into the correct position, mirror M is replaced by the cutting head and the aluminum honeycomb is cut to shape. The complete reflector is supported on an accurate air bearing and rotates under the template with the cutting head. With experience, Leighton has been able to improve his method. The fourth reflector produced by this method has achieved an accuracy of $\sim$ 10-$\mu$m rms. Note that, by virtue of this method, this accuracy includes panel manufacture, panel measurement, panel setting, and total reflector measurement.

Findlay has devised and tested two methods that have a basic, but ingenious, simplicity (Fig. 14). The first is a variation on the spherometer, widely used in checking optical mirror surfaces (Fig. 14a). The local curvature of the surface is measured with a cart that moves on the surface. The arc length $l$ is derived from an encoder on one of the wheels, while a probe on the

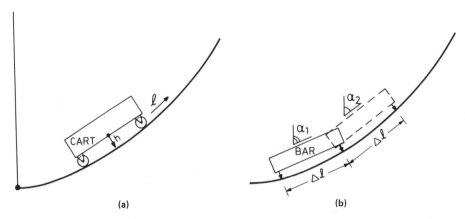

FIG. 14 Two methods applied by Findlay for measuring the reflector contour. (a) The spherometer cart measures the path along the surface ($l$) and the depth from cart to surface ($h$), delivering the surface contour after two integrations. (b) The stepping bar measures inclination angles $\alpha_n$ for consecutive positions of the bar in length increments $\Delta l$.

cart measures the depth $h$ to the surface. By double integration the reflector profile can be derived, provided the two integration constants have been determined. Payne, Hollis, and Findlay (1976) moved the cart along a set of radii and achieved an accuracy of $\approx 40\ \mu$m rms. Findlay's second method employs a bar of precisely known length $\Delta l$ equipped with an accurate inclinometer (Fig. 14b). The bar is stepped along the radius of the reflector and the angle is measured. An accuracy of $25\ \mu$m rms has been achieved (Findlay and Ralston, 1977).

A British group has devised a method using a laser interferometer (Shenton and Hills, 1976; Fig. 15). A cart is moved continuously along a radial arm, while the interferometer measures the *changes* in the coordinates along the radius ($U$) and toward the reflector ($V$). The position of the cart is maintained in relation to a laser beam and is therefore independent of the bending of the arm. Tests in the laboratory indicate an accuracy of $20\ \mu$m rms.

At the MPIfR, work is being done on a modified and improved version of a modulated laser ranger previously used by Payne (1973). The modulation frequency is 1.75 GHz, and an accuracy of $30\ \mu$m over 18 m has been demonstrated in the laboratory. The laser beam is sent through a specially modified, accurate theodolite (Kern DKM3) toward a target on an arm that moves tangentially along the surface (Fig. 15b). The target contains a corner reflector (C) for the ranger and a linear diode array (D) that measures vertical displacements with an accuracy of about $10\ \mu$m. This system,

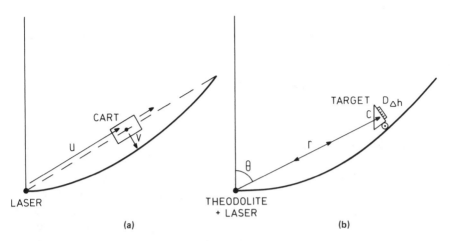

(a) (b)

FIG. 15 (a) The U – V method employs a radially moving cart with a laser interferometer, continuously measuring the *changes* in U and V and hence determining the reflector contour. (b) The r, θ, Δh method uses a modulated laser to measure r *absolutely,* whereas Δh is determined with a diode array on the tangentially moving–stepping target cart.

described by Greve and Harth (1980), allows a measuring accuracy of 40 μm for a 30-m reflector. A very similar system is being used by a Tokyo group for the 45-m antenna, with a quoted accuracy of 100 μm (Kaifu, personal communication).

Other methods are being studied, involving, e.g., a tangentially moving spherometer cart or a measurement from the center of curvature (A. Delannoy, IRAM, personal communication).

Two factors are of particular importance in obtaining high surface accuracy of millimeter telescopes. First, the method must be implemented at a time when thermal effects cannot cause deformations in the structure. For open-air telescopes without a stringent temperature control, the measuring time is limited to several hours at night. Second, as discussed in Section III, the effects of gravitational deformations on the support structure can be minimized by setting the surface at an intermediate elevation angle. If this setting can be accompanied by a profile measurement at this angle, a further increase in accuracy is possible because uncertainties in deformation calculations are accounted for. For practical reasons the measurement of an inclined reflector must be "noncontacting."

A very powerful method has been developed in recent years by Anderson and co-workers at the University of Sheffield (Bennett *et al.,* 1976; Godwin *et al.,* 1981). The method is called *radio holography.* A hologram is formed by scanning the antenna through a point source and combining the output with that from a reference antenna. From the hologram the aperture phase distribution is derived and phase departures are identified with reflector surface errors. The Sheffield group has employed ground-based transmitters, but recently performed a measurement of a 25-m antenna using a satellite beacon at 14 GHz with an accuracy of 35-$\mu$m rms (Godwin et al., 1981).

For radio astronomers it is natural to look at the sky for the source. Indeed, the very strong, pointlike, water vapor maser sources at 22.3 GHz are well suited for our purpose. The great advantage of these sources is their varying elevation angle with time of day, which enables measurement at several elevations and study of the structural deformations.

Scott and Ryle (1977) used a radio source and an interferometer of two 13-m-diam antennas to measure the surface of one of the antennas to an accuracy of 0.1 mm in a few hours' time. Shenton and Hills (1976) and von Hoerner (1978) have proposed a variation of this method, in which a moving feed samples the focal field of the reflector. The Fourier transform relation is used to derive the aperture field. This method is under development at NRAO. IRAM and MPIfR plan to measure the 30-m millimeter telescope using a small ($D \lesssim 2$ m) reference antenna with an expected accuracy of < 50-$\mu$m rms and a resolution of $\approx 1$ m on the surface in 2 h measuring time, using the strongest $H_2O$ maser sources.

Finally, how shall we measure and set future submillimeter telescopes to an accuracy of $\lesssim 10$ $\mu$m? Measuring and setting individual panels in the outside air appears extremely problematic. An integrated support structure with surface appears to offer a solution. The whole reflector, or at least a large section of it (between one-fourth and one-half), could be manufactured, measured, and adjusted in the shop. Leighton (1978) has demonstrated the feasibility of this approach. Other measuring methods are available, such as the measuring machine and optical methods. For instance, the surfaces of these telescopes might be good enough to allow the application of classical tests of optical surfaces in the "light" of a 10-$\mu$m-wavelength $CO_2$ laser (Kwon et al., 1980).

## X. Pointing and Tracking — Servo Control

After the reflector surface quality, the accuracy and stability of the pointing is the most critical parameter of large millimeter telescopes. A pointing error of 0.2 half-power beamwidth causes, in addition to positional uncertainty, a loss in sensitivity of 10%, which is similar to the effect of a $\lambda/40$ surface error. The large telescopes have a beamwidth well under 1 min$^{-1}$, reaching a value of less than 10 sec$^{-1}$ at the highest operational frequency. Thus a pointing accuracy and repeatability of 1 to 2 $\sec^{-1}$ is required under operational conditions.

In the diffraction-limited operation of radio telescopes, the requirements imposed on tracking and on absolute pointing are the same. This is in contrast to the usual situation with optical telescopes, where the absolute pointing accuracy is often poor, on the order of 1 $\min^{-1}$. This is acceptable because there is always a sufficiently bright star of known position in the field of view to use as a position calibrator. A star is also used as a guiding star for tracking, allowing a tracking accuracy on the order of 0.1 $\sec^{-1}$.

The absolute pointing of a radio telescope is influenced by flections in the structure. The usual procedure is to construct correction curves or tables by observation of a set of radio sources with known positions, which are then incorporated into the control computer (Schraml, 1969; Stumpff, 1972; Ulich, 1981).

Variable pointing and tracking errors occur as the result of imperfections in the drive system and the effects of temperature and wind. The stiffness of the telescope structure, the drive system, and the layout of the servo-control system must be such that under the specified operation conditions a pointing accuracy of 1 to 2 $\sec^{-1}$ is achieved. The control of wind effects is the major problem, and is particularly difficult for an exposed instrument on a high mountain site. The structure must be designed to exhibit relatively high eigenfrequencies, typically above 3 Hz. This, together with a servo system with a sufficiently large bandwidth, allows for a quick reaction to unpredictable wind gusts. Also, it is important to read the axis positions to a resolution significantly smaller than the specified pointing accuracy. Both absolute encoders — normally of the inductosyn type with up to 23 bits (0.15 $\sec^{-1}$) resolution — and incremental encoders are being applied.

The reaction to wind gusts (a step function in the input signal to the system) is probably the most sensitive test of the quality of the servo system. As an example, we describe the control system of the MPIfR 30-m telescope. The basic block diagram of a traditional *cascade controller* with position- and velocity-control loops is shown in Fig. 16a. Disturbances ($D$) on the antenna (e.g., wind moments) cause variations in one or more of the control parameters: the elevation encoder reading $\varphi_E^*$ and its velocity $\dot{\varphi}_E$, the motor axis velocity $\dot{\varphi}_M$, and the angle of the support house $\varphi_H$. Comparison with the required elevation angle $\varphi_R$ yields the required motor moment $M_M$. In Fig. 17 the reaction of the system to a wind gust (a step function from 0 to 12 m/sec wind speed) is shown. The telescope shows a deflection of 7″ and takes more than 1 sec to settle within the specified value of 0.5 $\sec^{-1}$; it does not return to zero error. This is unsatisfactory behavior.

As an alternative approach, the recently developed method of a *state-*

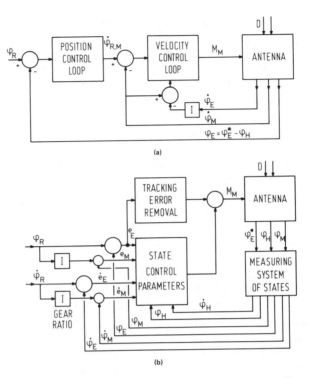

FIG. 16 Block diagrams of servo controllers: (a) classical cascade controller with velocity- and position-control loops; (b) simplified state controller, as used in the MPIfR millimeter-telescope design. The six input states for the elevation drive are the angular position and velocity of the elevation axis, drive-motor axis, and support house.

*space controller with dynamic pole assignment* [e.g., Luenberger (1979)] has been applied to the 30-m telescope. The block diagram of a simplified state controller is shown in Fig. 16b. The simplification consists in reduction of the number of required system states from 10 to 6 as a result of simulation studies. The 6 states are the angle and its time derivatives (velocity) of the elevation axis $\varphi_E$, motor axis $\varphi_M$, and support house $\varphi_H$. These must be measured to high accuracy in the cycle time of the servo system (about 10 msec). This measuring task is far from trivial. The resulting reaction to a wind gust is also shown in Fig. 17. The improvement is significant, and the system response remains within the specified value. Considerably more hardware is needed for a state controller than for a cascade controller. For the 30-m telescope, the improved performance was considered well worth the additional effort and cost.

The position-control system can be realized in other ways. Leighton

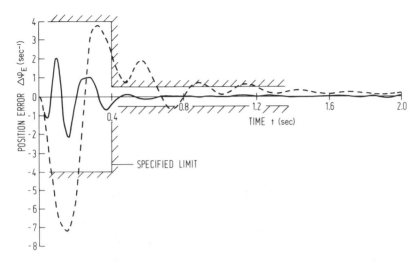

FIG. 17 Reaction of the elevation structure of the MPIfR millimeter telescope to a step function in wind speed from 0 to 12 m/sec. The state controller (——) keeps the position error within the specified value; the cascade controller (– – –) shows a large overshoot, a decay time of more than 1 sec, and a small nonzero tracking error.

(1978) has described in some detail the solution for the Caltech 10-m antenna. The Japanese 45-m telescope applies an optical *master collimator,* which should provide a pointing accuracy of 3 sec$^{-1}$ under all operation conditions. It can be said, however, that for the newest instruments with a beamwidth of 10 sec$^{-1}$ or less the accuracy of the servo-control system significantly determines the usefulness of the telescope. Radome-enclosed antennas are definitely in an advantageous position in this respect.

## XI. Electromagnetic and Operational Aspects

Strictly speaking, these subjects are outside the scope of this chapter, but because the telescope is supposed to collect electromagnetic radiation as effectively as possible and the user should be able to exploit this to obtain unambiguous astronomical data, a few remarks on these subjects are in order.

All newer millimeter telescopes employ a Cassegrain-type two-reflector system, thereby avoiding the need to locate the intricate and experimental receiver in a small and relatively inaccessible prime-focus box. Because observations in the millimeter and submillimeter range are extremely sensitive to variations in the weather, it is desirable that a quick changeover between receivers at several frequencies be possible. The space needed for a set of receivers is often not available in the central hub of the reflector support. Thus, several telescopes employ additional mirrors to direct the beam to a larger room behind the elevation axis (a so-called Nasmyth focus; Nos. 11–13 in Table I). In the Japanese telescope, between 6 and 10 additional reflections through a beam–waveguide system are used to direct the beam to a set of receivers in a large cabin located in the azimuth tower. The *quasi-optical* techniques used in these systems have been well surveyed by Goldsmith (1982).

A serious problem in the observation of spectral lines from astronomical sources arises from multipath reflections in the antenna–receiver combination; this is the "baseline ripple" problem (Allen, 1969; Penzias and Scott, 1968). Significant improvement can be obtained by choosing the correct geometry for the subreflector and its support legs to avoid multipath reflections (Morris 1978). To this end, the 30-m telescope of MPIfR/IRAM employs support legs with a sharp edge toward the main reflector and a subreflector with a conical central part and an edge with slight deviations from roundness. These should result in a more than 10 times weaker baseline ripple. The ogival shape of the support legs in ESSCO antennas is also beneficial in this respect.

It should also be remarked that the experimental character of many present-day millimeter and submillimeter receivers renders easy access to the receiver hardware essential. Thus, one might expect easier and more effective operation from telescopes that employ a Nasmyth or Cassegrain focus arrangement.

For the radio astronomer, the calibration of the gain of the telescope has always been a special problem. Over the years, techniques have been developed for using astronomical sources for the determination of antenna parameters. In the centimeter-wavelength range there is now a set of standard sources with accurately known *absolute flux density* with which antenna gain can be measured (Baars *et al.,* 1977; Findlay, 1966). Using sources of different angular extents enables additional antenna parameters to be determined (Baars, 1973).

The present situation at short millimeter wavelengths is far from satisfactory. Sources routinely used for pointing and gain measurements at centimeter wavelengths are now too weak for the relatively insensitive millimeter systems. Fortunately the planets are strong emitters at millimeter wavelengths, although, because of their finite angular size and irregular path in the sky, they are less convenient to use. Ulich, Davis, Rhodes, and Hollis (1980) have performed absolute measurements of planetary flux densities, which are very useful for antenna calibration.

Yet another problem is posed by the calibration of the intensity scale of observations of molecular spectral lines. Both the influence of the atmosphere and the large angular size of the molecular clouds render accurate calibration difficult. Ulich and Haas (1976) have described these problems. Some inconsistencies in the results of different authors now appear to be resolved, and Kutner and Ulich (1981) have presented a comprehensive method of spectral-line calibration, which — some remaining imperfections notwithstanding — deserves general acceptance so that a unified calibration scale might be achieved.

## XII. Conclusion

With the emergence of telescopes with a diameter of more than 10 m and a shortest usable wavelength of 0.5 mm or less, the technology of reflector antennas has evolved to an unsurpassed level with regard to reflector size (measured in wavelengths), reflector profile accuracy, and pointing accuracy. Along with the achievements in receiver technology, the sensitivity of radio telescopes in the millimeter range will dramatically increase in the near future. Also, special telescopes for the submillimeter windows usable from the earth will be built. The main design problems of these telescopes are reflector accuracy and measurement and thermal deformations. For the submillimeter telescope planned by the Max-Planck-Institut für Radioastronomie, the use of low-thermal-expansion carbon fiber material for both the reflector and its support structure is envisaged. An integrated support and panel structure should alleviate the measurement problem, and a surface accuracy of $\approx 10 \ \mu$m should be achieved through the use of Pyrex molds, figured by optical shop techniques to an accuracy of $\approx 2 \ \mu$m (Parks, personal communication). A similar accuracy is the goal of Caltech's submillimeter telescope.

### ACKNOWLEDGMENTS

Much of the more interesting work in the field treated in this chapter has not been published. I am very grateful to many colleagues for communicating very openly with me and for the permission to use material. I particularly thank Dr. Leighton and Dr. Moffet (Caltech), Dr. Morimoto and Dr. Kaifu (Tokyo), Dr. Hills (Cambridge), Dr. Cohen (ESSCO, Concord), and Dr. Findlay and Dr. Wong (NRAO) for fruitful discussions and for providing material. I could not have written this chapter without the close cooperation of my colleagues at the Max-Planck-Institut für Radioastronomie, Dr. Hooghoudt, Dr. Greve, Dr. Morris, and Dr. Schraml, and of the design team of Krupp and M. A. N., Dr. Eschenauer, Dr. Mäder, and Dr. Kärcher. Naturally I remain solely responsible for its content.

### REFERENCES

Allen, R. J. (1969). *Astron. Astrophys.* **3**, 316–322.
Altenhoff, W. J., *et al.* (1980). *Astron. J.* **85**, 9–12.
Baars, J. W. M. (1973). *IEEE Trans. Antennas Propag.* **AP-21**, 461–474.

Baars, J. W. M. (1980). *Proc. Int. URSI-Symp. Electro. Waves* (*Munich*), p. 143A.

Baars, J. W. M. (1981). *Mitteilungen Astron. Gesellschaft* (*Germany*) **54**, 61–69.

Baars, J. W. M., and Hooghoudt, B. G. (1980). *Proc. Opt. Infrared Telescopes 1990s* (*Tucson*), pp. 458–465.

Baars, J. W. M., and Rusch, W. V. T. (1983). "Aperture-blocking in Radio Telescopes" (in preparation).

Baars, J. W. M., Genzel, R., Pauliny-Toth, I. I. K., and Witzel, A. (1977). *Astron. Astrophys.* **61**, 99–106.

Baars, J. W. M., Greve, A., and Hooghoudt, B. G. (1983). "Thermal Control of Accurate Reflector Antennas" (in preparation).

Bennett, J. C., Anderson, A. P., McInnes, P. A., and Whitaker, A. J. T. (1976). *IEEE Trans. Antennas Propag.* **AP-24**, 295–303.

Carleton, N. P., and Hoffmann, W. F. (1978). *Phys. Today* **31**(9), 30–37.

Cheung, A. C., Rank, D. M., Townes, C. H., Thornton, D. D., and Welch, W. J. (1968). *Phys. Rev. Lett.* **21**, 1701–1705.

Cheung, A. C., Rank, D. M., Townes, C. H., Thornton, D. D., and Welch, W. J. (1969). *Nature* **221**, 626–628.

Chu, T. S., Wilson, R. W., England, R. W., Gray, D. A., and Legg, W. E. (1978). *Bell Syst. Tech. J.* **57**, 1257–1288.

Cogdell, J. R., *et al.* (1970). *IEEE Trans. Antennas Propag.* **AP-18**, 515–529.

Eschenauer, H., Brandt, P., and Schütz, K. (1977). *Tech. Mitteilungen Krupp-Forschungs-Ber.* **35**, 17–31.

Findlay, J. W. (1966). *Ann. Rev. Astron. Astrophys.* **4**, 77–94.

Findlay, J. W. (1971). *Ann. Rev. Astron. Astrophys.* **9**, 271–292.

Findlay, J. W., and Ralston, J. N. (1977). "Testing the Stepping Method on the Test Track and the 140-Foot Telescope." NRAO MM-Telescope Memos, Nos. 94–95.

Gardner, F. F., Batchelor, R. A., McCulloch, M. G., Simons, L. W., and Whiteoak, J. B. (1978). *Proc. Aust. Soc. Astron.* **3**, 264–266.

Godwin, M. P., Whitaker, A. J. T., Bennett, J. C., and Anderson, A. P. (1981). *IEE Conf. Antennas Propag. Publ.* **195**, 232–236

Goldsmith, P. F. (1982). In "Infrared and Millimeter Waves," vol. 6 (K. J. Button, ed.), pp. 277–343. Academic Press, New York.

Greve, A., and Harth, W. (1980). *1980 Eur. Conf. Opt. Syst. Appl. Proc. Soc. Photo-Opt. Inst. Eng.* **236**, 110–112.

Hachenberg, O., Grahl, B. H., and Wielebinski, R. (1973). *Proc. IEEE* **61**, 1288–1295.

Kaufmann, P., *et al.* (1976). *Rev. Bras. Tecnol.* **7**, 81–88.

Kutner, M. L., and Ulich, B. L. (1981). *Astrophys. J.* **250**, 341–348.

Kwon, O., Wyant, J. C., and Hayslett, C. R. (1980). *Appl. Opt.* **19**, 1862–1869.

Leighton, R. B. (1978). "A 10-Meter Telescope for Millimeter and Sub-Millimeter Astronomy." Technical Report, California Inst. Tech.

Luenberger, D. G. (1979). "Introduction to Dynamic Systems." Wiley, New York.

Menzel, D. H. (1976). *Sky Telescope* **52**, 240–242.

Morris, D. (1978). *Astron. Astrophys.* **67**, 221–228.

National Radio Astronomy Observatory Staff (1975, 1977). "A 25-Meter Telescope for Millimeter Wavelengths," Volumes I and II. Nat. Radio Astron. Obs., Green Bank, West Virginia.

Nobeyama Radio Observatory Staff (1981). "The 45-Meter Telescope." Nobeyama Radio Observatory Technical Report No. 6, 1–7

Payne, J. M. (1973). *Rev. Sci. Instrum.* **44**, 304–306.

Payne, J. M., Hollis, J. M., and Findlay, J. W. (1976). *Rev. Sci. Instrum.* **47**, 50–55.

Penzias, A. A., and Scott, E. H. (1968). *Astrophys. J.* **153**, L7-9.

Ruze, J. (1966). *Proc. IEEE* **54**, 633-640.

Schraml, J. (1969). *Int. Rep. Nat. Radio Astron. Obs.,* Green Bank, West Virginia.

Scott, P. F., and Ryle, M. (1977). *Mon. Not. R. Astron. Soc.* **178**, 539-545.

Shenton, D. B., and Hills, R. E. (1976). "A proposal for a United Kingdom Millimetre Wavelength Astronomy Facility." Science Research Council, London.

Stumpff, P. (1972). *Kleinheubacher Berichte* **15**, 431-437.

Ulich, B. L. (1981). *Int. J. Infrared Millimeter Waves* **2**, 293-310.

Ulich, B. L., and Haas, R. W. (1976). *Astrophys. J.* (Suppl.) **30**, 247-258.

Ulich, B. L., Davis, J. H., Rhodes, P. J., and Hollis, J. M. (1980). *IEEE Trans. Antennas Propag.* **AP-28**, 367-376.

von Hoerner, S. (1967a). *Astron. J.* **72**, 35-47.

von Hoerner, S. (1967b). *J. Struct. Div. Proc. Am. Soc. Civil Eng.* **93**, 461-485.

von Hoerner, S. (1969). *In* "Structures Technology for Large Radio and Radar Telescope Systems" (J. W. Mar and H. Liebowitz, eds.), pp. 311-333. MIT Press, Cambridge.

von Hoerner, S. (1975). *Astron. Astrophys.* **41**, 301-306.

von Hoerner, S. (1977a). *Vistas Astron.* **20**, 411-444.

von Hoerner, S. (1977b). "140-Ft. Pointing Errors After Thermal Shielding." NRAO Eng. Div. Int. Rept. No. 107.

von Hoerner, S. (1978). *IEEE Trans. Antennas Propag.* **AP-26**, 857-860.

von Hoerner, S., and Wong W. Y. (1975). *IEEE Trans. Antennas Propag.* **AP-23**, 689-695.

## CHAPTER 6

# A Gyrotron Study Program

*G. Boucher, Ph. Boulanger, P. Charbit, G. Faillon, A. Herscovici,*
*E. Kammerer, and G. Mourier*

*Thomson-CSF*
*Division Tubes Électroniques*
*Boulogne Billancourt, France*

## I. Single-Cavity Oscillators at 35 GHz

The characteristics of these single-cavity oscillators are as follows:

| | |
|---|---|
| Operating frequency | $35 \pm 0.1$ GHz |
| Peak output power (minimum) | 150 kW |
| rf output | $TE_{02}$ mode, circular waveguide |
| Pulse duration (limited by modulator) | 6 $\mu$sec |
| Beam voltage | 80 kV |
| Beam current | 8 A |
| Heater voltage | 7 V |
| Heater current | 10–11 A |
| Modulating anode voltage with respect to cathode (approximate) | 40 kV |
| Magnetic field (approximate) | 1.35 T |
| Superconducting electromagnetic current | 25 A |
| Power supply for electron-gun coil | 7/150 A/V |
| Liquid-helium consumption | 0.3 L/h |

Tests using long pulse durations of approximately 100 msec are in progress. The structure of the electron gun is compatible with continuous-wave operation.

FIG. 1  Single-cavity oscillator tube without electromagnets (overall length 94 cm).

A photograph of the tube without electromagnets is shown in Fig. 1. Figure 2 is a plot of beam current output ($I$) and power ($P$) versus cathode voltage ($U_k$), showing that a voltage stability on the order of 1% is usually sufficient. Figure 3 shows that a power of 230 kW, and probably more, can be obtained. This result corresponds to the desired oscillation in the $TE_{02}$ mode at 35.10 GHz. If the magnetic field is varied, however, the oscillation remains in the desired mode over a range of 13,250–14,000 G (Fig. 4). For values higher or lower than this range, discontinuities occur in the power–magnetic field curve and different oscillation frequencies are measured.

It has been possible to provide an output horn covered with a sheet of liquid crystal. Temperature differences on the sheet appear as different colors, so that a map of the rf electric field **E** (more exactly, of $|E|^2$) can be observed. Figure 5a shows the $TE_{02}$ mode operating at 13,700 G. If the field is lowered to 13,000 G, the frequency is 33.6 GHz, which corresponds to the $TE_{22}$ mode, and the pattern of Fig. 5b is observed. This pattern has not been clearly related to calculated values of $|E|$, but it corresponds very well to cold test measurements. It is known that the $TE_{22}$ mode forbids operation at low fields using very short pulses, and this may explain why the efficiency peaks at 32.5% for a calculated optimum value of 42% (Fig. 3). The extension of

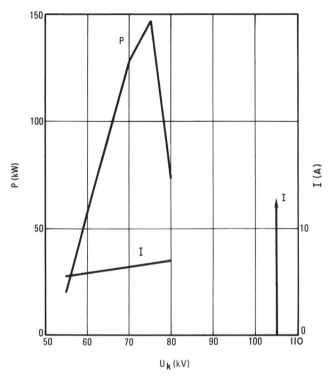

FIG. 2    Power $P$ and current $I$ vs voltage $U_k$ for $F = 35.10$ GHz and $B = 13,530$ G.

this work will include technological improvements necessary for operation at high average power and higher frequencies.

## II.    Self-Consistent Calculations of Beam–Waveguide Interaction

One way of obtaining complete solutions of Maxwell's equations for a waveguide structure excited by an electron beam of any thickness is to use an expansion of the transverse field components in eigenfunctions over the cross section of the waveguide (Charbit *et al.*, 1981; Felsen and Marcuvitz, 1973; Mourier, 1980). TM-like modes can only describe space charge. However, in many practical cases, only one mode is efficiently excited and, with a properly defined eigenfunction **e** and waveguide parameters $V$ and $I$, one can write the following for a TE mode:

$$\mathbf{E}_{\perp}(x,\, y,\, z,\, t) = V(z,\, t)\mathbf{e}(x,\, y), \tag{1}$$

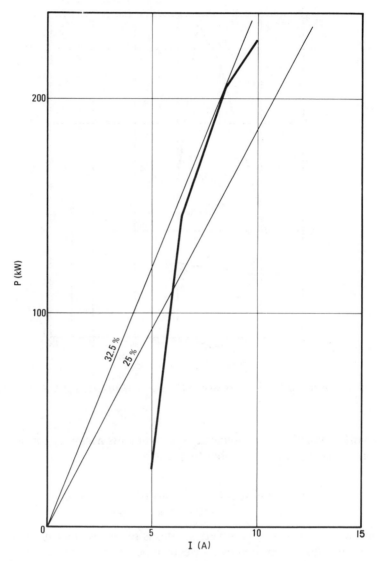

FIG. 3  Power $P$ vs current $I$ for $F = 35.10$ GHz, $V_k = 75$ kV, and $B = 13,540$ G.

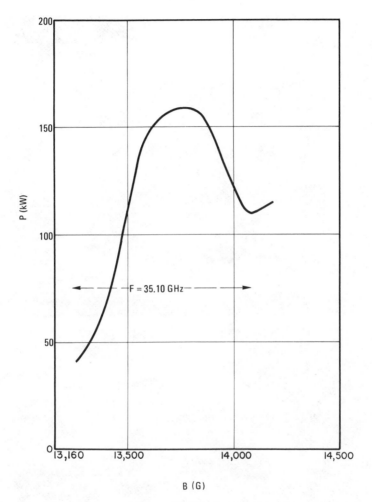

FIG. 4 Power $P$ vs magnetic field $B$ for $U_k = 77$ kV and $I = 8.1$ A: $TE_{02}$ mode.

$$\mathbf{H}_\perp(x, y, s, t) = I(z, t)\, \mathbf{w} \times \mathbf{e}(x, y), \qquad (2)$$

$$\partial V/\partial z + j\omega\mu I = 0, \qquad (3)$$

$$\partial^2 V/\partial z^2 + (\omega^2/c^2 - k_0^2)V = j\omega\mu \int_s \mathbf{e} \cdot \mathbf{J}\, ds, \qquad (4)$$

where $\mathbf{J}$ is the transverse electronic current and $s$ the cross section of the waveguide at a given $z$.

A solution obtained by relaxation has achieved an efficiency of 59%

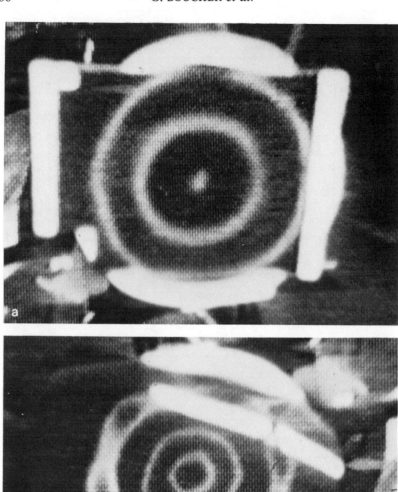

FIG. 5   The rf electric field **E** for the $TE_{02}$ mode at (a) 13,700 and (b) 13,000 G.

including the interaction in the horn (Charbit and Mourier, 1981; Charbit *et al.*, 1981). (An optimized calculation with prescribed field in a cylindrical cavity yields a maximum value of 42%.) In this solution, the calculated efficiency was 36.44% at the entrance of the horn, with prescribed electric field [$\sin(\pi z/L)$ variation], and 41.22% at the end of the horn. The horn fields are calculated from Eq. (4), with $k_0 \sim 1/z$ as in spherical wave equations and $J = 0$ (for a prescribed field) or $J = J_{el}$ (for self-consistent solutions) obtained from a calculation of electron trajectories. Thus, it appeared that the interaction in the horn might increase the efficiency if calculated non-self-consistently, but that a larger increase could be obtained with a self-consistent calculation that allows the phases of the fields to be altered by the interaction of the electrons. This solution, however, is not matched to the characteristic impedance of the horn and is therefore not typical.

Much better convergence has been achieved for the following case.

(1) Electron beam: 80 kV, 10 A, $\alpha = 1.94$.

(2) Cavity (Fig. 6): $TE_{02}$ mode, beam centered on first field maximum. The cavity is composed of a cylinder with $L_0 = 4\lambda_0$ followed by a 6° half-angle cone.

(3) Reference frequency: $f_0 = (f_{cut}^2 + c^2/4L_0^2)^{1/2}$.

(4) Magnetic field: $eB/2\pi m_0 f_0 = 1.08$.

The procedure used is as follows:

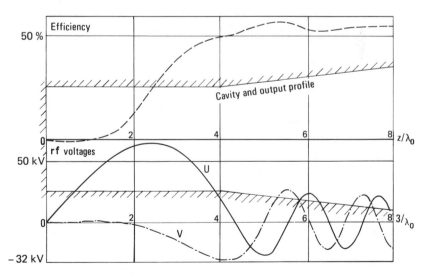

FIG. 6   Energy conversion efficiency and rf voltages along the cavity axis.

TABLE I

RESULTS FOR THREE OSCILLATION FREQUENCIES

| Frequency $f/f_0$ | Power $P_4$ (kW) | Power $P_8$ (kW) | Power[a] $P_2$ (kW) | Efficiency[a] $\eta$ (%) | Impedance $Z^a$ ($\Omega$) |
|---|---|---|---|---|---|
| 1 | 292 | 330 | 480 | 60 | $456 - 315i$ |
| 0.9985 | 296 | 312 | 455 | 57 | $525 - 14i$ |
| 0.9983 | 296 | 308 | 446 | 56 | $552 + 25i$ |

[a] For $z = 8 \lambda_0$.

(1) $J_0 = 0 \rightarrow U_0$ (in phase), $V_0$ (out of phase) $= 0$ [from Eq. (4)],
(2) $U_0, V_0 \rightarrow J_1$ (from motion equations),
(3) $J_1 \rightarrow U_1$ (in phase), $V_1$ (out of phase) [from Eq. (4)],
(4) $U_1, V_1 \rightarrow J_2$ (from motion equations),
(5) $J_2 \rightarrow U_2$ (in phase), $V_2$ (out of phase) [from Eq. (4)].

Table I summarizes the results for three different oscillation frequencies. When solving Eq. (4), the initial values $U(0)$ and $V(0)$ were taken to be zero and the maximum value of $U$ inside the cavity was kept at 66,200 V. It was found that for $z = 8\lambda_0$ the interaction has practically ceased owing to the Doppler effect. The table gives the values $P_4$ and $P_8$ of the power lost by the electrons at the end of the cylinder at $z = 4\lambda_0$ and $z = 8\lambda_0$, the power $P_2 = \frac{1}{2}$ Re $V_2 I_2^*$ carried by the only mode considered (TE$_{02}$) at $z = 8\lambda_0$, the efficiency $\eta$ calculated from $P_2$, and the circuit impedance $Z$ at $z = 8\lambda_0$ needed to match the solution.

The characteristic impedance in the horn at $z = 8\lambda_0$ is 54 $\Omega$. To obtain that value for the impedance with a prescribed field amplitude in the cavity, the values of the beam current and frequency must be adjusted. A good match is obtained for $I = 10$ A and $f/f_0 = 0.9984$. Figure 6 shows the efficiency calculated from the loss energy of the electrons along the axes as well as the values of $U_2$ and $V_2$.

It is apparent that in this case the interaction in the horn does not substantially increase the power according to a non-self-consistent calculation (compare $P_4$ and $P_8$). On the other hand, a self-consistent calculation predicts a large power increase in the horn, and a final efficiency of 56%. These calculations cannot be considered to be optimized, for neither the cavity shape nor the magnetic field has been varied.

Another way of increasing the efficiency of single-cavity gyrotron oscillators is to taper the magnetic field (Chu *et al.*, 1980; Sprangle, 1979). Efficiencies of up to 76% have been calculated by Chu *et al.* Their parameter

$N_c = L/\beta_\parallel \lambda_0$ would be $4/0.23 = 17.4$ in our calculation, and with $\alpha = 1.94$ their calculation also predicts an efficiency of somewhat less than 60%.

The two methods are similar in that they both alter the relative phases of the electron bunches and the rf field. They are therefore probably not cumulative. It appears, in any case, that non-self-consistent solutions cannot be accurate.

### III. Space-Charge Effects in Traveling-Wave Gyrotron Amplifiers (Small-Signal Theory)

There is ample evidence in the Russian literature that in operating a gyrotron the beam voltage and current must be chosen so as to avoid space-charge effects. We have discussed this fact in the light of some simple considerations (Charbit et al., 1981), some of which are based on a comparison of the perveances of different high-power microwave tubes.

Early calculations on space-charge effects in a relativistic electronic plasma (Mourier, 1978a,b) demonstrated the existence of an electron instability in a plasma slab almost irrespective of the geometric structure in which it is contained. The calculations used a tensorial conductivity or permittivity, as is often done in studies of propagation in plasmas. A relativistic permittivity was used. The calculations were later extended to propagating waves, and some results were presented (Charbit and Mourier, 1981). We have now obtained more detailed results, some of which should be, and indeed are, identical to those of other theories (Sprangle and Drobot, 1977) if space-charge effects are neglected.

Space-charge effects, of course, consist mainly of a modification of the electromagnetic field in the plane transverse to the direction of propagation and constant magnetic field. They can be dealt with by using a mode expansion. However, a description of the beam by its permittivity leads simple to a single second-order equation with variable coefficients and does away with field discontinuities. It is simpler to describe the electron cloud — assuming it has no velocity spread — in its own frame of reference, where the longitudinal velocity is zero. It is then very easy to introduce into the numerical code a Lorentz transform for the quantities of interest, primarily the angular frequency $\omega$ of the fields and their propagation constant $k$.

We shall begin by studying the interaction of a propagating wave with a population of electrons that rotate in a magnetic field parallel to the direction of propagation. The modes of interest are TE modes, which have no longitudinal electric field. Although space charge will excite a weak longitudinal field, its effect will always be much smaller than that of transverse fields, because there is no electron resonance in the longitudinal direction.

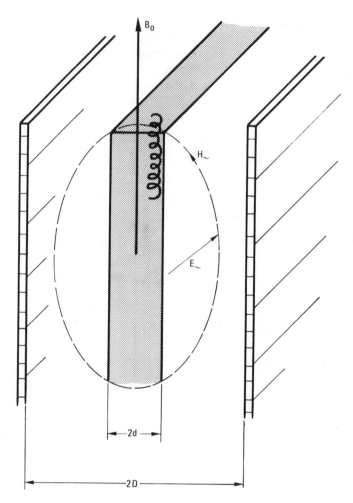

FIG. 7  Schematic diagram of a gyrotron showing field configurations.

We neglect the continuous-wave transverse electric field due to the beam. We assume that the electrons are close to resonance:

$$|\omega - \omega_c| \ll \omega \tag{5}$$

in the beam frame, where

$$\omega_c \equiv \Omega_0/\gamma \equiv eB/m_0\gamma \tag{6}$$

is the relativistic cyclotron frequency of the electrons. Equation (5) indeed justifies the neglect of rf longitudinal fields.

We then obtain a dyadic transverse electron susceptibility

$$\bar{\bar{\chi}} \equiv \chi \left\| \begin{array}{cc} 1 & j \\ -j & 1 \end{array} \right\|$$

where, in the beam frame ($v_\parallel = 0$),

$$\chi = \frac{\omega_p^2}{4\gamma\omega_c^2} \left[ \frac{\beta_\perp^2 \omega_c^2}{(\omega - \omega_c)^2} - \frac{2\omega_c}{\omega - \omega_c} \right]. \tag{8}$$

If we solve the field problem in the geometry of Fig. 7 for the lowest TE mode, we obtain the transcendental equation

$$\sqrt{\varepsilon/\mu} \tan (\omega d/c) \sqrt{\varepsilon\mu} = K \cot K(D - d)\omega/c, \tag{9a}$$

where

$$K^2 = 1 - k^2 c^2/\omega^2, \tag{10}$$

$$\varepsilon = K^2 + \chi/(1 + \chi), \tag{11a}$$

$$\mu = (1 + \chi)/[1 + \chi(1 + \omega\beta_\perp^2/2\omega_c)]. \tag{12a}$$

The quantity $\mu$ describes the magnetic moment of the orbits (Mourier, 1980). Equation (9a) can be solved numerically by approximation on a desk computer. If the beam density is low, some of the roots will closely correspond to modes propagating in the waveguide. We identify these by a subscript 0:

$$\omega_0^2 = k_0^2 c^2 + k_{cut}^2 c^2. \tag{13}$$

Some roots will lie close to the resonant frequency [Eq. (5)] and, after the Lorentz transformation, will fulfill the equation

$$\omega \simeq \omega_c + kv_\parallel. \tag{14}$$

They can be called cyclotron waves. We have calculated the roots after performing a Lorentz transformation from the beam frame ($v_\parallel = 0$) to the laboratory frame ($v_\parallel$ = electron axial velocity); we identify these roots in Eqs. (9a), (11a), and (12a).

It is possible to do away with space charge simply by neglecting $\chi + 1$ in the foregoing expressions. We then have (without space charge)

$$\varepsilon = K^2 + \chi, \tag{11b}$$

$$\mu = 1. \tag{12b}$$

Different approximations of Eq. (9a) can be made, resulting in Eqs. (9b)–(9d). If the beam is geometrically thin ($\omega d/c \ll 1$), we obtain, in the beam frame,

$$\varepsilon\omega d/c = K \cot K(D - d)\omega/c. \tag{9b}$$

If the current density in the beam is low, we can apply a first-order development of Eq. (9a) for the cold propagation case of Eq. (13),

$$\tan K_0(\omega d/c) = \cot K_0(\omega/c)(D - d),$$

with

$$K_0^2 = 1 - k_0^2 c^2/\omega_0^2. \tag{15}$$

The result, after a Lorentz transform from the beam frame to the laboratory frame, is

$$\omega^2 - k^2 c^2 - k_{\text{cut}}^2 c^2 = -\frac{\chi}{1 + \chi} \frac{2d}{D} \frac{1}{1 - \beta_{\parallel}^2} (\omega - kv_{\parallel})^2, \tag{9c}$$

where the cutoff frequency of the waveguide is $\omega_{\text{cut}} = k_{\text{cut}} c$ and

$$\chi = \frac{\omega_p^2}{4\gamma\omega_c^2} \left[ \frac{\beta_{\perp}^2 \omega_c^2}{(\omega - kv_{\parallel} - \omega_c)^2} - \frac{2\omega_c(1 - \beta_{\parallel}^2)}{\omega - kv_{\parallel} - \omega_c} \right], \tag{16}$$

in which the plasma $\omega_p$ is for any system, using the rest mass of the electron, and $\omega_c$ is for the realistic mass. The forms of (9c) and (16) without space charge [Eqs. (11b) and (12c)] are to be compared with Eq. (1) in Sprangle and Drobot (1977), which reads in our notation

$$\omega^2 - k^2 c^2 - k_{\text{cut}}^2 c^2$$
$$= -\frac{\omega_p^2}{\gamma} \frac{2d}{D} \frac{1}{4} \left[ \frac{\beta_{\perp}^2 (\omega^2 - k^2 c^2)}{(\omega - kv_{\parallel} - \omega_c)^2} - \frac{2[1 - (\omega_{\text{cut}}^2/\omega_c^2)\beta_{\perp}^2](\omega - kv_{\parallel})}{(\omega - kv_{\parallel} - \omega_c)} \right]. \tag{9d}$$

All solutions are based on the assumption of Eq. 14. Furthermore, the algebraic equations (9c) and (9d), either with or without space charge, are assumed to be restricted to the vicinity of waveguide propagation, but Eq. (9a) does not have this limitation. If these two assumptions are explicitly introduced in Eq. (9d), the difference between Eqs. (9c) and (9d) in the case without space charge essentially disappears.

We have calculated a number of solutions of Eq. (9a), (9c), and (9d) for real $\omega$ and obtained real roots or pairs of complex conjugate roots for $k$. The $\omega_{\text{Re}\,k}$ diagrams obtained from Eq. (9c) with and without space charge are shown in Figs. 8 and 9 for the set of parameters $\omega_p^2/\omega_c^2 = 2.11 \times 10^{-3}$, $\beta_{\perp} = 0.4866$, $\beta_{\parallel} = 0.23$, $d/D = 0.165$, $\omega_c/\omega_{\text{cut}} = 1.051$, and $\omega_p/\omega_{\text{cut}} = 0.048$. These values do not correspond to a practical amplifier; they were chosen to illustrate some features of the solutions.

Complex roots for $k$ are obtained, with or without space charge, in two regions:

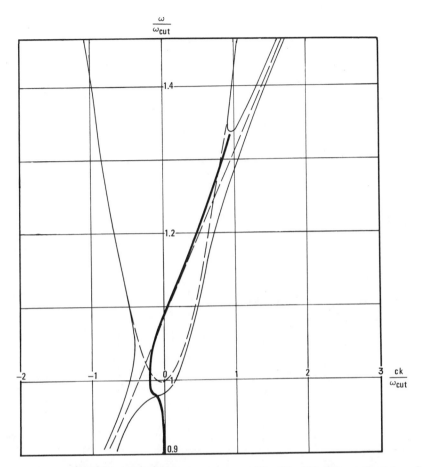

FIG. 8 Dispersion characteristics of a gyrotron amplifier [Eq. (9c) without space charge]: dashed lines, uncoupled beam and waveguide dispersion; thin solid lines, constant amplitude coupled dispersion; thick solid lines, amplified–attenuated coupled dispersion. $\beta_\perp = 0.4866$, $\beta_\parallel = 0.23$, $\omega_r/\omega_{cut} = 0.1$, $\omega_c/\gamma\omega_{cut} = 1.08$, $d/D = 0.165$.

(1) along the straight line of cyclotron resonance [Eq. (14)] inside the guided wave hyperbola [Eq.(13)], which corresponds to amplification from the beam–waveguide interaction;

(2) for $k \approx 0$ and $\omega/\omega_{cut} < 1$, which corresponds to cutoff in the waveguide with little effect from the electron beam except in the vicinity of the cutoff frequency, where the nature of the waves must be investigated.

The calculations with space charge predict amplified waves along the cyclotron resonance line on both sides of the beam–circuit amplification

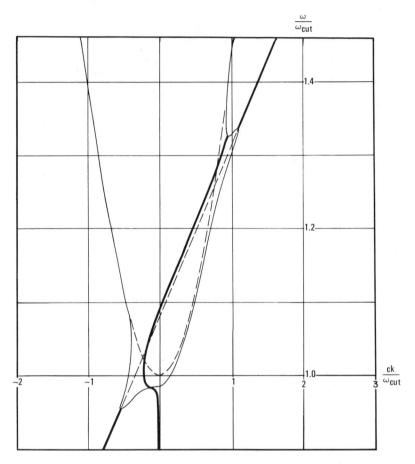

FIG. 9    Dispersion characteristics of a gyrotron amplifier [Eq. (9c) with space charge]: dashed lines, uncoupled beam and waveguide dispersion; thin solid lines, constant amplitude coupled dispersion; thick solid lines, amplified–attenuated coupled dispersion. $\beta_\perp = 0.4866$, $\beta_\parallel = 0.23$, $\omega_r/\omega_{cut} = 0.1$, $\omega_c/\gamma\omega_{cut} = 1.08$, $d/D = 0.165$.

region. These amplified cyclotron waves correspond to the instabilities obtained earlier in the beam frame for the special case $k = 0$ (Charbit *et al.*, 1981; Mourier, 1978a, 1978b, 1980). Some of the solutions given by Charbit *et al.* (1981) are erroneous, owing to a mistake in the numerical code.

The diagrams obtained from Eq. (9a) are very similar to those of Figs. 8 and 9. Equation 9(d), the most widely used in the literature, gives results very similar to those of our theoretical calculations without space charge. However, it has complex solutions with very small amplification for $\omega/\omega_{cut} < 0.9$, which to our knowledge have not been explained.

The imaginary part of $k$ as calculated from the algebraic equation (9c) with and without space charge has been plotted in Fig. 10 for the parameters $\omega_p/\omega_{cut} = 0.1$, $\beta_\perp = 0.5$, $\beta_\parallel = 0.23$, $d/D = 0.166$, and $\omega_p/\omega_{cut} = 1.08$. A beam instability appears below $\omega/\omega_{cut} = 0.93$ and above 1.28. In the central region, the beam–waveguide amplification is larger and less frequency dependent when space charge is taken into account.

In Table II the numerical values obtained from Eq. (9a), (9c), and (9d) are compared. The results of the algebraic equation (9c) are almost identical to those of Eq. (9a) everywhere, either with or without space charge, although Eq. (9c) is in principle valid only in the vicinity of waveguide propagation. The two algebraic equations without space charge, Eqs. (9c) and (9d), also

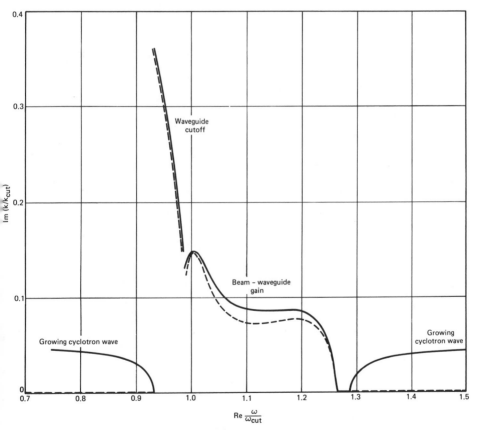

FIG. 10 Gain vs frequency in a gyrotron amplifier [Eq. (9a) with (———) and without (— —) space charge].

TABLE II

NUMERICAL VALUES OBTAINED FROM EQS. (9A), (9C), AND (9D)[a]

| Frequency $\omega/\omega_{wt}$ | Equation (9a)[b] | | Equation (9c)[c] | | Equation (9d)[d] |
|---|---|---|---|---|---|
| | With space charge | Without space charge | With space charge | Without space charge | |
| **Beam amplification** | | | | | |
| 1.40 | — | 0.0000 | 0.0430 | 0.0000 | 0.0149 |
| 1.38 | 0.0404 | 0.0000 | 0.0415 | 0.0000 | 0.0104 |
| 1.36 | 0.0382 | 0.0000 | 0.0393 | 0.0000 | 0.0000 |
| 1.34 | 0.0350 | 0.0000 | 0.0363 | 0.0000 | 0.0000 |
| 1.32 | 0.0301 | 0.0000 | 0.0314 | 0.0000 | 0.0000 |
| 1.30 | 0.0204 | 0.0000 | 0.0222 | 0.0000 | 0.0000 |
| 1.28 | 0.0000 | 0.0000 | 0.0000 | 0.0000 | 0.0000 |
| 1.26 | 0.0127 | 0.0287 | 0.0133 | 0.0290 | 0.0000 |
| **Beam-waveguide amplification** | | | | | |
| 1.24 | 0.0638 | 0.0611 | 0.0644 | 0.0616 | 0.0569 |
| 1.22 | 0.0798 | 0.0731 | 0.0805 | 0.0739 | 0.0715 |
| 1.20 | 0.0861 | 0.0769 | 0.0870 | 0.0779 | 0.0767 |
| 1.18 | 0.0876 | 0.0764 | 0.0887 | 0.0776 | 0.0770 |
| 1.16 | 0.0871 | 0.0742 | 0.0883 | 0.0755 | 0.0755 |
| 1.14 | 0.0864 | 0.0722 | 0.0876 | 0.0737 | 0.0740 |
| 1.12 | 0.0865 | 0.0717 | 0.0878 | 0.0732 | 0.0737 |
| 1.10 | 0.0880 | 0.0731 | 0.0893 | 0.0747 | 0.0749 |
| 1.08 | 0.0918 | 0.0773 | 0.0931 | 0.0789 | 0.0786 |
| 1.06 | 0.0994 | 0.0863 | 0.1008 | 0.0879 | 0.0864 |

| | | | | | |
|---|---|---|---|---|---|
| 1.04 | 0.1153 | 0.1046 | 0.1167 | 0.1062 | 0.1029 |
| 1.02 | 0.1411 | 0.1348 | 0.1425 | 0.1362 | 0.1316 |
| 1.00 | 0.1467 | 0.1463 | 0.1482 | 0.1478 | 0.1408 |
| Beam amplification | | | | | |
| 0.98 | (0.1847) 0.0000 | (0.1854) | (0.1844) 0.0000 | (0.1851) | (0.1842) |
| 0.96 | (0.2747) 0.0000 | (0.2748) | (0.2746) 0.0000 | (0.2747) | (0.2741) |
| 0.94 | (0.3381) 0.0000 | (0.3381) | (0.3380) 0.0000 | (0.3381) | (0.3376) |
| 0.92 | (0.3898) 0.0207 | (0.3898) | (0.3898) 0.0223 | (0.3898) | (0.3893) |
| 0.90 | (0.4343) 0.0302 | (0.4343) | (0.4343) 0.0315 | (0.4343) | (0.4339) |
| 0.88 | (0.4738) 0.0351 | (0.4738) | (0.4738) 0.0363 | (0.4738) | (0.4734) |
| 0.86 | (0.5093) 0.0383 | (0.5093) | (0.5093) 0.0394 | (0.5093) | (0.5090) |
| 0.04 | (0.5418) 0.0404 | (0.5418) | (0.5418) 0.0415 | (0.5418) | (0.5415) 0.0101 |
| 0.82 | (0.5717) 0.0420 | (0.5717) | (0.5717) 0.0431 | (0.5717) | (0.5714) 0.0148 |
| 0.80 | (0.5994) 0.0432 | (0.5994) | (0.5995) 0.0443 | (0.5995) | (0.5992) 0.0175 |

[a] Values in parentheses correspond to cutoff fields and not amplification.

[b] Charbit et al.

[c] Kammerer et al.

[d] Sprangle et al.

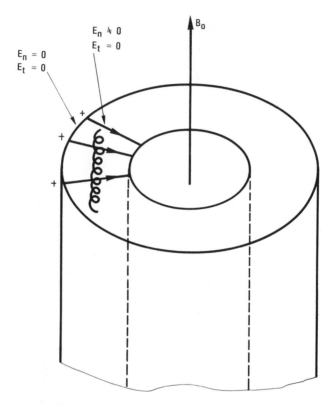

FIG. 11   Field map for a cylindrical system (cyclotron wave instabilities).

yield practically identical results, except for the weak beam instability predicted by Eq. (9d). Values in parentheses correspond to cutoff fields and not to amplification.

The agreement between our space-charge theory and other non-space-charge theories is of course not a proof of its validity. Our theory assumes, in any case, a beam thickness larger than the electron orbits and neglects the dc space-charge field. It has not yet been possible to compare other theories, with or without space charge.

We know of no experimental evidence of the existence or nonexistence of the predicted beam instabilities. The cyclotron beam instabilities that we find correspond very closely to the condition

$$\chi + 1 = 0. \tag{17}$$

In this case, it can be shown that the electric field parallel to the boundaries of the electron beam (Fig. 7) almost vanishes. In a cylindrical system

(Fig. 11), the electric field is almost radial and is confined inside the beam by surface charges due to the radial displacement of the electrons.

## IV. Proposed Scheme for an Infrared Electron Maser and Its Electron Source

At relativistic energies, the rotating frequency $\omega_c$ of an electron in a magnetic field is reduced by a factor $\gamma$ equal to the relativistic energy of the electron in mass–energy units ($\gamma = 1 + eV/m_0c^2 \simeq 1 + V_{\text{kev}}/512$), but its radiation occurs on all harmonics of $\omega_c$. The maximum of the radiation occurs around the harmonic of order $0.4\gamma^3$ (Ginzburg and Nusinovich, 1979; Schwinger, 1949). Thus, we have

$$\omega_c = eB/m = eB/\gamma m_0, \tag{18}$$

and the intensity of spontaneous emission will be distributed in frequency approximately as in Fig. 12.

Enhanced emission can be obtained by confining the radiated energy in a resonator consisting of a pair of mirrors, whereby the beam will be bunched at a frequency around $0.4\,\gamma^3\omega_{c0}$ selected by the resonator (Fig. 13). Such a radiation source would be fairly coherent and would constitute a maser, tunable in a range of at least a few percent by changing the magnetic field and the mirror spacing. The rf interaction requires a detailed analysis, but it seems that obtaining an appropriate electron beam is in no way trivial.

If, for instance, the magnetic field is 5 T, the low-energy cyclotron frequency $\omega_{c0} = eB/m_0$ is 140 GHz. Assuming a value of $\gamma = 10$ corresponding to $(\gamma - 1) \times 512 = 4.61$ MeV, radiation would occur at $0.4\gamma^2 \times 140$ GHz $= 5600$ GHz ($\gamma = 54\ \mu$m). The efficiency of such devices is expected to vary as $1/\gamma$ or $1/\gamma^2$. If an efficiency of 1% could be obtained with an electron beam current of 100 mA, a peak output power of a few kilowatts could be attained.

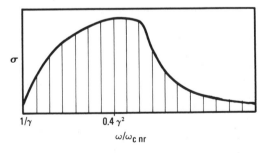

FIG. 12 Synchrotron radiation for various harmonics (nr: nonrelativistic).

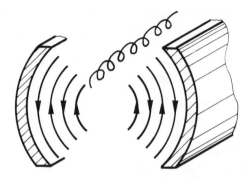

FIG. 13  A synchrotron maser.

What is needed, then, is a beam of electrons of well-defined energy spiraling in a magnetic field. For various reasons, it is desirable to keep the longitudinal velocity fairly low and also well defined. The most obvious way to impart rotational energy to the electron is by the cyclotron resonance. However, the variation in mass will produce a change of $\omega_c$ in a ratio of approximately $10:1$ for the values quoted, and if the magnetic field is tapered to compensate for this variation, a strong longitudinal force arises in the direction of the electron gun.

We propose to keep an arrangement with a growing magnetic field and accelerate the electrons longitudinally with a lower-frequency rf field (Fig. 14). If the electrons originally have an angular velocity and the process is slow enough, the angular momentum $mr^2\dot{\phi}$ remains constant and $m\dot{\phi}$ remains equal to $eB(z)$, so that the following relation holds between the state of the electron at any time and the state at time $t = t_1$:

$$m^2 r^2 \dot{\phi}^2 / m_1^2 r_1^2 \dot{\phi}_1 = B(z)/B(z_1). \qquad (19)$$

If we assume that $r\dot{\phi}$ approaches the velocity of light and $m_1$ is close to $m_0$, we have for the final value of $\gamma$

$$\gamma \simeq (r_1 \dot{\phi}_1 / c)\sqrt{B(z)/B(z_1)}. \qquad (20)$$

Note that $r^2 B$ is a constant, so that

$$\gamma \simeq (r_1 \dot{\phi}_1 / c)(r_1 / r). \qquad (21)$$

Thus, the linear compression in the case of a 5-T magnetic field may be greater than 10.

Consider again Fig. 14, where the accelerating structure is represented by a periodic structure. Because the angular momentum is a constant of the motion in the accelerator, it must be already present at the entrance. This can be achieved using a magnetron-type gun with an annular cathode and a

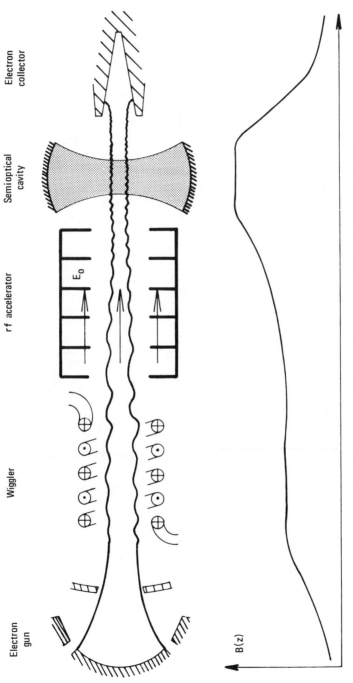

FIG. 14 Schematic representation of a synchrotron maser with an internal rf accelerator.

cusped magnetic field. The system shown is a Pierce-type gun, which produces a confined beam, followed by a wiggler. In the linear accelerator, the phase velocity should remain substantially equal to the longitudinal velocity $v_z$ or $v_\parallel$ $f$ $\beta_\parallel c$ of the electrons, and therefore the accelerator circuit should guide a slow wave; hence its periodic structure.

We shall now analyze the operation of the accelerator. Electrons enter at any phase $\alpha$ of the rf field and have different energies for the same position $z$, defined by functions $\gamma(z, \alpha)$. We shall examine first the case of electrons of one particular phase $\alpha_0$ and then the behavior of the other electrons.

One can assume that, for electrons of the phase $\alpha_0$, the longitudinal electrical field as well as $\beta_\parallel$ are constant along $z$. $B_z(z)$ is chosen so that the decelerating magnetic force balances the rf electric force, the work of the latter being used to increase the transverse velocity. Thus, we neglect the longitudinal force due to rf magnetic field of the accelerator; this can be done only if the rf frequency is low enough.

If $E_0$ is the field experienced by electrons of phase $\alpha_0$, the law for $\gamma$ is simply

$$\gamma(z, \alpha_0) = \gamma_1 + eE_0 z, \tag{22}$$

and we obtain from Eq. (19)

$$\frac{B(z)}{B_1} = \frac{\gamma^2(1 - \beta_{\parallel 1}^2) - 1}{\gamma_1^2(1 - \beta_{\parallel 1}^2) - 1} = \frac{(\gamma_1 + eE_0 z)^2(1 - \beta_{\parallel 1}^2) - 1}{\gamma_1^2(1 - \beta_{\parallel 1}^2) - 1}. \tag{23}$$

Assuming $\gamma_2 = 10$ and $\beta_\parallel = 0.2$, we have

$$\begin{aligned} B_2/B_1 &= 81.9 \quad &&\text{if} \quad \gamma_1 = 1.5 \\ &= 248 \quad &&\text{if} \quad \gamma_1 = 1.2. \end{aligned}$$

The convergence clearly becomes very high if the initial electron energy is low.

We examine the behavior of the other electrons by integrating the set of equations

$$\frac{dW}{dt} = ev_\parallel E(z, t, \alpha), \tag{24}$$

$$E(z, t, \alpha) = E_0 \cos(\omega t - kz + \alpha)/\cos \alpha_0, \tag{25}$$

$$\frac{dv_\parallel}{dt} = \left(1 - \frac{v_\parallel^2}{c^2}\right)\frac{e}{m}E + \frac{e}{m}v_\perp\left(-\frac{r}{2}\frac{dB}{dz}\right), \tag{26}$$

$$v_\perp = r(e/m)B, \tag{27}$$

$$\frac{dz}{dt} = v_{\parallel}, \tag{28}$$

$$W = mc^2 = m_0 c^2 [1 - (v_{\parallel}^2 + v_{\perp}^2)/c^2]^{-1/2}. \tag{29}$$

Figures 15–17 show significant results for initial values of $v_{\perp} = 0.5c$, $v_{\parallel} = 0.3c$, and $\gamma = 1.231$ (118 keV) and $\alpha_0 = \pi/3$, $E_0 = 0.05\ m_0\omega c/e$, and $z_{\max} = 15\ 2\pi c/\omega$. The magnetic field scales inversely with the transverse dimensions.

Figure 15 shows how the energy of a number of test particles increases along the axes. Figure 16 shows how many of the particles shift initially in phase to be accelerated at the same rate, in spite of some decaying oscilla-

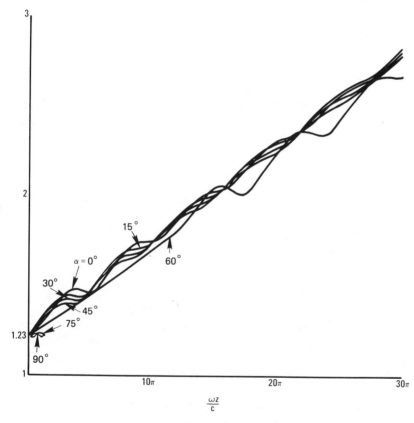

FIG. 15 Energy $\gamma$ vs normalized distance $\omega z/c$ in the accelerator.

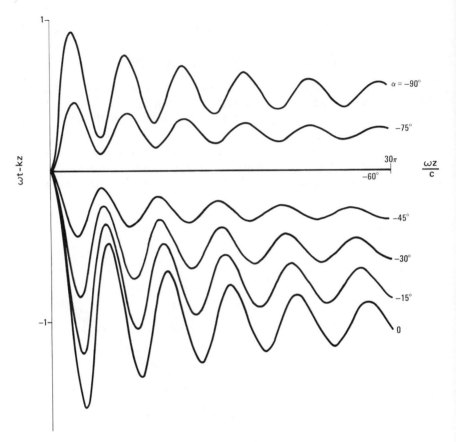

FIG. 16   Phase $\alpha$ vs normalized distance $\omega z/c$ in the accelerator for $\beta_{\perp 0} = 0.5$, $\beta_{\parallel 0} = v\gamma/c = \omega/kc = 0.3$, $eE/m_0\omega_c = 0.1$, and $\alpha_0 = \pi/3$.

tions around the stable phase. Figure 17 shows that particles with initial phases of between $-150°$ and $+60°$ have a stable acceleration. The other particles are reflected back. The particle taken as the initial test particle ($\alpha_0 = 60°$) has "indifferent" equilibrium, but the particle at $\alpha = 60°$ has the same theoretical acceleration and is stable. It is clear from Figs. 16–17 that the other stable particles bunch longitudinally around this particle. We can say nothing in this analysis about azimuthal bunching. Owing to the decaying character of the oscillations of the phases and of the energy, it is clear that the integration could have been carried out further with a linear increase in the energy.

The type of accelerator proposed seems to be feasible and capable of

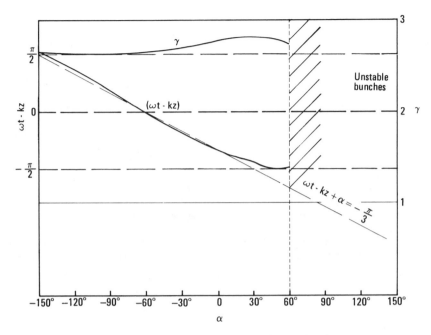

FIG. 17 Final energy $\gamma$ and phase $\alpha$ after acceleration.

handling large currents, although some problems are posed by the large variations along the axis of the magnetic field, of the Larmor radius, and of the accelerating structure. Detailed analyses of both the maser interaction and the accelerating section — with all terms included — are needed, and the device should be compared with a free-electron laser.

## V. Conclusion

We have built a 200-kW, 35-GHz gyrotron oscillator and tested it using short pulses. Tests using long pulses are in progress, and modifications for continuous-wave operation are planned, as well as extensions to higher frequencies.

The theoretical study of traveling-wave amplifiers has been completed and compared successfully to other theories by neglecting space charge. Growing cyclotron waves are predicted.

A tentative scheme for a synchrotron maser using a new type of accelerator has been proposed, and some of the properties of the accelerator — notably, its phase stability — have been numerically investigated.

REFERENCES

Charbit, P., and Mourier, G. (1981). *IEEE Int. Conf. Plasma Science, Santa Fe, New Mexico,* May 18–20. p. 123.

Charbit, P., Herscovici, A., and Mourier, G. (1981). *Int. J. Elec.* **51**(4), 303–330.

Chu, K. R., Read, M. E., and Ganguly, A. K. (1980). *IEEE Trans. Microwave Theory Tech.* **MTT-28**(4), 318–325.

Felsen, L. B., and Marcuvitz, N. (1973). *In* "Radiation and scattering of waves," chapter II. Prentice-Hall, London.

Ginzburg, N. S., and Nusinovich, G. S. (1979). *Izv. Vyz. Uch. Zav. Radiofizika* **6**, 754–753.

Mourier, G. (1978a). *Int. Symp. Heating Toroïdal Plasmas, Grenoble, France,* July 3–7.

Mourier, G. (1978b). *8th Eur. Microwave Conf. Paris,* Sept. 4–8. Microwave Exhibitions and Publishers Ltd., Kent, England.

Mourier, G. (1980). *Elec. Commun.* **34**(12), 473–484.

Schwinger, J. (1949). *Phys. Rev.* **75**(12), pp. 1912–1925.

Sprangle, P. (1979). "The non-linear theory of efficiency enhancement in the electron cyclotron maser." NRL Memorandum Report 3983.

Sprangle, P., and Drobot, A. T. (1977). The linear and self-consistent nonlinear theory of the electron cyclotron maser instability. *IEEE Trans. Microwave Theory Tech.* **MTT-25**(6), 528–544.

CHAPTER 7

# Multimode Analysis of Quasi-Optical Gyrotrons and Gyroklystrons

*Anders Bondeson**

Laboratory for Plasma and Fusion Energy Studies
University of Maryland
College Park, Maryland

*Wallace M. Manheimer*

Plasma Theory Branch, Plasma Dynamics Division
Naval Research Laboratory
Washington, D.C.

*Edward Ott*

Laboratory for Plasma and Fusion Energy Studies
University of Maryland
College Park, Maryland

Plasma Theory Branch, Plasma Dynamics Division
Naval Research Laboratory
Washington, D.C.

* Present Address: Institute for Electromagnetic Field Theory, Chalmers University of Technology, 412 96 Göteborg, Sweden.

## I. Introduction

Electron cyclotron masers (also called gyrotrons) have demonstrated efficient high-power generation of electromagnetic waves at centimeter wavelengths (Symons and Jory, 1981). For many applications it is of great practical interest to develop high-power generation capability also at *millimeter* wavelengths. (One example is the use of such waves for plasma heating in controlled thermonuclear fusion reactors.) Currently no high-power millimeter-wave source exists. Simple scaling of present-day gyrotrons to millimeter wavelengths implies a scaling down in size and power capability because of the upper limit on electric field and hence power density imposed by considerations of breakdown. To achieve high power ($\gtrsim 10^2$ kW), the dimensions of the resonant wave cavity must be large compared with the millimeter wavelengths of interest here. Resonant cavities with dimensions large compared to the wavelength yield a very dense spectrum of resonant frequencies. In fact, for a closed cavity of typical dimension $L$ the spacing $\Delta\omega$ between resonant frequencies scales like $\Delta\omega/\omega \sim (\lambda/L)^3$, where $\lambda$ is the wavelength. The small $\Delta\omega/\omega$ would be expected to make mode selection and the achievement of coherent operation impossible at millimeter wavelengths. Thus it has been proposed that a quasi-optical open-cavity configuration be used. Because higher-order transverse modes are suppressed by diffraction losses, the spectrum is essentially one dimensional and the mode density in $\omega$ is vastly reduced so that $\Delta\omega/\omega \sim \lambda/L$.

The configuration currently of most interest is shown in Fig. 1, where the electron beam propagation direction is transverse to the direction of electromagnetic wave propagation. Here the gyrating electron beam propagates along a magnetic field through a quasi-optical cavity formed by two curved

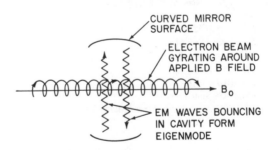

Fig. 1   Single-cavity quasi-optical gyrotron.

mirror segments. An experiment using this concept is presently under construction at the Naval Research Laboratory. This configuration has the advantage of being very insensitive to the thermal spread in pitch angle of the beam. Single-mode nonlinear analyses of the configuration in Fig. 1 have been carried out by Rapoport *et al.* (1967) and Sprangle *et al.* (1981). In these analyses a particular mode frequency and amplitude are selected and the orbits of the beam particles are solved for. The resulting efficiency of the device is then calculated by comparing the input particle energy flux to that after traversal through the interaction region. The single-mode analyses indicated the very encouraging result that good efficiency for such devices is possible. The major questions left unanswered by these analyses are whether a strongly overmoded device will indeed achieve single-mode operation, and, if so, which mode will be selected. The question of mode selection is extremely important, because the nonlinearly saturated single-mode efficiency is strongly dependent on the mode number. Furthermore, it is also possible that the saturated state may involve more than one mode, may be time dependent, or may even be highly turbulent.

This chapter has two purposes. The first is to remove the above-mentioned restrictions of previous analyses by performing a multimode time-dependent analysis. The second is to examine the possibility of improving device performance by utilizing a variant of the original scheme, here called the quasi-optical gyroklystron.

The gyroklystron is a two-cavity system, in which the first quasi-optical cavity is fed with low-power radiation and is utilized to give the beam particles a gyrophase-dependent energy increase (Fig. 2). Owing to the relativistic mass effect, the beam then ballistically bunches in gyration angle in a drift region between the two cavities. Thus, when the beam enters the second cavity it efficiently transfers energy to the high-power radiation there. There are several possible advantages to this scheme. First, the

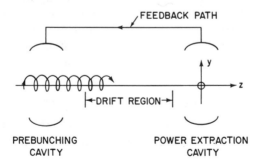

FIG. 2  Quasi-optical gyroklystron.

prebunching should give rise to higher efficiency. Second, one might expect a double-cavity system to be more mode selective than a single-cavity system. For instance, the two cavities could have different lengths so that they are simultaneously resonant at only the desired frequency. Gyroklystrons operating with ordinary microwave cavities have been investigated by Ergakov *et al.* (1979), Ganguly and Chu (1981), Symons and Jory (1977), and Jory *et al.* (1977). The field contouring efficiency enhancement schemes used by Sprangle *et al.* (1981) are also applicable in a quasi-optical gyroklystron. We do in fact find that maximum efficiency is obtained with nonuniform fields.

This chapter is organized as follows. In Section II we give the field equations and particle orbit equations including the case of operation at harmonics of the cyclotron frequency. In Section III we discuss the linear theory of the quasi-optical gyroklystron and show that the linear mechanism of the klystron is fundamentally different from that of the single-cavity gyrotron. The latter is a "negative mass" driving mechanism and operates only for frequencies above the relativistic cyclotron frequency (Sprangle and Manheimer, 1975). The former, however, is not a negative mass mechanism. Its efficiency maximizes at the relativistic cyclotron frequency and falls off sharply on either side. Linear theory also shows the klystron to be more frequency selective than the single-cavity gyrotron if the drift region is longer than the cavity size. Finally, it is shown that the quasi-optical klystron is relatively insensitive to beam thermal spread as long as the thermal spread is in pitch angle and not in energy. In fact, a thermal spread in pitch angle principally decreases the bandwidth over which the phase bunching is effective. However, a thermal spread in beam energy greatly weakens the power transfer from the beam to the radiation. In Section IV we present detailed nonlinear numerical results for different beam shapes and magnetic field profiles and derive an approximate scaling law for the efficiency. In Section V we develop the nonlinear multimode formulation. In Section VI we present the results of several multimode simulations. We find that in many instances the saturated state is dominated by one single mode. We also find that the contouring of the magnetic field controls the process of mode selection, and that, with the proper contouring, nonlinear mode competition favors modes that give high efficiency.

## II. Field and Particle Equations

In this section we shall develop the equations that form the basis of the subsequent analysis for both the double-cavity gyroklystron and the single-cavity gyrotron.

In each gyroklystron cavity or in the single cavity of the gyrotron, the high-frequency electric field $\mathbf{E}$ and magnetic field $\mathbf{B}$ of a single mode are given by

$$E_x(x, y, z, t) = E(x, y, z) \sin[ky + \alpha(x, y, z)] \cos \omega t \tag{1a}$$

and

$$cB_z(x, y, z, t) = E(x, y, z) \cos[ky + \alpha(x, y, z)] \sin \omega t, \tag{1b}$$

where $x$, $y$, $z$ are rectangular coordinates, $c$ is the speed of light, $\omega$ is the angular frequency, $k = \omega/c$,

$$E(x, y, z) = E_0[r_0/r_s(y)] \exp[-(x^2 + z^2)/r_s^2(y)],$$

$$\alpha = R^{-1}(y)(x^2 + z^2)\omega/2c - \tan^{-1}(y/y_R),$$

$r_0$ is the minimum spot size at $y = 0$, $r_s(y) = r_0(1 + y^2/y_R^2)$ is the spot size in the plane $y = \text{const}$, $y_R = r_0^2\omega/2c$ is the Rayleigh length, and $R(y) = y(1 + y_R^2/y^2)$ is the radius of curvature of the spherical wave front. Only the lowest-order transverse modes—those with an $x-z$ dependence $\exp[-(x^2 + z^2)/r_s^2]$—are included, because higher-order modes suffer larger diffraction losses and can thus be eliminated. In the analysis that follows, we envisage a beam localized near the center of the two cavities. As it travels through, it samples all $z$. However, if $y \ll y_R$, then

$$\alpha \simeq 0, \tag{2a}$$

$$E(x, y, z) \simeq E_0 \exp(-z^2/r_0^2). \tag{2b}$$

To simplify the analysis we make this approximation in all subsequent calculations.

We assume that there is an ambient magnetic field of magnitude $B_0$ in the $z$ direction and that a beam of electrons, each having a transverse momentum $p_\perp$ and parallel momentum $p_z$, traverses the cavity (gyrotron) or cavities (klystron) from a source at the left to a dump at the right (cf. Figs. 1 and 2). On traversing the power extraction cavity, the beam loses some of its energy flux, which is transferred to the radiation fields. To calculate this power loss to radiation, we integrate the particle equations from the source to the dump.

To describe the particle motion, we use as dependent variables $p_\perp$, $\psi$, $x_g$, and $y_g$, which are related to $p_x$, $p_y$, $x$, and $y$ by

$$p_x = p_\perp \cos \psi, \tag{3a}$$

$$p_y = p_\perp \sin \psi, \tag{3b}$$

$$y_g = y + p_x/m\Omega, \tag{3c}$$

$$x_g = x - p_y/m\Omega, \tag{3d}$$

where $\Omega$ is the nonrelativistic cyclotron frequency in the unperturbed field $\Omega = eB_0/m$ and $m$ the electron rest mass. The particle equations then reduce to

$$dp_\perp/dz = (-em\gamma/p_z) \cos \psi\, E_x, \tag{4}$$

$$d\psi/dz = (m\Omega/p_z) + (eB_z/p_z) + (m\gamma e \sin \psi\, E_x/p_\perp p_z), \tag{5}$$

$$dx_g/dz = (-p_x/p_z)(B_z/B_0), \tag{6}$$

$$dy_g/dz = (-e\gamma/p_z\Omega)[E_x + (p_y/m\gamma)B_z], \tag{7}$$

where $\gamma^2 = 1 + (p_\perp^2 + p_z^2)/m^2c^2$ and $E_x$ and $B_z$ are given in terms of r and $t$ by Eqs. (1) and (2). Here $z$ is regarded as the independent variable and $t$ is related to $z$ for each particle by

$$dt/dz = m\gamma/p_z. \tag{8}$$

Let us also note that for the polarization given in Eq. (1) $p_z$ is a constant of motion.

Equations (4)–(8) are much simpler to examine (analytically or numerically) if one exploits the fact that the coupling between the particles and the field is strong only when $\omega \simeq n\Omega/\gamma$ and $n$ is an integer. Then the right-hand sides of Eqs. (4)–(8) will be the sum of quantities that oscillate rapidly in $z$ and quantities that oscillate slowly in $z$. The key approximation is that to first order in $E/cB_0$ one can neglect the coupling between the rapidly and slowly varying terms, so that only terms oscillating slowly in $z$ are retained on the right-hand sides of Eqs. (4)–(8).

We shall now illustrate the calculation of the slowly varying part of the right-hand side of Eq. (4). Using $y = y_g - (p_\perp/m\Omega) \cos \psi$ in Eq. (4) and the identity

$$\sin k\left(y_g - \frac{p_\perp}{m\Omega} \cos \psi\right)$$

$$= \sin ky_g \left[J_0(Q) + 2 \sum_{l=1}^\infty (-1)^l J_{2l}(Q) \cos 2l\psi\right]$$

$$- 2 \cos ky_g \sum_{l=0}^\infty (-1)^l J_{2l+1}(Q) \cos (2l + 1)\psi,$$

where $Q = kp_\perp/m\Omega = \omega\gamma\beta_\perp/\Omega$, we find that

$$\frac{dp_\perp}{dz} = -\frac{em\gamma}{p_z} \cos \psi\, E(z) \cos \omega t$$

$$\times \left\{\sin ky_g \left[J_0(Q) + 2 \sum_{l=1}^\infty (-1)^l J_l(Q) \cos 2l\psi\right]\right.$$

$$\left. - 2 \cos ky_g \sum_{l=0}^\infty (-1)^l J_{2l+1}(Q) \cos (2l + 1)\psi\right\}. \tag{9}$$

Now, if we assume $\omega \simeq n\Omega/\gamma$ and take into account that to lowest order $d\psi/dt \simeq \Omega/\gamma$, the low-frequency term on the right-hand side of Eq. (9) gives rise to

$$\frac{dp_\perp}{dz} = -\frac{em\gamma}{p_z} E(z)J_n'(Q) \cos(\omega t - n\psi) \cos(ky_g - n\pi/2). \quad (10)$$

A similar calculation yields

$$\frac{d\psi}{dz} = \frac{m\Omega}{p_z} - \frac{eE(z)m\gamma}{p_z p_\perp} \frac{nJ_n(Q)}{Q}\left(1 - \frac{p_\perp}{m\gamma c}\frac{Q}{n}\right)$$
$$\times \sin(\omega t - n\psi) \cos(ky_g - n\pi/2) \quad (11)$$

and

$$\frac{dy_g}{dz} = \frac{e\gamma E(z)}{p_z\Omega} \cos(\omega t - n\psi)\left(\frac{n\Omega}{\gamma\omega} - 1\right)J_n(Q) \sin(ky_g - n\pi/2). \quad (12)$$

Because, in the approximation specified by Eq. (2), the electric field is independent of $x$, an equation for $x_g$ is not needed. As is apparent from Eqs. (10)–(12), there is a strong wave particle interaction whenever $\omega$ is near a harmonic of the cyclotron frequency. However, if we wish to consider a gyrotron working at high efficiency, the relativistic cyclotron frequency of the particles experiences a change of order $\Omega(1 - 1/\gamma_0)$, where $\gamma_0$ is the input $\gamma$. Thus, the separation of time scales is valid only if

$$n(\gamma_0 - 1) \ll 1. \quad (13)$$

In the remainder of this chapter we assume $n = 1$ and $p_\perp/mc \ll 1$. If we introduce the slowly varying phase angle

$$\theta = \psi - \omega t, \quad (14)$$

Eqs. (10)–(12) reduce to

$$dp_\perp/dz = (-em\gamma/2p_z)E(z) \sin ky_g \cos \theta, \quad (15)$$
$$d\theta/dz = (m/p_z)(\Omega - \gamma\omega) + (m\gamma eE(z)/2p_\perp p_z) \sin ky_g \sin \theta, \quad (16)$$
$$dy_g/dz \simeq 0, \quad (17)$$

where we have neglected terms of order $E(z)(\omega - \Omega/\gamma)$. Equations (15)–(17) are the equations we shall analyze in the remainder of this chapter.

## III. Linear Theory of the Quasi-Optical Gyroklystron

In a conventional gyrotron (Rapoport *et al.*, 1967; Sprangle *et al.*, 1981; Symons and Jory, 1981), the power extraction occurs in the same cavity as the phase bunching. For the fields to gain energy from the phase-bunched beam, $\omega$ must be somewhat larger than $\Omega/\gamma_0$ so that the gyrophases of the particles lag the wave phase. This causes the slow phase $\theta$ of the bunches to be less than $\pi/2$ [its value at exact resonance; see Eqs. (15) and (16)]; hence, most particles lose energy. If $\omega < \Omega/\gamma_0$, the bunches occur at $\theta > \pi/2$ and the particles gain energy. Thus the linearized efficiency is asymmetric in the frequency mismatch $\omega - \Omega/\gamma_0$ (Sprangle, Vomvoridis, and Manheimer, 1981). In the klystron, we obtain the necessary phase relation by controlling the phase difference between the prebunching and power extraction cavities, and the linearized efficiency maximizes at exact resonance $\omega - \Omega/\gamma_0 = 0$. In what follows we shall use subscripts 1 and 2 to denote the first (prebunching) and second (power extraction) cavities (Fig. 2).

## A.   LINEAR THEORY WITHOUT THERMAL SPREAD

Throughout this section we shall use linear theory for the traversal of the electrons through each of the two cavities. However, the nonlinearity due to phase bunching in the drift region can be dealt with analytically, and we shall include this nonlinearity to demonstrate the importance of optimizing the prebunching field.

To investigate the phase bunching, we first calculate $\Delta p_\perp$, the change in $p_\perp$ that a particle experiences on traversing the first cavity. An unperturbed orbit has

$$\theta = \theta_0 + (m/p_z)(\Omega - \gamma\omega)z,$$

where $\theta_0$ is the slow phase at $z = 0$, the center of the first cavity. Assuming a low electric field amplitude in the first cavity, Eq. (15) yields

$$\begin{aligned}
\Delta p_\perp &= -\frac{em\gamma}{2p_z} \int_{-\infty}^{\infty} E(z) \sin ky_g \\
&\quad \times \cos\left[\theta_0 + \frac{m}{p_z}(\Omega - \gamma\omega)z\right] dz \\
&= -\frac{\sqrt{\pi}em\gamma r_{01}}{2p_z} E_{01} \sin ky_g \cos \theta_0 \\
&\quad \times \exp\left[-\frac{(\Omega - \gamma\omega)^2 r_{01}^2 m^2}{4p_z^2}\right].
\end{aligned} \tag{18}$$

The change in $p_\perp$ leads to a change in $\gamma = [1 + (p_\perp^2 + p_z^2)/m^2c^2]^{1/2}$, and the particles gyrate at different frequencies in the drift region. As the beam propagates, the more rapidly gyrating particles ($\Delta p_\perp < 0$) tend to catch up with the more slowly gyrating particles and *phase* bunching occurs. The idea of the gyroklystron is to adjust the amplitude and phase of the prebunching field so that the beam enters the second cavity strongly bunched at the right phase angle and thereby loses power efficiently to the fields.

In the drift region, $E = 0$ and Eq. (16) yields

$$\frac{d\theta}{dz} = \frac{m}{p_z}(\Omega - \gamma\omega) \simeq \frac{m}{p_z}\left(\Omega - \gamma_0\omega - \frac{p_\perp \Delta p_\perp}{\gamma_0 m^2 c^2}\omega\right), \qquad (19)$$

where $\gamma_0$ is the relativistic factor of the particle before it enters the first cavity. If the drift region has length $L$ and a uniform magnetic field, the slow phase in the second cavity will be $\theta_0 + \Delta\theta$, where

$$\Delta\theta = [m(\Omega - \gamma_0\omega)/p_z](L + z) + q \cos\theta_0 \sin ky_g \qquad (20)$$

and

$$q = \frac{\sqrt{\pi}\, \omega p_\perp e E_{01} r_{01} L}{2 p_z^2 c^2}\exp\left[-\frac{(\Omega - \gamma_0\omega)^2 r_{01}^2 m^2}{4 p_z^2}\right]. \qquad (21)$$

In Eq. (20) we have redefined the variable $z$ so that $z = 0$ is now at the center of the second cavity, Eq. (21) we have neglected the $z$ dependence of the phase bunching, which is permissible if $L \gg r_{02}$.

The energy that the particle transfers to the wave fields on traversing the second cavity is

$$\Delta W = e \int v_x E_x \, dt = e \int (p_x/p_z)\, E_x \, dz. \qquad (22)$$

The integral in Eq. (22) in general requires solution of the nonlinear particle orbit equations. Analytical results can, however, be obtained for small field amplitudes. In this case

$$p_x = p_\perp \cos\psi = p_\perp \cos(\omega t + \theta_0 + \Delta\theta),$$

with $\Delta\theta$ given by Eq. (20). Substituting this $p_x$ and

$$E_x = E_{02} \sin ky \exp(-z^2/r_{02}^2) \cos(\omega t + \phi_0),$$

where $\phi_0$ is the temporal phase shift between the two cavities, into Eq. (22) and averaging over the gyromotion, we obtain the single-particle energy loss

$$\Delta W = \frac{e r_{02}\sqrt{\pi}\, E_{02} p_\perp}{2 p_z}\exp\left[-\frac{(\Omega - \gamma_0\omega)^2 r_{02}^2 m^2}{4 p_z^2}\right]\sin ky_g$$

$$\times \cos\left[\frac{m(\Omega - \gamma_0\omega)}{p_z}L - \phi_0 + \theta_0 + q \sin ky_g \cos\theta_0\right]. \qquad (23)$$

The average energy loss for a uniform distribution in $\theta_0$ and an arbitrary distribution $f_y(ky_g)$ in $y_g$ becomes

$$\langle \Delta W \rangle = \frac{er_{02}\sqrt{\pi}\,E_{02}p_\perp}{2p_z} \exp\left[-\frac{(\Omega - \gamma_0\omega)^2 r_{02}^2 m^2}{4p_z^2}\right]$$

$$\times \sin\left[\phi_0 - \frac{m(\Omega - \gamma_0\omega)}{p_z}L\right]F(q), \qquad (24a)$$

where

$$F(q) = \int J_1(q\sin\xi)\sin\xi\, f_y(\xi)\,d\xi \qquad (24b)$$

Specifically, a pencil beam located at a field maximum gives $F(q) = J_1(q)$, whereas a sheet beam in the $y$ direction gives $F(q) = J_0(q/2)J_1(q/2)$. We see that each beam profile has a specific optimum bunching field at which $F(q)$ is maximum; for the sheet beam $F_{max} = F(1.96) = 0.339$ and for the pencil beam $F_{max} = F(1.84) = 0.582$.

Equation (24b) illustrates how the output in the linear regime depends on the prebunching. In the nonlinear calculations, we shall still be using linear theory in the first cavity [i.e., Eq. (20)], because $E_{01}$ is presumed small. However, it will be necessary to optimize the parameters $\phi_0$ and $q$ numerically, because the power lost is no longer a harmonic function of the entrance angle to the second cavity.

## B. THERMAL EFFECTS

We shall now consider the effect of a velocity spread in the beam. It is characteristic of gyrotron electron guns that the velocity spread is due primarily to a spread in pitch angle rather than in energy. Thus, with $p_\perp = p\sin\alpha$ and $p_z = p\cos\alpha$, where $p = mc(\gamma^2 - 1)^{1/2}$, our main concern will be with a spread in $\alpha$, although we shall also consider a spread in $\gamma$. It will be shown that the prebunching is relatively insensitive to a spread in $\alpha$ if the frequency mismatch $\omega - \Omega/\gamma_0$ is properly chosen. This is in marked contrast to the conventional traveling-wave tube gyrotron, where the presence of a $k_z v_z$ term in the resonance condition $\omega = \Omega/\gamma + k_z v_z$ makes the efficiency sensitive to a spread in pitch angle. (For the configuration here, $k_z \equiv 0$.)

The power loss of a thermal beam can be calculated by averaging Eq. (24) over a distribution in $\alpha$ and $\gamma$. The dominant $\alpha$ and $\gamma$ dependence in Eq. (24a) is in the sine factor, because its argument is related to the change in the slow phase of the particles over the whole drift region $L$, whereas the exponential factor results from the evolution of $\theta$ over a much shorter interval, the radiation waist of the second cavity. Thus, the linearized efficiency of a gyroklystron with a thermal beam can be found simply by averaging the sine factor in Eq. (24a). In addition to thermal spreads, we also wish to consider the effect of $B$ field nonuniformity in the drift region. To that end we calculate the contributions to $d\theta/dz$ from variations of $B$ over the drift region using Eq. (19). Let $B(z) = B_0 + \delta B(z)$, where $\langle \delta B \rangle = L^{-1} \int_0^L \delta B \, dz = 0$, $\Omega = \Omega_0 + \delta\Omega$, etc. Using the conservation of the magnetic moment $v_\perp^2/B$, we obtain

$$\delta v_z = (-v_{\perp 0}^2/2v_{z0})(\delta B/B_0).$$

Hence, in the absence of prebunching, we have to second order in $\delta B$

$$\Delta\theta(z = L) = \int_0^L \left(\omega - \frac{\Omega}{\gamma}\right) \frac{dz}{v_z} \simeq \frac{L}{v_{z0}}\left(\omega - \frac{\Omega_0}{\gamma}\right) + \int_0^L \frac{\delta\Omega}{\gamma} \frac{\delta v_z}{v_{z0}^2} \, dz$$

$$= \frac{\omega L}{v \cos \alpha}\left[\frac{\omega - \Omega_0/\gamma}{\omega} - \left\langle\left(\frac{\delta B}{B_0}\right)^2\right\rangle \frac{1}{2}\tan^2 \alpha\right], \qquad (25)$$

where terms of order $\langle \delta B^2 \rangle$ $(\omega - \Omega_0/\gamma)$ have been dropped.

Let us now assume a Gaussian distribution in pitch angle $\alpha = \alpha_0 + \delta\alpha$ and energy $\gamma = \gamma_0 + \delta\gamma$,

$$f(\delta\alpha, \delta\gamma) = \frac{1}{2\pi a g} \exp\left[-\frac{(\delta\alpha)^2}{2a^2}\right] \exp\left[-\frac{(\delta\gamma)^2}{2g^2}\right], \qquad (26)$$

and expand $\Delta\theta$ to second order in $\delta\alpha$ [because $\partial(\Delta\theta)/\partial\alpha$ can become equal to zero] and to first order in $\delta\gamma$,

$$\Delta\theta = \Delta\theta(\alpha_0, \gamma_0) + A_1\delta\alpha + \tfrac{1}{2} A_2(\delta\alpha)^2 + C\delta\gamma. \qquad (27)$$

The expansion coefficients are given by

$$A_1 = \frac{L\omega}{v} \frac{\sin \alpha_0}{\cos^2 \alpha_0}\left[\frac{\omega - \Omega_0/\gamma_0}{\omega} - \left\langle\left(\frac{\delta B}{B_0}\right)^2\right\rangle\left(1 + \frac{3}{2}\tan^2 \alpha_0\right)\right], \qquad (28a)$$

$$A_2 = \frac{L\omega}{v} \frac{1}{\cos \alpha_0}\left[\frac{\omega - \Omega_0/\gamma_0}{\omega} (1 + 2\tan^2 \alpha_0)\right.$$

$$\left. - \left\langle\left(\frac{\delta B}{B_0}\right)^2\right\rangle\left(1 + \frac{13}{2}\tan^2 \alpha_0 + 6\tan^4 \alpha_0\right)\right], \qquad (28b)$$

$$C = \frac{\omega L}{v} \frac{1}{\gamma_0 \cos \alpha}\left[1 - \frac{\gamma_0^2}{\gamma_0^2 - 1}\left(\omega - \frac{\Omega_0}{\gamma_0}\right)\right] \simeq \frac{\omega L}{\gamma_0 v \cos \alpha_0}. \qquad (28c)$$

Averaging $\sin[\phi_0 - \Delta\theta(z = L)]$ over the distribution function of Eq. (26) and using Eq. (27), we obtain the average power loss per particle

$$\langle\Delta W\rangle = \tfrac{1}{2} e r_0 \sqrt{\pi} E_{02} \tan \alpha_0 \exp[-(\omega - \Omega/\gamma_0)^2 r_{02}^2/4v^2]$$

$$\times F(q) \sin(\phi_0 - \bar{\theta}) \exp(-\tfrac{1}{2}C^2g^2)$$

$$\times (1 + a^4 A_2^2)^{-1/4} \exp[-\tfrac{1}{2}A_1^2a^2/(1 + a^4A_2^2)], \qquad (29)$$

$$\bar{\theta} = \Delta\theta(\alpha_0, \gamma_0) + \tfrac{1}{2}\tan^{-1}A_2 a^2 - \tfrac{1}{4}A_1^2 A_2 a^4/(1 + a^4 A_2^2).$$

For the bunching to be effective, we need $Cg \lesssim 1$, i.e.,

$$L\omega/v \lesssim (\gamma_0/g)\cos\alpha_0. \tag{30}$$

If the energy spread is due to the variation of electrostatic energy within the beam, a 60-kV annular beam with a 10% spread in the radial location of the gyrocenters and $\alpha_0 = \pi/3$ will have $g < 5 \times 10^{-4}$ for currents of less than 20 A. This means that we must have $L \lesssim 80\lambda$.

We see from Eqs. (28a–c) that one can make $A_1 = 0$ by a judicious choice of the frequency mismatch $\omega - \Omega_0/\gamma_0$, so that $\Delta\theta$ becomes independent of $\alpha$ to first order. The requirement on $a$, the thermal width in $\alpha$, is $|A_1 a| \lesssim 1$ and $|a^2 A_2| \lesssim 1$, from which we find

$$\left| \frac{\omega - \Omega_0/\gamma_0}{\omega} - \left\langle\left(\frac{\delta B}{B_0}\right)^2\right\rangle\left(1 + \frac{3}{2}\tan^2\alpha_0\right)\right| \lesssim \frac{v}{L\omega a}\frac{\cos^2\alpha_0}{\sin\alpha_0}, \tag{31a}$$

$$\left\langle\left(\frac{\delta B}{B_0}\right)^2\right\rangle \lesssim \frac{v}{3L\omega a^2}\frac{\cos^5\alpha_0}{\sin^2\alpha_0}. \tag{31b}$$

If we take $L\omega/v = 500$, $a = 0.1$, and $\alpha_0 = \pi/3$, we see that the bandwidth over which the prebunching is effective is about 1.2%; thus the klystron mechanism is fairly frequency selective. With the same parameters, we require $\langle(\delta B/B_0)^2\rangle^{1/2} \lesssim 0.06$, which should raise no experimental difficulties.

Although the criteria (30) and (31) were derived using linear theory in the power extraction cavity, they should also be roughly applicable in the nonlinear regime, because they are related to phase-mixing of the prebunched beam. If these criteria are not satisfied, the device will operate essentially as a single-cavity gyrotron, which is highly insensitive to thermal spreads in pitch angle.

## IV. Single-Mode Nonlinear Computations

### A. Numerical Procedures

In most applications, efficiency is a main figure of merit for a radiation source, e.g., in rf heating of magnetic fusion plasmas. To calculate and optimize efficiency, we need to compute particle orbits in a highly nonlinear regime. Because the radiation field and guide magnetic field are $z$ dependent, analytical solution is impossible and numerical computation is necessary.

For the first harmonic and small beam voltages, $\gamma - 1 \ll 1$, we use the small $\beta_\perp$ approximation in Eqs. (15) and (16), including also the changes in $p_\perp$ and $p_z$ due to the nonuniform magnetic field. The particle orbits have been integrated from $z = -2r_{02}$ to $z = 2r_{02}$ for different distributions in initial phase, y position, and pitch angle. For the purposes of this chapter, to include both the gyroklystron and the single-cavity gyrotron in the same discussion, $r_{02}$ and $E_{02}$ will refer to either the output cavity of the gyroklystron or the $r_0$ and $E_0$ of the single-cavity gyrotron.

As shown in Section III, a moderate spread in pitch angle is not very harmful. We have vertified this numerically in the nonlinear regime by averaging over a distribution $\alpha$. In what follows we shall only consider beams with no thermal spread. The efficiency here is defined as the difference in beam energy flux entering and leaving normalized to the entering energy flux,

$$\eta = \Delta\bar{\gamma}/(\bar{\gamma}_0 - 1), \qquad \Delta\bar{\gamma} = \bar{\gamma}_0 - \bar{\gamma}(z = 2r_{02}), \qquad (32)$$

where $\gamma_0 = \gamma(z = -2r_{02})$ and the overbar denotes an average weighted by the current density

$$\bar{\gamma}(z) = \int \gamma f(\mathbf{v}_\perp, v_z, y_g, z)\, v_z\, d^3v\, dy_g \Big/ \int f(\mathbf{v}_\perp, v_z, y_g, z)\, v_z\, d^3v\, dy_g,$$

where $f$ is the beam particle distribution function.

For a single-cavity device (i.e., with no prebunching), the distribution in initial phase angle $\theta_{in} = \theta(z = 2r_{02})$ is uniform. For the klystron, we use linear theory in the first cavity. Ignoring spreads in pitch angle and energy, we obtain from Eq. (20)

$$\theta_{in} = \theta_0 - \phi_0 + \varepsilon \sin ky_g \cos \theta_0, \qquad (33)$$

where $\varepsilon$ is proportional to the prebunching field in cavity 1 and $\phi_0$ is the phase difference of the fields in the two cavities. We now assume a uniform distribution in $\theta_0$, calculate $\theta_{in}$ from Eq. (33), and use interpolation over precalculated orbits (uniformly distributed in $\theta_{in}$) to find the energy extracted from the particle. The efficiency of the klystron is then optimized with respect to $\varepsilon$ and $\phi_0$. This procedure avoids recalculation of the particle orbits for different $\varepsilon$ and $\phi_0$. For the klystron case, about 100 integrations of Eqs. (15) and (16) with different initial phases are required to find well-defined efficiency maxima. In the single-cavity case, accurate results require only about 30 initial phases.

The distribution in $y_g$ is determined by the shape of the beam emanating from the electron gun. As one example, we consider a sheet beam with an extent $L_b$ in the y direction, which is an integer multiple of half a wavelength, $kL_b = n\pi$. In this case we take $f_y(ky_g) = 2/\pi$ for $0 \le ky_g \le \pi/2$. (Only the interval $0 \le ky_g \le \pi/2$ need be considered if the initial phase is averaged from 0 to $2\pi$.) We have used eight different y positions for the averaging in y. For the case of an annular beam, we have attributed weights $w(j)$ to each y position, proportional to the arc length of the circle $x_g^2 + y_g^2 = r_b^2$ for which

$$\sin(j - 1)\, \pi/16 < |\sin ky_g| < \sin j\, \pi/16, \qquad j = 1,...,8.$$

The choice of beam radius was motivated by linear single-cavity theory, i.e., so as to maximize

$$\langle \sin^2 ky_g \rangle = \tfrac{1}{2}\, [1 \pm J_0(2kr_b)], \tag{34}$$

for a beam centered at a field maximum (+) or node (−). At $\lambda = 2$ mm, a beam with $kr_b = 5.087$ (corresponding to the second minimum of $J_0$) has a diameter of 3.3 mm, which is experimentally tractable (Kim, private communication). Finally, we have considered a pencil beam (possibly multiple), centered at the field maxima so that $|\sin ky_g| = 1$ for all the particles, and no averaging in $y_g$ is required.

## B. RELATION TO PHYSICAL PARAMETERS

In Section IV.A we described the calculation of efficiency by following single-particle orbits through a radiation field of given amplitude $E_{02}$ and spot size $r_{02}$. The actual power delivered to the cavity also involves the current and voltage of the beam and the parameters of the cavity in the following simple way. If the beam has voltage $V$ and current $I$ and the efficiency of the energy transfer is $\eta$, the power input to the cavity is $\eta IV$. The power loss from the cavity field energy is $(\omega/Q)W$, where $Q$ is the cavity quality factor and $W$ is the electromagnetic energy stored in the cavity,

$$W = \tfrac{1}{2} \int (\varepsilon_0 E^2 + B^2/\mu_0)\, d^3r = (\pi\varepsilon_0/8)E_{02}^2 r_{02}^2 L_y,$$

where $L_y$ is the length of the cavity in the $y$ direction. In steady state, the power loss of the cavity must be equal to the energy supplied by the beam, so that

$$(\omega/Q)(\pi\varepsilon_0/8)E_{02}^2 r_{02}^2 L_y = \eta IV. \tag{35}$$

Thus, if $E_{02}$, $r_{02}$, and $V$ are specified and $\eta$ is calculated from Eq. (32) by integrating the particle orbits through the radiation fields, the remaining parameters of the beam and cavity are specified by the single quantity $IQ/L_y$. It is also evident that if $Q$, $L_y$, $I$, and $V$ are known, single-mode equilibria can be found by drawing a line $\eta \propto E_{02}^2$ as expressed by Eq. (35) in a diagram of $\eta$ vs $E_{02}$ obtained from Eq. (32). Any intersection of the two curves represents a single-mode equilibrium. (The stability of such an equilibrium can be determined by a multimode simulation; see Section VI.)

To calculate $Q$ we assume that one of the mirrors constituting the cavity has a power reflection coefficient $1 - Y$, so that a fraction $Y$ is transmitted through or diffracted around this mirror as output radiation. (For simplicity, we neglect dissipation in the mirrors.) A simple calculation then yields

$$Q = 2L_y\omega/Yc. \qquad (36)$$

In a time-dependent calculation, this output corresponds to a damping rate of the electric field

$$\Gamma = \tfrac{1}{2}\,\omega/Q = \tfrac{1}{4}Yc/L_y. \qquad (37)$$

## C. Dependence on Magnetic Field Profile

In general, we fix all parameters except the electric field $E_{02}$ in the second cavity and use Eq. (32) to calculate the efficiency $\eta$ as a function of $E_{02}$ for the single-cavity gyrotron and the klystron, optimizing the prebunching as described in the preceding section. Four typical results are shown in Fig. 3, where results obtained using four different ambient magnetic fields $B_0(z)$ are compared. In all these cases $\gamma_0 = 1.118$ (60-kV beam), $\alpha = 1$ ($p_\perp/p_\parallel = 1.56$),

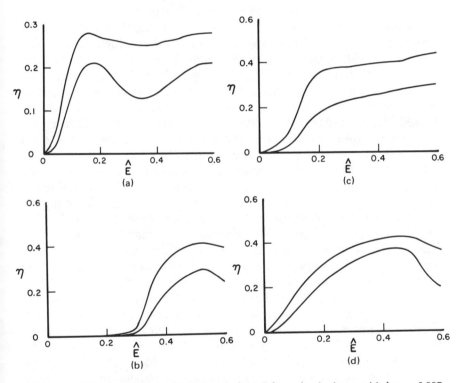

FIG. 3 Efficiency $\eta$ vs normalized electric field $E$ for a circular beam with $k_y r_b = 5.087$, $\alpha = 1$, $r_{02}/\lambda = 4.8$, and $\gamma_0 = 1.118$. In (a), $\hat{\Omega} = 1.095$, $\Delta\hat{\Omega} = 0$; in (b), $\hat{\Omega} = 1.075$, $\Delta\hat{\Omega} = 0$; in (c), $\hat{\Omega}(-2r_0) = 1.135$, $\Delta\hat{\Omega} = 0.12$ ($B_0$ decreasing); in (d), $\hat{\Omega}(-2r_{02}) = 1.04$, $\Delta\hat{\Omega} = -0.1$ ($B_0$ increasing). Upper curves, gyroklystron; lower curves, gyrotron.

$r_{02}/\lambda = 4.8$, and the beam is annular. The normalized electric field is defined by

$$\hat{E} = e \int_{-\infty}^{\infty} E_x \, dz/m_0 c^2 = \sqrt{\pi} r_{02} e E_{02}/m_0 c^2.$$

The magnetic field in cavity 2 is taken to be a linear function of $z$, and we have introduced the notation

$$\hat{\Omega}(z) = \Omega(z)/\omega, \qquad \hat{\Omega}_0 = \hat{\Omega}(-2r_{02}), \qquad \Delta\hat{\Omega} = \hat{\Omega}(-2r_{02}) - \hat{\Omega}(2r_{02}).$$

In Fig. 3a and 3b $B_0$ is uniform; in Fig. 3a $\hat{\Omega} = 1.095$ and in Fig. 3b $\hat{\Omega} = 1.075$. For small electric fields, the case shown in Fig. 3a is more efficient, as predicted by linear theory (Sprangle, Vomvoridis, and Manheimer, 1981). However, much higher nonlinear efficiency at large electric fields is obtained for a lower value of $\Omega/\omega$ as in Fig. 3b. The efficiency of the single-cavity gyrotron has an oscillatory dependence on $\hat{E}$ because of particles bouncing in the potential well of the wave.

Figure 3c shows a case where the magnetic field decreases throughout the cavity and $\hat{\Omega}$ varies from 1.135 at $z = -2r_{02}$ to 1.015 at $z = 2r_{02}$. The maximum efficiency is slightly higher than in Fig. 3b, and high efficiency can be obtained with considerably lower electric fields. Furthermore, the efficiency is almost monotonically increasing with $\hat{E}$. The advantage of a decreasing $B_0$ is easily understood; the electrons can remain trapped at resonance $\Omega \simeq \gamma\omega$ throughout the interaction region. Because $\gamma$ is decreased as a result of the interaction, it is necessary to decrease $\Omega$ to extract more energy from the particles. The effect of the prebunching is quite noticeable; at $\hat{E} = 0.25$, $\eta$ is increased from 20 to 38%.

Finally, Fig. 3d shows a case where $B_0$ is increasing and $\hat{\Omega}$ is going from 1.04 to 1.14. This configuration turns out to have a very good efficiency, even for a single cavity, whereas the klystron does not lead to a substantial improvement. In this case, the electrons are at resonance $\Omega = \gamma\omega$ only at the end of the interaction region. The first half, $z < 0$, mainly serves to phase-bunch the beam so that the energy extraction in the second half becomes very efficient. This also explains why external prebunching does not increase the efficiency very much. Figure 4 shows the average $\gamma$ as a function of $z$ for the two different magnetic field profiles for the single-cavity gyrotron.

The maximum efficiency of a single-cavity gyrotron with $B_0$ increasing is almost as large as for the klystron with $B_0$ decreasing. It might seem, therefore, that the klystron, with its increased degree of complexity, is not a particularly useful configuration. However, as we shall show in Section VI, multimode simulations demonstrate that mode competition is very much affected by the contouring of $B_0$. In fact, the efficient high-frequency modes, which yield the optimistic single-cavity results of Fig. 3d, never get excited when $B_0$ is increasing. The multimode analysis shows that a decreasing $B_0$ indeed gives higher efficiency, even for a single-cavity gyrotron.

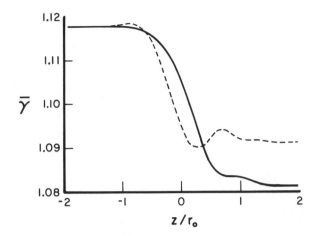

FIG. 4 Average $\gamma$ of the beam vs position for decreasing $B_0$ (---) and increasing $B_0$ (——). $\hat{E} = 0.28$ in both cases.

## D. DEPENDENCE ON BEAM SHAPE

In the quasi-optical gyrotron, the wave vector of the radiation fields is perpendicular to the beam direction, which has the advantage of strongly reducing the effects of a thermal spread in pitch angle. On the other hand, electrons with different $y$ coordinates will see a different electric field, and those at the nodes of the electric field do not interact at all. Moreover, because of the transverse mode structure, parameters such as the guide field profile and the amplitude and phase of the prebunching field can be optimized only in an average sense. To illustrate the importance of the beam

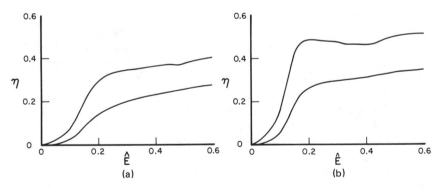

FIG. 5 Efficiency vs normalized electric field for (a) a beam uniform in $y$ and (b) a pencil beam at a field maximum. $\alpha = 1$, $r_{02}/\lambda = 4.8$, and $\gamma_0 = 1.118$. Upper curves, gyroklystron; lower curves, gyrotron.

shape, Fig. 5 shows the efficiency for a beam uniform in $y$ (Fig. 5a) and concentrated at the field maxima (Fig. 5b) for the same parameters as in Fig. 3c. For the uniform beam, the efficiency at $\hat{E} = 0.3$ is about 19% for the single cavity and 33% for the klystron, whereas the annular beam gives 22 and 38%. A pencil beam at the field maxima gives 28 and 48%, respectively. We conclude that the shape of the beam is of considerable importance. Furthermore, the klystron can greatly improve the efficiency; $\eta$ is increased by a factor $\geq 1.7$ in all three cases.

## E. SCALING LAWS

Owing to the large number of parameters relevant to the gyrotron, it is not an easy task to find the global maximum in efficiency. Thus, it is of great value to find at least approximate scaling laws to reduce the number of parameters. To derive such scaling laws, we assume a large pitch angle, $p_\perp^2 \gg p_z^2$, and a low beam voltage, $\gamma - 1 \ll 1$, and approximate $\gamma$ by 1 everywhere in Eqs. (15) and (16) except in the frequency detuning term, where we set $\gamma \simeq 1 + p_\perp^2/2m^2c^2$. We also normalize $z$ with respect to the radiation beam radius $r_0$ and find the set of equations

$$dp/d\zeta = -\mathcal{E}e^{-\zeta^2}\cos\theta\sin ky_g,$$
$$d\theta/d\zeta = \Delta - \kappa p^2 + (\mathcal{E}/p)e^{-\zeta^2}\sin\theta\sin ky_g, \tag{38}$$

where

$$\rho = p_\perp/p_{\perp 0}, \qquad \zeta = z/r_0. \tag{39}$$

The parameters $\mathcal{E}$, $\Delta$, and $\kappa$ are given by

$$\mathcal{E} = \frac{eEr_{02}}{2\beta_z\beta_{\perp 0}mc^2}, \qquad \Delta = \frac{r_{02}(\Omega - \omega)}{c\beta_z}, \qquad \kappa = \frac{\omega r_{02}\beta_{\perp 0}^2}{c\beta_z}. \tag{40}$$

Equation (40) shows that three parameters replace the six we had originally: $E$, $r_{02}$, $\beta_{\perp 0}$, $\beta_z$, $\omega$, and $\Omega$. Earlier in this section we discussed the dependence on $\mathcal{E}$ and $\Delta$. To cover a three-dimensional parameter space, we need to vary one additional parameter, e.g., $r_{02}$, which has been kept fixed at $\omega r_{02}/c = 30$. We have varied $r_{02}$ and found that the efficiency, after optimization with respect to $\Delta$, is very insensitive to the value of $r_{02}$ as long as $10 < \omega r_{02}/c < 45$. The optimum value of $\omega r_{02}/c$ is approximately 15, i.e., half of the value used here. However, the increase in efficiency for our case is small, typically 1% or less.

The dependence on $\sin ky_g$ in Eq. (38) is already taken care of by considering different beam shapes. If we are interested only in single-particle orbits, the $\sin ky_g$ factor can be incorporated in $\mathcal{E}$.

## V.  Multimode Formulation

In this section we derive equations for the time-dependent response of the quasi-optical gyrotron when many modes are excited and interact in a self-consistent manner. This allows us to see whether a particular configuration will evolve into a steady state with one single mode excited, and, if so, which mode is selected and how the mode selection is influenced by various parameters. From the single-mode analysis in Section IV we know that mode selection is very important for the efficiency. Generally, low-frequency modes ($\omega$ only slightly larger than $\Omega/\gamma$) have high linear growth rate but poor maximum efficiency, whereas higher-frequency modes have small linear growth rate but good efficiency for large electric field.

Numerical studies of multimode gyrotrons have been carried out recently by Vomvoridis and Sprangle (1981) and Dialetis and Chu (1983). However, these studies were only concerned with a relatively small number of interacting modes in traveling-wave tube gyrotrons and differ considerably from the present treatment. Dialetis and Chu demonstrate that, in a two-mode system, the growth of a linearly unstable mode can be quenched by the presence of another mode already excited at large amplitude. Vomvoridis and Sprangle reach the same conclusion from a dynamical two-mode simulation. In addition to including a large number of modes, our simulations differ from those of Vomvoridis and Sprangle (1981) by making use of the separation in time scales corresponding to the frequency separation between different modes, $\Delta\omega$, and the inverse time scale over which a mode amplitude changes, $\delta\omega$, the inequality $\delta\omega \ll \Delta\omega$ being satisfied when the single-pass gain of the radiation is small.

The axisymmetric modes described in Eqs. (1) and (2) are characterized only by their longitudinal mode number and have approximately equal spacing in frequency. Thus, we can write

$$E_x = \mathrm{Re} \sum_{l=-M}^{M} \tilde{E}_l(t) \exp\left[i\omega_0\left(1 + \frac{l}{N}\right)t\right] \exp\left(-\frac{z^2}{r_{02}^2}\right) \sin\left(ky - \frac{l\pi}{2}\right), \quad (41)$$

where $\tilde{E}_l = E_l \exp(i\phi_l)$ is slowly varying compared with $\exp(i\omega_0 t)$; $N$ is twice the number of wavelengths between the mirrors and $M \ll N$, so that the frequency band is narrow. We have neglected the $l$ dependence of $k$, which is permissible if the beam extends only a few wavelengths in the $y$ direction, $kL_b M/N \ll 1$. We have also assumed that the beam is centered with respect to the mirrors, so that half of the modes have maxima where the others have nodes and vice versa.

The change in the amplitudes $\tilde{E}_l$ over a time $\tau = 2\pi N/\omega_0 = 2L_y/c$ is assumed to be negligible; i.e., we assume that the band-width of each mode is small compared with the frequency spacing $\Delta\omega = \omega_0/N$. We exploit this by noting that with constant $\tilde{E}_l$ the multimode field (41) is periodic in time with a period $\tau$. To advance the amplitudes, we integrate particle orbits over a time $\tau$, calculate $d\tilde{E}_l/dt$ from the currents, and step forward in time by a time step that is not related to $\tau$, but is restricted only by the requirement that the change in $\tilde{E}_l$ be "small." Thus, by simulating an interval $\tau = 2\pi/\Delta\omega$ we can take a time step of order $1/\delta\omega$, where $\delta\omega$ is the frequency width of the modes. For a cavity where the reflection losses at the mirrors are small, this implies a large reduction in computation time.

The equation of motion for each particle is a straightforward generalization of the single-mode equation

$$\frac{dp_\perp}{dz} = -\frac{em\gamma}{2p_z} \sum_l E_l \cos\left(\theta - \omega_0 \frac{l}{N} t - \phi_l\right) \exp\left(-\frac{z^2}{r_{02}^2}\right) \sin\left(ky_g - \frac{l\pi}{2}\right), \quad (42)$$

$$\frac{d\theta}{dz} = \frac{m}{p_z}(\Omega - \gamma\omega_0) + \frac{m\gamma e}{2p_z p_\perp} \sum_l E_l \sin\left(\theta - \omega_0 \frac{l}{N} t - \phi_l\right)$$

$$\times \exp\left(-\frac{z^2}{r_{02}^2}\right) \sin(ky_g - l\pi/2), \quad (43)$$

where

$$dt/dz = \gamma m/p_z \quad (8)$$

and

$$\theta = \psi - \omega_0 t. \quad (44)$$

The single-mode theory is recovered by considering only the $l = 0$ term in the summation. An important simplification of the single-mode theory is that all quantities depend only on the relative phase between the gyromotion and one single wave, $\theta$, and there is no other explicit time dependence. Thus, to initialize the particle orbits, we only had to initialize a single phase variable $\theta$. For a multimode computation this simplification no longer applies. As the particle enters the second cavity, its phase relation to the whole wave spectrum is now characterized by two variables: $\theta$, its phase with respect to the central wave in the spectrum, and $t_0$, the time at which it enters the cavity. Because the wave spectrum is periodic with period $\tau = 2\pi N/\omega_0$, $t_0$ varies between 0 and $\tau$. Thus, in doing the single-cavity gyrotron simulations, we have initialized particles uniformly in $\theta$ and $t_0$, typically using 30 intervals between $t_0 = 0$ and $t_0 = \tau$ and 30 intervals between $\theta_{in} = 0$ and $\theta_{in} = 2\pi$. This implies a total of 900 particles per time step. [We see from Eqs. (42) and (43) that it is not necessary for $N$ to be an integer, but the mode frequencies must satisfy $\omega_l = (1 + l/N)\omega_0$ within the frequency band under consideration.] The extension to the klystron case is straightforward; the uniform distribution in $\theta_{in} = \theta \, (z = -2r_{02})$ is modified as described by Eq. (42).

Finally, we need an equation for the electric fields generated by the particles. From Maxwell's equations,

$$c^2\nabla^2\mathbf{A} - \frac{\partial^2}{\partial t^2}\mathbf{A} = -\frac{1}{\varepsilon_0}\mathbf{J}. \tag{45}$$

We now write $\mathbf{A}$ as a superposition of normal modes,

$$\mathbf{A} = \text{Re}\sum_l \mathbf{F}_l(\mathbf{r})\tilde{a}_l(t)\exp(i\omega_l t), \tag{46}$$

where $\omega_l = (1 + l/N)\omega_0$ and the spatial parts $\mathbf{F}_l$ are assumed to satisfy

$$c^2\nabla^2\mathbf{F}_l + \omega_l^2\mathbf{F}_l = 0 \tag{47}$$

for an imagined lossless resonator. Substituting Eqs. (46) and (47) into Eq. (45), multiplying by $F_l^*\exp(-i\omega_l t)$, and integrating in space and over a time period $\tau = 2\pi N/\omega_0$, we obtain

$$-i\omega_l\frac{d\tilde{a}_l}{dt} = -\frac{1}{\varepsilon_0\tau}\int d^3r\int_0^\tau \mathbf{J}\cdot\mathbf{F}_l^*\exp(-i\omega_l t)\,dt, \tag{48}$$

where we have used the normalization $\int |\mathbf{F}_l|^2\,d^3r = 1$. For the modes we are concerned with,

$$\mathbf{F}_l = 2\hat{x}(\pi L_y r_{02}^2)^{-1/2}\exp[-(x^2 + z^2)/r_{02}^2]\sin(ky - l\pi/2), \tag{49}$$

where $L_y$ is the cavity length. Because $\mathbf{E} = -\partial\mathbf{A}/\partial t$, the complex amplitudes of Eqs. (41) and (46) are related by

$$\tilde{E}_l = -i\omega_l 2(\pi L_y r_{02}^2)^{-1/2}\tilde{a}_l.$$

If we also include a damping factor $\Gamma$, due to reflection and diffraction losses [cf. Eq. (46)], which will be approximately the same for all these modes, we obtain

$$\left(\frac{d}{dt} + \Gamma\right)\tilde{E}_l = -\frac{4}{\pi\varepsilon_0 L_y r_{02}^2}\int d^3r\frac{1}{\tau}\int_0^\tau J_x\exp(-i\omega_l t)$$
$$\times\exp[-(x^2 + z^2)/r_{02}^2]\sin(ky - l\pi/2)\,dt. \tag{50}$$

Finally, we substitute

$$J_x = -\sum_j \delta^3(\mathbf{r} - \mathbf{r}_j)ev_{\perp j}\cos\psi_j, \tag{51}$$

where $j$ labels the particles, average over the gyromotion, and change the variable of integration from $t$ to $z$ to obtain

$$\left(\frac{d}{dt} + \Gamma\right)\tilde{E}_l = \frac{2e}{\pi\tau\varepsilon_0 L_y r_{02}^2} \sum_j \int_{-\infty}^{\infty} \frac{p_{\perp j}}{p_{zj}} \exp\left[i\left(\theta_j - \frac{l\omega_0 t}{N}\right)\right]$$

$$\times \exp(-z^2/r_{02}^2) \sin(ky_g - l\pi/2) \, dz, \tag{52}$$

where the sum refers to beam particles that enter the cavity during one periodicity time $\tau$. It may be noted that Eqs. (42) and (52), together with the expression for the wave energy

$$W = (\pi\varepsilon_0/8)L_y r_{02}^2 \sum_l |\tilde{E}_l|^2, \tag{53}$$

conserve energy.

The sum over particles in Eq. (52) represents the averaging in initial phase $\theta$ and time $\omega_0 t_0/N$ as described earlier in this section, and, if appropriate, also averages in $y_g$, pitch angle, and energy. [Note that in Eqs. (50) and (51) the integration interval in actual time is from 0 to $\tau$ and the entrance time $t_0$ varies from $-\infty$ to $+\infty$, whereas in Eq. (52) the integration intervals have been interchanged. However, because the integrand of Eq. (50) remains unchanged under a translation of both times by a multiple of $\tau$, $(t,t_0) \rightarrow (t + n\tau, t_0 + n\tau)$, we have $\int_0^\tau dt \int_{-\infty}^{\infty} dt_0 = \int_{-\infty}^{\infty} dt \int_0^\tau dt_0$ and the integral over $t_0$ becomes a summation over entry times during a single period $\tau$ (Sprangle et al., 1980)].

Equation (52) is used to advance the electric field in time. We emphasize that when more than two modes are present, phase correlations are important. Even though we neglect the frequency shifts $\Delta\omega_l = d\phi_l/dt$ in the particle equation of motion, their time-integrated effect [included in Eq. (52)] is to introduce phase correlations between the waves. In all numerical examples described in the following subsection, we use a normalized time $\hat{t}$ defined by

$$t = \hat{t}\frac{mc^2}{e}\frac{\varepsilon_0 L_y}{2I} = \hat{t}\tau\frac{339}{I},$$

where $\tau = 2L_y/c$ is the periodicity time and I is measured in amperes. For a 50-cm cavity and a 10-A beam, the time unit is about 0.11 $\mu$s.

## VI. Multimode Simulations

### A. Illustrative Examples for a Single-Cavity Gyrotron

The three configurations shown in Fig. 3 (no tapering, decreasing $B_0$, and increasing $B_0$) were studied in multimode simulations. Three different runs were set up for a single-cavity gyrotron with a 4-A pencil beam at the field

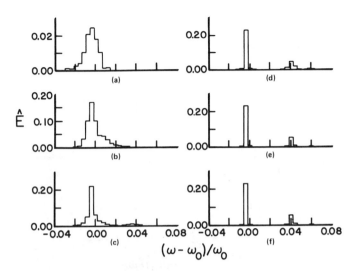

FIG. 6   Normalized electric field amplitudes $E_l = \sqrt{\pi} e r_0 E_l / mc^2$ at different values of normalized time $\hat{t}$ for a single-cavity quasi-optical gyrotron. $\hat{\Omega} = \Omega/\omega_0$ increases linearly from $\hat{\Omega}(-2r_{02}) = 1.05$ to $\hat{\Omega}(2r_{02}) = 1.15$ and the beam is a 4-A pencil beam. (a), $\hat{t} = 2$, $\eta = 0.019$; (b), $\hat{t} = 10$, $\eta = 0.213$; (c), $\hat{t} = 25$, $\eta = 0.216$; (d), $\hat{t} = 50$, $\eta = 0.216$; (e), $\hat{t} = 75$, $\eta = 0.215$; (f), $\hat{t} = 100$, $\eta = 0.218$.

maxima and a power loss of 2% at one of the mirrors. Each of these runs used 2500 particles per time step and 25 modes, with a frequency separation of $(\omega_{l+1} - \omega_l)/\omega_0 = 1/N = 4 \times 10^{-3}$. At time $\hat{t} = 0$ all the modes were given an amplitude of $10^{-4}$ (in normalized variables) and randomized phases.

The case of increasing $B_0$ is shown in Fig. 6. Initially, a broad band of frequencies grows up, and the fastest growing mode tends to suppress the others. Later, a sideband with about 4% higher frequency is destabilized and the final state has one large mode (the initially most unstable one) and a single high-frequency sideband with small amplitude. The large mode is about 1.5% lower in frequency than the nonlinearly most efficient mode, and the steady-state efficiency is moderate, 22%.

The case where $B_0$ decreases with $z$ is shown in Fig. 7. Here, the linearly fastest growing mode cannot suppress the higher-frequency sidebands, and the peak of the spectrum drifts toward higher frequency as a result of the nonlinearity. Eventually, the system settles down to an equilibrium with one single mode and an efficiency of 28%. The frequency is 2.4% higher than that of the linearly most unstable mode and is indeed close to the optimum for high efficiency. Thus, we see a striking contrast between the two ways of contouring the magnetic field. When $B_0$ decreases with $z$, higher-frequency modes are favored in the nonlinear regime and the steady-state efficiency of

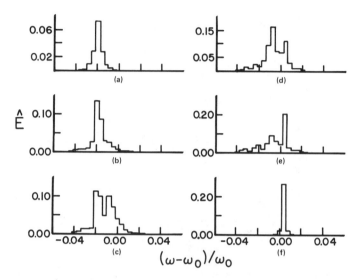

FIG. 7   Normalized electric field amplitudes $\hat{E}_l = \sqrt{\pi} er_0 E_l/mc^2$ at different values of normalized time $\hat{t}$ for a single-cavity quasi-optical gyrotron, with $\hat{\Omega}$ decreasing linearly from 1.134 to 1.014. (a), $\hat{t} = 5$, $\eta = 0.063$; (b), $\hat{t} = 10$, $\eta = 0.117$; (c), $\hat{t} = 15$, $\eta = 0.172$; (d), $\hat{t} = 25$, $\eta = 0.245$; (e), $\hat{t} = 35$, $\eta = 0.233$; (f), $\hat{t} = 50$, $\eta = 0.275$. Note the overall upshift in frequency and the corresponding increase in efficiency.

an overmoded gyrotron is actually higher than when $B_0$ is increasing. In the latter case, the very efficient high-frequency modes are not nonlinearly competitive.

The nonlinear interaction of the different cavity modes is obviously a very complicated process. For instance, we have seen that coherent wave interactions do play a role; occasionally the entire system, or groups of modes, phaselock. Nevertheless, we believe that the most important nonlinear effects are quasi-linear in nature. In particular, one effect always present is a nonlinear destabilization of higher-frequency modes. We recall that for a single-cavity gyrotron the growth rate as a function of frequency peaks at $\omega$ slightly larger than $\Omega/\gamma_0$. If the gyrotron already operates at high efficiency, the average $\gamma$ at some point within the interaction region will be less than $\gamma_0$, so that modes with higher $\omega$ will be favored. The overall upshift in frequency in Fig. 7 and the high-frequency sideband in Fig. 6 are examples of this nonlinear destabilization. An even clearer example will be seen in a simulation with constant $B_0$.

Some of the differences between the two runs discussed so far can be explained at least qualitatively by considering the point in space where the particles are at resonance, $\Omega(z)/\bar{\gamma}(z)\omega \simeq 1$. Figure 8 illustrates how $\Omega/\gamma_0\omega$

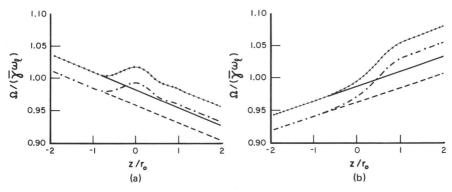

FIG. 8  Variation of $\Omega(z)/\gamma_0\omega_l$ for the linearly most unstable mode (———), $l = l_a$, and a nonlinearly efficient mode (———), $l = l_b$. Also shown is $\Omega(z)/\bar{\gamma}(z)\omega_l$ for the linearly most unstable mode (· · ·) and the nonlinearly efficient mode (– · –) when $E_{l_b} = 0.3$ and all other modes have zero amplitude. Figure 8a corresponds to the decreasing $B_0$ of Fig. 7; with $\omega_0$ as in Fig. 7, $(\omega_{l_a} - \omega_0)/\omega_0 = -0.02$ and $(\omega_{l_b} - \omega_0)/\omega_0 = 0.004$. Figure 8b corresponds to the increasing $B_0$ of Fig. 6; with $\omega_0$ as in Fig. 6, $(\omega_{l_a} - \omega_0)/\omega_0 = -0.002$ and $(\omega_{l_b} - \omega_0)/\omega_0 = 0.022$.

varies with $z$ for the linearly most unstable mode and the mode with highest nonlinear efficiency in the two simulations. In both cases, the mode with the largest linear growth rate has $\Omega/\gamma_0\omega \simeq 0.98$ at the center of the cavity. However, when $B_0$ is decreasing, the higher-frequency modes are at resonance with the particles before they reach the midpoint. If a high-frequency mode has an amplitude large enough to trap particles, it causes the average $\gamma$ to keep decreasing throughout the interaction region and so prevents the particles from reaching the resonance with the lower-frequency modes. This implies that lower-frequency modes become nonlinearly blocked. Obviously, such a mechanism is not possible if $B_0$ increases with $z$. In this case, the electrons first encounter resonances with the low-frequency modes. Before the beam is in resonance with the high-frequency mode, it will have acquired considerable spreads in energy and lost most of its phase coherence, and therefore interacts poorly with the high-frequency mode. Consequently, if $B_0$ is increasing, a large-amplitude low-frequency mode can nonlinearly block the more efficient high-frequency modes. We have verified that an equilibrium with a single high-frequency mode is unstable and eventually evolves to the equilibrium in Fig. 6.

For comparison, one run was made with a uniform magnetic field (Fig. 9). Initially ($\hat{t} = 2$ and $\hat{t} = 10$), one observes a slight drift of the spectrum toward higher frequency. At $\hat{t} = 20$, a broad band of modes at a frequency about 2% higher than the main peak has been excited. At the end of the run, the system approaches a state with only two modes. We emphasize that the

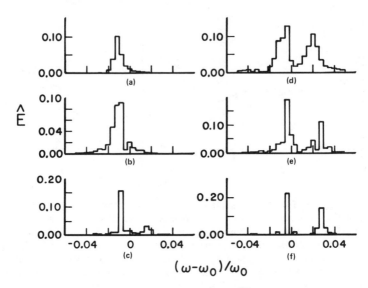

FIG. 9   Normalized electric field amplitudes $\hat{E}_l = \sqrt{\pi} e r_0 E_l / mc^2$ at different values of normalized time $\hat{t}$ for a single-cavity quasi-optical gyrotron, with $\hat{\Omega} = 1.096$ a constant in space. (a), $\hat{t} = 2$, $\eta = 0.069$; (b), $\hat{t} = 5$, $\eta = 0.071$; (c), $\hat{t} = 10$, $\eta = 0.104$; (d), $\hat{t} = 20$, $\eta = 0.229$; (e), $\hat{t} = 40$, $\eta = 0.222$; (f), $\hat{t} = 75$, $\eta = 0.263$. The equilibrium has two modes with large amplitudes. The high-frequency band has been nonlinearly destabilized.

second (i.e., high-frequency) mode has been nonlinearly destabilized by the large-amplitude lower-frequency mode; in the linear regime the high-frequency mode is strongly stable. The efficiency is remarkably high, over 26%. From a single-mode calculation for the first mode one would expect an efficiency of 14%. Thus, multimoding is very beneficial in this particular case. Figure 10a shows how the beam loses energy to the two modes as a function of space. In the first part of the interaction region, the electrons lose about 2.8% of their rest energy to the first (low-frequency) mode and 0.3% to the second (high-frequency) mode. The average $\gamma$ has now decreased by 3.1%, thus enhancing the interaction with the high-frequency mode. In the second part of the interaction region, the particles lose energy to mode 2 (0.6%) but also gain energy (0.5%) from mode 1. Figure 10b shows the energy deposition profile with the high-frequency mode removed. The electrons oscillate in the potential well of the wave and recover a substantial part of their energy (1.6%) in the second part of the interaction region. Evidently, in addition to extracting energy from the particles, the high-frequency mode also destroys the coherence of the bounce motion in the low-frequency wave. The amount of energy recovered by the particles from wave 1 is much decreased by the presence of wave 2.

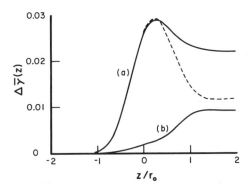

FIG. 10 Fraction of the rest energy of the electrons that is lost to the low-frequency mode (a) and high-frequency mode (b) of Fig. 9 when both are excited (——) and when the high-frequency mode is artifically removed (–––).

The way in which these runs have been initialized (with small wave amplitudes, but with the full beam current and voltage on at $t = 0$) does not very well represent an actual experiment. Rather than trying to model the time dependence of the beam current and voltage during startup, we have investigated the dependence of the final state on the initial condition and tested the stability of different single-mode equilibria. For decreasing $B_0$, the final state is always single mode, if the current is not too small. The frequency range within which these single-mode equilibria are stable is very narrow ($< 0.5\%$). Clearly, if the mode density is very high, there may be more than one possible final state; however, they all have approximately the same frequency and efficiency.

After completion of this work, we became aware of the analytical work of Nusinovich (1981), who derives the time evolution of the modes from a four-wave interaction treatment. Even though our calculations involve a large number of modes and are certainly too nonlinear to be described by four-wave interaction (the transition from the linear to the nonlinear regime is roughly at $\hat{E} = 0.01$), it is interesting to note that many of the effects predicted and discussed by Nusinovich do occur in the simulations.

We hope that the three examples discussed will have given the reader some understanding of the multitude of effects that occur in a simulation of this kind. We do not claim that we fully understand all that has been observed so far, nor have we covered a sufficient region in parameter space to have observed everything that might happen. Nevertheless, simulations are helpful in allowing us to isolate, diagnose, and study in detail particular aspects of a multimode gyrotron that may not be easily isolated in an experiment.

### B. NUMERICAL RESULTS

To study how the efficiency of a strongly overmoded gyrotron depends on parameters such as beam current and beam shape, and to assess the effects of prebunching, a series of runs was made with beam currents of 2, 4, and 8 A using either a sheet beam, an annular beam with $kr_b = 5.087$, or a pencil beam at a field maximum. The magnetic field was decreased linearly by 11% from $z = -2r_{02}$ to $z = 2r_{02}$, which seemed favorable from a single-mode analysis. The radius of the radiation beam was 4.8 times the wavelength, and $\Delta\omega/\omega_0 = 4 \times 10^{-3}$. For the pencil beam, the particles are located at the nodes of half of the modes, and these modes do not interact at all. Thus, for the pencil beam, half of the modes are eliminated, so that $\Delta\omega/\omega = 8 \times 10^{-3}$. All runs included 17 modes, enough to cover the whole spectrum that could be excited. Table I gives the steady-state efficiencies of the different configurations discussed in this section.

The prebunching was always made at one single frequency. The field in the prebunching cavity was assumed to be provided by linear feedback from the power extraction cavity. The amplitude and phase relation between the two cavities was chosen by searching for single-mode equilibria using the efficiency optimization described in Section V. With a judicious choice of prebunching frequency, single-mode operation at substantially increased efficiency could be obtained. Figure 11 shows the time history of a klystron run with a 4-A annular beam. (Note the almost complete suppression of half of the modes with this beam.) On the other hand, multimoding occurs if the prebunching frequency is not high enough. In this case, the prebunched

TABLE I

GYROTRON AND GYROKLYSTRON EFFICIENCIES FOR
DIFFERENT BEAM CURRENTS AND SHAPES[a]

| Beam current (A) | Device | Efficiency (%) | | |
|---|---|---|---|---|
| | | Sheet beam | Annular beam | Pencil beam |
| 16 | single cavity | 27 | 29 | 34 |
| 8 | klystron | 34 | 38 | 38 |
| | single cavity | 27 | 24 | 28 |
| 4 | klystron | 31 | 37 | 45 |
| | single cavity | 22 | 19 | 28 |
| 2 | klystron | 28 | 35 | 48 |
| | single cavity | — | — | 25 |

[a] The magnetic field in all cases decreases by 11% across the interaction region (from $z = -2r_{02}$ to $z = 2r_{02}$).

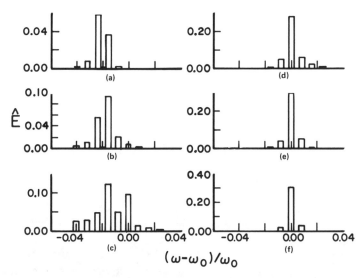

FIG. 11    Time history for a klystron with a 4-A circular beam, with $k_y r_b = 5.087$ and $\hat{\Omega}$ decreasing from 1.134 to 1.014. The prebunching is done at frequency $\omega_0$. (a), $\hat{t} = 2, \eta = 0.034$; (b), $\hat{t} = 5, \eta = 0.061$; (c), $\hat{t} = 10, \eta = 0.123$; (d), $\hat{t} = 15, \eta = 0.359$; (e), $\hat{t} = 25, \eta = 0.367$; (f), $\hat{t} = 50, \eta = 0.370$. The modes that interact weakly with the beam are almost completely suppressed.

mode can nonlinearly destabilize higher-frequency modes with which it cannot successfully compete. For the klystron, substantial multimoding degrades the efficiency. Usually the frequency of the prebunched mode has to be taken so high that the klystron is not linearly unstable, which is the case in Fig. 11.

For both the pencil and the annular beam, the klystron configuration is a significant improvement, in particular at low beam currents. For instance, at 4 A the efficiency is increased from 28 to 45% for the pencil beam, from 19 to 37% for the annular beam, and from 22 to 31% for the sheet beam. The highest efficiency observed in any of these multimode runs was about 48%, which was accomplished with a 2-A pencil beam.

When the beam is uniform in $y_g$, the final state for the single-cavity gyrotron has one dominant mode, and the two nearest neighbors have about half the amplitude of the center mode (Fig. 12). More distant sidebands have smaller amplitudes. The efficiency of this equilibrium is higher than the single-mode prediction, because the two large sidebands interact strongly with the particles at the nodes of the center mode. In fact, the efficiency of a single-cavity gyrotron is often larger for a sheet beam than for an annular beam.

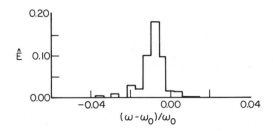

FIG. 12   Coherent, multimode equilibrium for a single-cavity gyrotron with a 4-A beam uniform in $y$.

The multimoding final state for a sheet beam is a coherent equilibrium, the amplitudes are constant, and all the phases are locked, so that $d\phi_l/dt$ is a linear function of the mode index $l$. We also found that if the frequency spacing is decreased to less than about $1 \times 10^{-3}\omega_0$, the shape of the equilibrium, i.e., the amplitude vs mode number graph, becomes independent of $N$, so that the actual bandwidth decreases with $\Delta\omega$.

When the beam current is 2 or 4 A, the sheet beam klystron achieves single-mode operation if the prebunching frequency is high enough (in the 4-A case actually higher than the frequency with optimum efficiency). It appears impossible to obtain single-mode conditions in the 8-A klystron.

For the annular and pencil beams, the final state is generally single mode. The system first evolves toward a state where one mode is completely dominant and has one satellite on each side. These sidebands then decay extremely slowly, and it is not computationally economical to follow the system to a presumed single-mode equilibrium. In a few of these runs, we have restarted the computation after reducing the amplitudes of the sidebands and observed stable single-mode operation. The computer analysis thus predicts a single-mode steady state under these conditions. The motivation for a multimode calculation is then to see which mode will be selected. For both the single cavity and the klystron, the pencil beam is superior, giving up to 28 and 48% efficiency, respectively, whereas the circular beam efficiencies reach 24 and 37%. For the single cavity, the efficiency can be increased by using even larger currents; a set of 16-A runs gave 35% for the pencil beam, 29% for the annular beam, and 27% for the sheet beam.

## VII.   Conclusion

We have explored the physics of quasi-optical gyrotrons and gyroklystrons in strongly overmoded cavities by self-consistent, time-dependent multimode simulations. The mode spectrum has been assumed dense (typically 5–10 modes are linearly unstable) but one dimensional by virtue of the quasi-optical resonator. Mode competition almost always leads to coherent steady states, which are usually single moded, despite the broad band of linearly unstable modes. However, in some cases the equilibrium involves more than one mode; in particular, if different cavity modes interact with different sections of the beam, a phase-locked state with several modes excited may result.

It is well known that the efficiency of a saturated single-mode equilibrium is strongly dependent on the mode frequency; maximum efficiency is achieved at a positive value of the frequency mismatch $\omega - \Omega/\gamma$, and the optimum mismatch increases with an increasing degree of nonlinearity (and hence beam current). We have demonstrated that the nonlinear effects in a gyrotron generally favor higher-frequency modes, because of a quasi-linear decrease of the beam energy. However, the mode competition is also strongly influenced by the contouring of the dc magnetic field in the interaction region. If the ambient magnetic field decreases along the propagation direction of the beam, high-frequency (and nonlinearly efficient) modes are particularly favored, and single-mode operation at high efficiency is possible. If the field contouring is strong enough, the final mode will be higher in frequency than all the linearly unstable modes.

We have proposed and analyzed a new device, the quasi-optical gyroklystron, where the beam is premodulated in a low-power cavity before it reaches the power extraction cavity. Because of the zero parallel wave number of the configuration discussed here, the prebunching is not sensitive to a thermal spread in pitch angle for the beam particles. The prebunching has been shown to lead to significantly improved efficiency, in both single-mode and multimode simulations. In particular, it allows very efficient operation with smaller electric fields (and beam currents) in the power extraction cavity than does the single-cavity gyrotron.

### ACKNOWLEDGMENTS

In carrying out this research we have benefited from discussions with V. Granatstein, D. Kim, B. Levush, M. Read, P. Sprangle, and M. Q. Tran. This work was supported by the Office of Naval Research at the University of Maryland and by the Department of Energy at the Naval Research Laboratory.

### REFERENCES

Dialetis, D., and Chu, K. R. (1983). *In* "Infrared and Millimeter Waves," vol. 7 (K. J. Button, ed.), pp. 537–581. Academic Press, New York.
Ergakov, V. S., Moiseev, M. A., and Erm, R. E. (1979). *Radiophys. Quantum Electron.* **22,** 700.
Ganguly, A. K., and Chu, K. R. (1981). *Int. J. Elec.* **51,** 503–520.

Jory, H. R., Friedlander, F., Hegji, S. J., Shively, J. F., and Symons, R. S. (1977). *Digest IEEE Int. Elec. Dev. Meeting,* Washington, D.C., 234–237.

Nusinovich, G. S. (1981). *Int. J. Elec.* **51,** 457–474.

Ott, E., and Manheimer, W. M. (1975). *IEEE Trans.* **PS-3,** 1–5.

Rapoport, G. N., Nemak, A. K., and Zhurakhovsky, V. A. (1967). *Radiotekh. Elektron.* **12,** 633–649.

Sprangle, P., and Manheimer, W. M. (1975). *Phys. Fluids* **18,** 224–230.

Sprangle, P., Tang, C. M., and Manheimer, W. M. (1980). *Phys. Rev.* **A21,** 302–318.

Sprangle, P., Vomvoridis, J. L., and Manheimer, W. M. (1981). *Phys. Rev.* **A23,** 3127–3138.

Symons, R. S., and Jory, H. R. (1977). *Proceedings of the 7th Symposium on Energy Problems of Fusion Research,* Knoxville, 1111–1115.

Symons, R. S., and Jory, H. R. (1981). *Adv. Elec. Elect. Phys.* **55,** 1–75.

Vomvoridis, J. L. (1983). *Int. J. Infrared Millimeter Waves,* forthcoming.

Vomvoridis, J. L., and Sprangle, P. (1981). Naval Research Laboratory Report 4426.

# INDEX